Alphaviruses

Current Biology

Edited by

Suresh Mahalingam

Lara J. Herrero

Institute for Glycomics
Griffith University
Gold Coast, QLD
Australia

and

Belinda L. Herring

School of Biomedical Sciences, Faculty of Health
Queensland University of Technology
Brisbane, QLD
Australia

Caister Academic Press

Caister Academic Press
Norfolk, UK

www.caister.com

British Library Cataloguing-in-Publication Data
A catalogue record for this book is available from the British Library

ISBN: 978-1-910190-15-9 (hardback)
ISBN: 978-1-910190-16-6 (ebook)

Cover images courtesy of Weiqiang Chen: chikungunya virus infection of bone cells in murine proximal tibial epiphysis.

Contents

Current Books of Interest

Microalgae: Current Research and Applications	2016
Arboviruses: Molecular Biology, Evolution and Control	2016
Shigella: Molecular and Cellular Biology	2016
Aquatic Biofilms: Ecology, Water Quality and Wastewater Treatment	2016
Thermophilic Microorganisms	2015
Flow Cytometry in Microbiology: Technology and Applications	2015
Probiotics and Prebiotics: Current Research and Future Trends	2015
Epigenetics: Current Research and Emerging Trends	2015
Corynebacterium glutamicum: From Systems Biology to Biotechnological Applications	2015
Advanced Vaccine Research Methods for the Decade of Vaccines	2015
Antifungals: From Genomics to Resistance and the Development of Novel Agents	2015
Bacteria-Plant Interactions: Advanced Research and Future Trends	2015
Aeromonas	2015
Antibiotics: Current Innovations and Future Trends	2015
Leishmania: Current Biology and Control	2015
Acanthamoeba: Biology and Pathogenesis (2nd edition)	2015
Microarrays: Current Technology, Innovations and Applications	2014
Metagenomics of the Microbial Nitrogen Cycle: Theory, Methods and Applications	2014
Pathogenic *Neisseria*: Genomics, Molecular Biology and Disease Intervention	2014
Proteomics: Targeted Technology, Innovations and Applications	2014
Biofuels: From Microbes to Molecules	2014
Human Pathogenic Fungi: Molecular Biology and Pathogenic Mechanisms	2014
Applied RNAi: From Fundamental Research to Therapeutic Applications	2014
Halophiles: Genetics and Genomes	2014
Molecular Diagnostics: Current Research and Applications	2014
Phage Therapy: Current Research and Applications	2014
Bioinformatics and Data Analysis in Microbiology	2014
The Cell Biology of Cyanobacteria	2014
Pathogenic *Escherichia coli*: Molecular and Cellular Microbiology	2014
Campylobacter Ecology and Evolution	2014
Burkholderia: From Genomes to Function	2014
Myxobacteria: Genomics, Cellular and Molecular Biology	2014
Next-generation Sequencing: Current Technologies and Applications	2014
Omics in Soil Science	2014

Full details at www.caister.com

Contributors

Weiqiang Chen
Emerging Viruses and Inflammation Research Group
Institute for Glycomics
Griffith University
Queensland
Australia

javier.chen@griffith.edu.au

Lark L. Coffey
Center for Vectorborne Diseases and Department of Pathology
Microbiology and Immunology
School of Veterinary Medicine
University of California
Davis, CA
USA

lcoffey@bloodsystems.org

John K. Fazakerley
The Pirbright Institute
Surrey
UK

john.fazakerley@pirbright.ac.uk

Suan Sin Foo
Emerging Viruses and Inflammation Research Group
Institute for Glycomics
Griffith University
Queensland
Australia

j.foo@griffith.edu.au

Naomi L. Forrester
Institute for Human Infections and Immunity
Center for Biodefense and Emerging Infectious Diseases
Department of Pathology
University of Texas Medical Branch
Galveston, TX
USA

nforrester@utmb.edu

Lara J. Herrero
Institute for Glycomics
Griffith University
Queensland
Australia

l.herrero@griffith.edu.au

Belinda L. Herring
School of Biomedical Sciences
Faculty of Health
Queensland University of Technology
Brisbane, QLD
Australia

belinda.herring@qut.edu.au

Linda Hueston
Arbovirus Emerging Diseases Unit
Centre for Infectious Diseases and Microbiology Services
PathologyWest-ICPMR
Westmead Hospital
Westmead
Australia

linda.hueston@health.nsw.gov.au

Natkunam Ketheesan
Faculty of Medicine, Health and Molecular Sciences
James Cook University
Townsville
Australia

n.ketheesan@jcu.edu.au

William B. Klimstra
Department of Microbiology and Molecular Genetics
Center for Vaccine Research
University of Pittsburgh
Pittsburgh, PA
USA

klimstra@pitt.edu

Alain Kohl
MRC-University of Glasgow Centre for Virus Research
Glasgow
UK

alain.kohl@glasgow.ac.uk

Brett A. Lidbury
John Curtin School of Medical Research
The Australian National University
Canberra
Australia

brett.lidbury@anu.edu.au

Kenneth Lundstrom
PanTherapeutics
Lutry
Switzerland

lundstromkenneth@gmail.com

Melanie McFarlane
MRC-University of Glasgow Centre for Virus Research
Glasgow
UK

melanie.mcdonald@glasgow.ac.uk

Clive S. McKimmie
Institute of Infection, Immunity and Inflammation
College of Medical, Veterinary and Life Sciences
University of Glasgow
Glasgow
UK

clive.mckimmie@glasgow.ac.uk

Suresh Mahalingam
Emerging Viruses and Inflammation Research Group
Institute for Glycomics
Griffith University
Queensland
Australia

s.mahalingam@griffith.edu.au

Lisa F.P. Ng
Laboratory of Microbial Immunity
Singapore Immunology Network
Agency for Science, Technology and Research (A*STAR); and
Institute of Infection and Global Health
University of Liverpool
Liverpool
UK

lisa_ng@immunol.a-star.edu.sg

Lynden Roberts
Townsville Hospital Health Service
Townsville Hospital; and
Faculty of Medicine, Health and Molecular Sciences
James Cook University
Townsville
Australia

lynden.roberts@jcu.edu.au

Pierre Roques
Center for Immunology of Viral Infections and Autoimmune Diseases
IDMIT Infrastructure
CEA–IMETI Division of Immuno-Virology
Université Paris Sud
France

pierre.roques@cea.fr

Nestor E. Rulli
Emerging Viruses and Inflammation Research Group
Institute for Glycomics
Griffith University
Queensland
Australia

ezequiel.rulli@gmail.com

Kate D. Ryman
Department of Microbiology and Molecular Genetics
Center for Vaccine Research
University of Pittsburgh
Pittsburgh, PA
USA

ryman@pitt.edu

Kuo-Ching Sheng
Emerging Viruses and Inflammation Research Group
Institute for Glycomics
Griffith University
Queensland
Australia

kcsheng@scientist.com

Adam Taylor
Emerging Viruses and Inflammation Research Group
Institute for Glycomics
Griffith University
Queensland
Australia

a.taylor1@griffith.edu.au

Scott C. Weaver
Institute for Human Infections and Immunity
Center for Biodefense and Emerging Infectious Diseases
Department of Pathology
University of Texas Medical Branch
Galveston, TX
USA

sweaver@utmb.edu

Preface

Alphaviruses are small, enveloped, single-stranded positive-sense RNA viruses that are typically transmitted by arthropods, especially mosquitoes. A number of alphaviruses are known as clinically important human pathogens (e.g. Venezuelan equine encephalitis virus, chikungunya virus). Key areas of alphavirus research are focused on treatment and prevention of disease, but there is also considerable potential for harnessing alphaviruses as vaccine vectors and in gene therapy (e.g. Semliki Forest virus). In addition, alphaviruses have also been very valuable model systems for a wide range of basic studies in diverse areas such as cell biology, virology and immunology.

The book presents the current status of alphavirus research across a wide range of disciplines, but has a particular focus on molecular biology, pathogenesis and host interactions. The book, which is the first of its kind to focus specifically on alphaviruses, has been written in response to a clear need within the alphavirus research community for a comprehensive and up-to-date overview of the field.

Each chapter has been written by leaders in that particular research field. Chapter topics covered include genome structure and replication; viral evolution; laboratory diagnosis and detection; interaction with the interferon system; antiviral responses in mosquitoes; animal models of alphavirus-induced inflammatory disease; clinical manifestations of arthritogenic alphaviruses; encephalitic alphaviruses; the application of alphavirus vectors for gene therapy; and chikungunya virus pathogenesis and the development of control strategies.

Finally, we thank all contributors for their hard work in providing chapters describing the latest developments in their fields of expertise and for their patience while waiting for the book to be published.

Professor Suresh Mahalingam

Alphavirus Genome Structure and Replication

1

Belinda L. Herring, Lara J. Herrero, Weiqiang Chen, Kuo-Ching Sheng, Nestor E. Rulli and Suresh Mahalingam

Abstract

The family *Togaviridae* consists of two genera, *Alphavirus* and *Rubivirus* (Fields *et al.*, 2013). The organization of their genomes is similar, but phylogenetic analysis shows that alphaviruses and rubiviruses are quite divergent (Koonin and Dolja, 1993). This book focuses on the genus *Alphavirus*, which consists of small (45–75 nm in diameter), enveloped viruses, typically transmitted by arthropod vectors, including mosquitoes, mites and ticks (Powers *et al.*, 2001). Approximately 30 species of Alphavirus have been identified and differentiated into seven groups based on antigen similarity determined by serum neutralization. These antigenic groups are the Barmah Forest complex, the Ndumu complex, the Middelburg complex, Semliki Forest complex, Western equine encephalitis complex, Eastern equine encephalitis complex, and Venezuelan equine encephalitis complex (Powers *et al.*, 2001). All seven serotypes share similarity in both genome organization (having two open reading frames encoding both the structural and non-structural proteins) and viral replication cycle (which includes virus binding, entry, replication and release). This chapter will focus on the alphaviral genome and the events of alphaviral replication.

Alphavirus structure and genomic organization

Alphaviruses are small, enveloped RNA viruses measuring approximately 70 nm in diameter, with a molecular mass of 52×10^6 Da and a buoyant density in sucrose of 1.22 g per cm^3 (Cheng *et al.*, 1995; Paredes *et al.*, 1992; Strauss and Strauss, 1994). The alphavirus is composed of highly organized layers of structural proteins arranged in a $T=4$ icosahedral protein lattice, with 240 copies of each protein E1, E2 and capsid protein C (Paredes *et al.*, 1993). A mature alphavirus particle contains an outer protein shell of E1 and E2 heterodimeric envelope glycoproteins, arranged as trimers, embedded in a host cell-derived lipid bilayer, followed by an underlying nucleocapsid, composed of capsid protein interacting with a single copy of viral genomic RNA (Paredes *et al.*, 1992) (Fig. 1.1).

Alphaviruses have a single-stranded (ss), positive-sense 49S RNA approximately 11.7 kb in length, capped at the 5′ terminus and containing a 3′ polyadenosine tail (Simmons and Strauss, 1972). The viral genome segregates into two large open-reading frames (ORF): the first ORF, containing the non-structural genes, covers the 5′ two-thirds of the genome (7.6 kb) and the second, containing the structural genes (4.1 kb), covers the 3′ terminal third of the genome (Simmons and Strauss, 1972) (Fig. 1.2). The translation of 49S genomic RNA gives rise to four non-structural proteins (nsP1–4), which are involved in viral replication (Foo *et al.*, 2011). After viral RNA synthesis, transcription of 49S negative-strand RNA template results in a 5′ capped and 3′ polyadenylated subgenomic 26S RNA which is similar in sequence to the 3′ terminal non-structural ORF (Faragher *et al.*, 1988; Simmons and Strauss, 1972; Strauss and Strauss, 1994). The subgenomic 26S RNA is translated as a polyprotein and generates the five structural proteins (C, E3, E2, 6K/transframe protein (TF) and E1) that are required for viral structural assembly, virus budding and RNA encapsidation (Foo *et al.*, 2011; Zhang *et al.*, 2002).

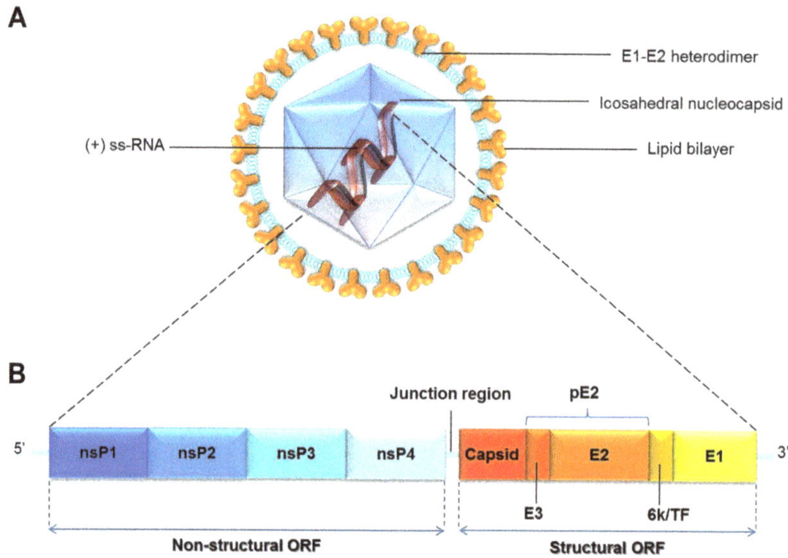

Figure 1.1 Alphavirus genome structure. (A) The alphavirus virion contains a positive single-stranded genomic (49S) RNA encapsulated within the nucleocapsid and enveloped in a glycoprotein-embedded lipid bilayer. (B) The genomic RNA of alphaviruses encodes two ORFs: non-structural and structural ORFs. The non-structural ORF is found at the 5′ end, which codes for four non-structural proteins: nsP1, nsP2, nsP3 and nsP4. The structural ORF encodes six structural proteins, including capsid, E1, E2, E3, 6K. ORF, open reading frame; TF, transframe.

Alphaviral non-structural region

nsP1

Studies of temperature-sensitive (ts) mutants led to the discovery that nsP1 is specifically required for (–) RNA strand synthesis (Hahn *et al.*, 1989; Sawicki *et al.*, 1981; Wang *et al.*, 1991). This activity is associated with nsP1 acting as both a methyltransferase and guanyltransferase in the process of 5′ methyl-guanosine cap formation, capping the genomic and subgenomic RNAs, (Mi and Stollar, 1991; Scheidel *et al.*, 1987). In 2007, the palmitoylation of nsP1 was reported to help the 3D structure of nsP1 and assist its interaction with other components of the viral replicase complex (Zusinaite *et al.*, 2007). Furthermore, it has been shown that nsP1 can regulate the activity of the nsP2 protease as the presence of nsP1 significantly reduced cleavage between nsP2 and nsP3 in Sindbis virus (SINV) by nsP2 (de Groot *et al.*, 1990).

nsP2

The nsP2 protein of alphaviruses is the largest of the non-structural proteins and possesses multiple enzymatic activities. It contains two structural and functional domains: the N-terminal and the C-terminal (Russo *et al.*, 2006).

The N-terminal domain has been shown to possess ATPase, GTPase (Rikkonen *et al.*, 1994) and RNA 5′-triphosphatase activity (Vasiljeva *et al.*, 2000) with nucleoside triphosphate (NTP) binding domains. In addition it serves as an RNA helicase (Gomez de Cedron *et al.*, 1999) to unwind the RNA-RNA duplex during RNA replication and transcription. The RNA helicase activity belongs to a superfamily of enzymes called the DEAD-box proteins, which includes the well-characterized RecBCD helicase of *E. coli* (Gorbalenya *et al.*, 1989). The nucleoside triphosphate (NTP) binding domains are characterized by two motifs, represented by the sequences GSGKS (starting at residue 189) and DEAF (starting at residue 252) (Gorbalenya *et al.*, 1988, 1990; Hodgman, 1988). Furthermore, it is believed that the N-terminal region plays a critical role in the folding of the nsP replicase complex, which is required for the synthesis of the 26S subgenomic mRNA (De *et al.*, 1996; Hahn *et al.*, 1989; Sawicki and Sawicki, 1985, 1993).

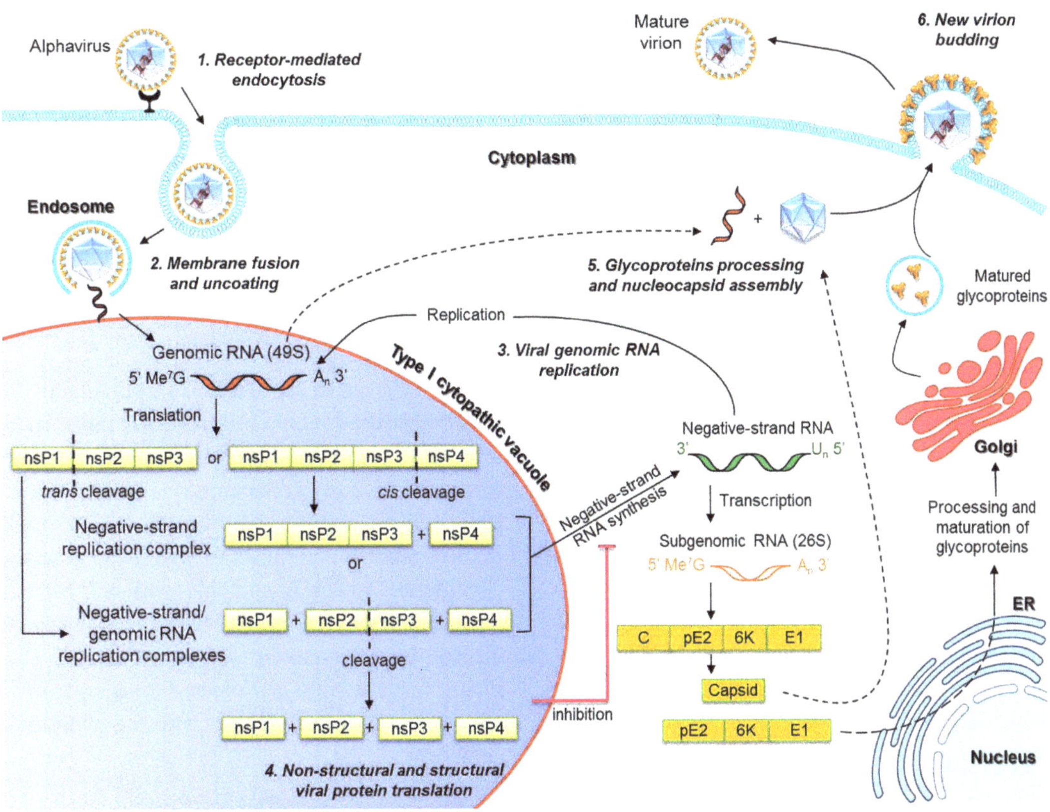

Figure 1.2 Replication of alphaviruses. Alphavirus replication consists of the following steps: (1) Alphavirus enters the host cell through clathrin-dependent endocytosis. (2) In the endosome, membrane fusion allows the viral genome to be released into cytoplasm, where replication occurs. (3) Downstream processing of the polyproteins enables the synthesis of both genomic and negative-strand RNA. Replication of genomic RNA from negative-strand RNA synthesized allows translation of structural proteins, which will be processed co-translationally and post-translationally. (4) Translation of the viral genome gives rise to two possible non-structural polyproteins, being P1234 and P123. (5) Glycoproteins are translocated to the ER, followed by downstream processing in the Golgi apparatus. Mature glycoproteins are transported to the host plasma membrane for budding. (6) Association of glycoproteins with the nucleocapsid allows the newly synthesized virus particle to bud out from the host membrane, completing the final assembly of the virion.

The C-terminal domain of the protein has been shown to have proteinase activity and is associated with proteolytic processing of the alphavirus non-structural polyprotein. nsP2 is believed to be a cysteine protease with residues 475 to 728 being essential for proteolytic activity (Hahn *et al.*, 1989). This was later confirmed when the crystal structure of nsP2 from Venezuelan equine encephalitis virus (VEEV) was described identifying a previously unrecognized methyltransferase (MTase)-like domain in the C-terminus (Russo *et al.*, 2006). This led to further work by Mayuri *et al.* (2008), who suggested that different regions of the nsP2 MTase-like domain were able to differentially modulate host defence mechanisms. It has also been shown that the C terminal domain is involved in a number of other functions including regulating 26S subgenomic RNA synthesis (Suopanki *et al.*, 1998), down-regulating minus-strand RNA synthesis late in infection (Sawicki *et al.*, 2006), targeting nsP2 for nuclear transport (Peranen, 1991; Peranen *et al.*, 1990) including nuclear export via the CRM1-dependent nuclear export pathway (Montgomery and Johnston, 2007), and RNA 5′-triphosphatase activity (Vasiljeva *et al.*, 2000). Recently, Akhrymuk *et al.* (2012) showed that the nsP2 of SINV, Semliki Forest virus (SFV) and chikungunya virus

(CHIKV) can inhibit host cellular protein transcription by degradation of Rpb1, a key subunit of the RNAPII complex for the host RNA synthesis. The sequence which determines these functions is located in the C-terminal domain but outside the helicase and protease domains (Garmashova *et al.*, 2006).

nsP2 has been shown to play an important role in suppression of the type I interferon response in alphavirus infected cells (see Chapter 4 for details of interferon evasion by alphaviruses). The infection of NIH 3T3 cells with a strain of SINV with the mutation nsP2-P726G causes higher levels of IFN secretion and the activation of 170 cellular genes that are induced by IFN and/or virus replication (Frolova *et al.*, 2002). Additionally, a large proportion of nsP2 of SFV has been found to localize in the nucleus. If the nuclear localization sequence in SFV is inactivated, the resulting virus was found to induce a significantly more robust IFN response in infected cells compared with parental virus (Breakwell *et al.*, 2007), further supporting a role for nsP2 in IFN antagonism.

nsP3

The role of the nsP3 in alphavirus replication is not fully understood. The nsP3 protein comprises three domains, the first being highly conserved among many viruses including alphaviruses, coronaviruses, hepatitis E virus and rubella virus (Koonin and Dolja, 1993). The second domain is conserved among alphaviruses and the third C-terminal domain is hypervariable (Strauss and Strauss, 1994). nsP3 is a phosphoprotein with the phosphorylation sites mapped to serine and threonine residues located predominantly in the hypervariable domain (Li *et al.*, 1990; Peranen, 1991; Vihinen and Saarinen, 2000). The phosphorylation of this domain is thought to play a role in negative-strand RNA synthesis (De *et al.*, 2003), with nsP3 having been shown to play a role in subgenomic 26S and negative-strand RNA synthesis (LaStarza *et al.*, 1994; Lemm and Rice, 1993; Wang *et al.*, 1994). It has been proposed that it acts in concert with nsP1 to mediate association of the replication complex with cytoplasmic membrane structures (Peranen, 1991). It is also known to affect the cleavage specificity of the nsP2 proteinase (de Groot *et al.*, 1990).

Several studies have demonstrated that the hypervariable region of nsP3 might modulate replication of SINV *in vitro* (LaStarza *et al.*, 1994) as well as virulence and pathogenesis of SFV in the mouse model. (Vihinen *et al.*, 2001). It was suggested that for the New World alphavirus VEEV, the hypervariable domain may reflect an adaptation to growth in different hosts or vectors species (Oberste *et al.*, 1996).

nsP4

nsP4 is believed to act as the RNA polymerase of the alphaviruses. It contains the GDD motif characteristic of viral RNA polymerases (Kamer and Argos, 1984), and its concentration is tightly regulated in most alphaviruses by both read-through of the opal codon at the end of nsP3 and efficient degradation by the N-end rule pathway (Strauss and Strauss, 1994). It has been shown that a small fraction of nsP4 remains stable in infected cells, suggesting that nsP4 is protected from rapid degradation while it is associated with the replicative complex.

Alphaviral structural region

Capsid protein

The subgenomic 26S RNA of the alphavirus translates to a single polyprotein precursor, which is cleaved to produce structural or virion proteins in the order of capsid-PE2 (E3 + E2)-6K/TF-E1 (Melancon and Garoff, 1987; Strauss and Strauss, 1994; Zhang *et al.*, 2002) in the cytoplasm of infected cells. The first structural protein to be translated is the capsid protein, whose size ranges from 258 to 275 residues (Strauss and Strauss, 1994). The capsid protein is shown to be located at the N-terminal of the polyprotein and functions as an autoprotease, with His131, Asp163 and Ser215 forming a serine protease catalytic triad essential for the rapid cleavage from the nascent polyprotein (Aliperti and Schlesinger, 1978; Boege *et al.*, 1981; Hahn and Strauss, 1990). Following autoproteolysis, the C-terminal residue, Trp264, remains in the protease active-binding site, thus preventing further protease activity (Strauss and Strauss, 1994).

The alphavirus nucleocapsid is made up of an internal core of viral RNA and a protein region composed of capsid proteins, arranged in a $T=4$ icosahedral lattice (Choi *et al.*, 1991). Capsid proteins bind to viral genomic RNA, including the large ribosomal subunit, and this binding process is important in nucleocapsid assembly or disassembly (Geigenmuller-Gnirke *et al.*, 1993; Wengler and Wurkner, 1992). The binding of capsid protein and viral genomic RNA is facilitated by a conserved sequence element (residues 81–113) on the surface of alphavirus capsid proteins (Choi *et al.*, 1991; Wengler and Wurkner, 1992). Deletion analysis also identified a 32-amino-acid recognition element (residues 76–107) in capsid proteins that govern specific binding to viral genomic RNA (Geigenmuller-Gnirke *et al.*, 1993). In the process of nucleocapsid assembly, a specific 132 nucleotide sequence in SINV genomic RNA is believed to promote the encapsidation of RNA into the nucleocapsid core by binding to the recognition sequence on capsid proteins (Geigenmuller-Gnirke *et al.*, 1993; Weiss *et al.*, 1989, 1994).

Envelope glycoproteins

Following capsid cleavage into the cytoplasm, the envelope structural polyprotein translocates across the endoplasmic reticulum (ER) via an N-terminal signal sequence on E3 glycoprotein (Bonatti *et al.*, 1984; Liljestrom and Garoff, 1991). The structural polyprotein is cleaved by cellular enzyme signalase within host ER to yield pE2, 6K and E1 proteins. E1–pE2 heterodimers are assembled and three heterodimers are oligomerized in the ER, forming 80 immature trimeric spikes (Lobigs *et al.*, 1990; Mulvey and Brown, 1996). The glycoprotein E3 is then cleaved from pE2 by a furin-like protease in the Golgi apparatus, rendering the 80 trimeric spikes fusogenic with the nucleocapsid (de Curtis and Simons, 1988).

The alphavirus E2 is a 432-amino-acid long transmembrane protein primarily responsible for receptor binding and entry into infected cells (Jose *et al.*, 2009; Mukhopadhyay *et al.*, 2006). E2 from SINV has been shown to carry major neutralizing epitopes (Dalrymple *et al.*, 1976). In addition, the C-terminal region of E2 glycoprotein has been reported to act as a signal sequence for 6K protein in SFV (Liljestrom and Garoff, 1991). A 33-amino-acid endodomain of E2, also known as the cytoplasmic domain (cdE2), is known to associate with the assembled nucleocapsid through a hydrophobic cleft in the capsid protein (Lee and Brown, 1994; Owen and Kuhn, 1997). Recently, Jose *et al.* (2012) reported that the mutations of the C-terminal cdE2 signal sequence in SINV can adversely affects virus budding.

E1 is a 439-amino-acid transmembrane protein which is involved in mediating viral and cellular membrane fusion (Dubuisson and Rice, 1993; Garoff *et al.*, 1980). The alphavirus E1 also plays a role in converting viral surface proteins into ion-permeable pores at the endosomal membrane during viral entry into infected cells (Wengler *et al.*, 2003). These pores allow exchange of endosomal proteins into the cytoplasm, leading to a low-pH environment, facilitating nucleocapsid disassembly (Glomb-Reinmund and Kielian, 1998).

E3 is a small, extracellular protein of approximately 64 amino acids that plays a critical role in folding of pE2, formation of pE2–E1 complex, and in mediating functional spike formation during virus maturation (Parrott *et al.*, 2009; Snyder and Mukhopadhyay, 2012). When E3 was replaced with an ER signal sequence, the heterodimerization of E1–E2 spike formation was inhibited (Lobigs *et al.*, 1990). Furthermore, E1 and E2 are retained in the ER, suggesting that E3 is crucial in pE2 folding and in E1–E2 spike assembly. E3 has been hypothesized to exhibit enzymatic functions in the process of virus assembly but there has been limited evidence available (Parrott *et al.*, 2009). Interestingly, E3 is non-covalently associated with the spike E1–E2 heterodimers in SFV, but is not a component of other alphavirus such as CHIKV, SINV and WEEV (Simizu *et al.*, 1984).

6K protein

6K is a small, cysteine-rich protein, having a molecular mass of 6000 Da, generated by the proteolytic cleavage of structural polyprotein by signalase (Strauss and Strauss, 1994). It is comprised of approximately 60 hydrophobic amino acid residues, some of which are acylated, conferring its hydrophobicity and ability to interact with the host cell membrane. Unlike mature glycoproteins E1 and E2, which are mostly incorporated into newly synthesized virion, 6K is incorporated in sub-stoichiometric amounts, approximately 7–30 copies per virion (Gaedigk-Nitschko and Schlesinger, 1990; Lusa *et al.*, 1991). 6K is primarily involved in protein translocation, membrane permeability, virus assembly and budding. The C-terminal domain of 6K serves as signal sequence for translocation of glycoprotein E1 (Strauss and Strauss, 1994) and pE2 (Lusa *et al.*, 1991) to the cell surface. However, it has been shown that the translocation of glycoprotein E1 by 6K is independent of the signal sequence on E2.

6K is also a viroporin (viral proteins that enhance membrane permeability to ions or small molecules through membrane destabilization contributed by the presence of basic amino acids). 6K proteins have been implicated in the formation of ion channels on planar lipid bilayers (Gonzalez and Carrasco, 2003). The ability of 6K proteins to form ion channels may result in disruption of host cell membrane, enhancing the entry of virions.

More importantly, 6K has been reported to facilitate virus assembly and budding. A site directed mutagenesis study performed on Sindbis virus has shown that mutations at regions between E2 and 6K, and 6K and E1 lead to defects in proteolytic cleavage by signalase, which results in delayed release of newly synthesized virus particles (Gaedigk-Nitschko and Schlesinger, 1991). Recently, the deletion of 6K from SFV resulted in a mutant with altered spike structure, which decreased its ability to fuse with liposomes, subsequently affecting the viral budding processes (McInerney *et al.*, 2004).

Alphavirus replication

Attachment and entry

Direct microscopic examination provides a fast and effective means of viewing the physical interaction between the virus and host cells. While early electron microscopy studies demonstrated that alphaviruses reside in the endocytic compartment and cytopathic vacuoles of host cells (Froshauer *et al.*, 1988; Marsh *et al.*, 1983) current microscopic technology does not allow accurate dissection of how alphaviruses enter and exit the plasma membrane during infection. (Haywood, 2010; Kononchik *et al.*, 2011a,b). The exact pathway by which alphaviruses attach and penetrate the host cell and the subsequent release of the viral genome from its capsid remain somewhat controversial (Kononchik *et al.*, 2011a) despite the fact that the process has been extensively studied.

Receptor recognition of alphaviruses

To initiate infection, alphaviruses scan the cell surface and probe for the site of attachment. A range of cell surface molecules/receptors have been reported to be involved in alphavirus binding including C-type lectins [dendritic cell-specific ICAM3-grabbing non-integrin (DC-SIGN) and liver/lymph node-specific ICAM3-grabbing non-integrin (L-SIGN)] (Klimstra *et al.*, 2003), heparan sulfate (Heil *et al.*, 2001; Klimstra *et al.*, 1998), major histocompatibility antigen (MHC)-class I (Helenius *et al.*, 1978), integrin a1b1 (La Linn *et al.*, 2005), heat shock protein 70 (Ryan *et al.*, 1992) and the high-affinity laminin receptor (Wang *et al.*, 1992). The viral factor that mediates binding of the virion to the cell surface is a protein sequence located on E2 (Kielian *et al.*, 2010; Kononchik *et al.*, 2011a). The N-linked glycosylated nature of E2 has been reported to influence receptor-mediated host responses. This strategy has been proposed to allow virus to achieve effective infection and subsequent propagation in host cells. For instance, mosquito cell-derived alphaviruses are rich in high-mannose structures on envelope proteins (e.g. E2) and consequently have an increased affinity for DC-SIGN/L-SIGN. *In vitro* infection with these viruses has been shown to dampen the mammalian host cell immune defence mechanism and in turn facilitate viral survival/replication (Klimstra *et al.*, 2003; Shabman *et al.*, 2007). Taken together, alphaviruses are able to persist in the transmission cycle between different animal species by using a broad spectrum of receptor recognition molecules enabling then to infect a variety of cells and possibly by modulating the host antiviral response.

Alphavirus entry: the membrane fusion model

The classical model for alphavirus entry is host receptor-mediated endocytosis via a clathrin-dependent pathway, followed by low-pH-triggered membrane fusion between host and virus membranes in the endosomal compartment, akin to the model described for influenza entry and uncoating. (Bernard *et al.*, 2010; DeTulleo and Kirchhausen, 1998; Helenius *et al.*, 1980; Kielian, 1995; Kielian *et al.*, 2010; Kielian and Helenius, 1985; Marsh and Helenius, 1980; Smit *et al.*, 1999; Strauss and Strauss, 1994; White and Helenius, 1980). The pH dependence suggests that alphaviral particles initially reside in early endosomes, which are mildly acidic (pH 5.3–6.2). In contrast to E2, which is involved in receptor binding, the viral E1 protein is involved in the membrane fusion reaction (Boggs *et al.*, 1989; Garoff *et al.*, 1980; Kielian and Helenius, 1985; Omar and Koblet, 1988). E1 appears to be a β-sheet which contains a central domain termed domain I (DI). This domain connects on one side to domain II (DII) by a hydrophobic fusion loop and on the other side of the sheet connects to domain III (DIII) via a linker protein. In the model of membrane fusion (Gibbons *et al.*, 2004; Li *et al.*, 2010; Roussel *et al.*, 2006; Voss *et al.*, 2010) following receptor binding and uptake of the virion the E1–E2 heterodimer undergoes a conformational change within endosomes (pH <6) resulting in dissociation of the E1–E2 dimer (Fields and Kielian, 2013; Wahlberg *et al.*, 1989, 1992). While E2 trimers are likely to be disrupted in the low pH environment (Cao and Zhang, 2013) the E1 fusion loop is then exposed. This conformational change has been shown to correspond to the rearrangement of the viral spike, which is a trimer of heterotrimers $(E1/E2/E3)_3$ or heterodimers $(E1/E2)_3$ depending on the type of virus. Subsequently, monomeric E1 is inserted into the target membrane through a cholesterol-dependent mechanism (Kielian and Helenius, 1984). E1 monomers then refold through DI and DII and form a stable post-fusion homotrimeric structure (Gibbons *et al.*, 2004; Wahlberg *et al.*, 1992). DIII and the stem region are folded back against the core trimer. This is followed by distortion of the target membrane leading to hemifusion where two outer leaflets meet and merge. Finally, the force of this cooperative action results in the formation of a fusion pore, with the fusion loops and transmembrane domains on the same side of trimer (reviewed in Sanchez-San Martin *et al.*, 2009). Membrane fusion is believed to be the process by which alphaviruses release their capsid and genomic material into the cytoplasm to gain the access to host's molecular machinery.

Blocking acidification of endosomes with inhibitors such as dynamin, eps15 and rab-5 has been the primary method to study the endocytic pathway of viral uptake (DeTulleo and Kirchhausen, 1998; Sieczkarski and Whittaker, 2002, 2003). Similarly, weak bases like ammonium chloride, chloroquine, the ionophore monensin, and concanamycin A, a vacuolar proton pump inhibitor, can all block the endocytic pathway (Glomb-Reinmund and Kielian, 1998; Helenius *et al.*, 1980, 1982; Irurzun *et al.*, 1997; Kielian *et al.*, 1986; Marsh *et al.*, 1982). However, data from these studies are somewhat questionable as most inhibitors are not specific to the endocytic pathway and may also affect other pathways in viral replication. For example, ammonium chloride can inhibit steps in the non-structural protein processing at the initial phase of RNA translation and chloroquine has been shown to directly affect viral production (Coombs *et al.*, 1981; Hernandez *et al.*, 2001; Kononchik *et al.*, 2011a). Therefore, data generated from these studies should be interpreted with caution as a combination of methodologies may be required to delineate endosomal function in viral replication with confidence.

Alternative pathway to membrane fusion

In contrast to the conventional view of alphavirus entry as previously described, increasing evidence suggests that other mechanisms for viral entry exist. For instance, SINV does not require exposure to low pH to enter mosquito cells (Cassell *et al.*, 1984; Coombs *et al.*, 1981; Hernandez *et al.*, 2001) and fuses rapidly with receptor-free liposomes in mammalian/mosquito cells (Smit *et al.*, 1999). The vacuolar proton pump inhibitor bafilomycin A1, which blocks endocytosis does not prevent alphaviral entry (Hunt *et al.*, 2011). Only minimal blockage of CHIKV infection can be achieved when dominant-negative mutants of Esp15, Rab5 and chemical inhibitors are used (Bernard *et al.*, 2010). CHIKV can enter host cells by bypassing the endocytic pathway (Solignat *et al.*, 2009). The presence of cholesterol and sphingolipid, which form 'rafts' in the plasma membrane, is required for membrane fusion and raft domains have been shown to promote alphavirus fusion (Bron *et al.*, 1993; Corver *et al.*, 1995; Kielian and Helenius, 1984; Moesby *et al.*, 1995; Nieva *et al.*, 1994; Phalen and Kielian, 1991; Smit *et al.*, 1999; White and Helenius, 1980; Wilschut *et al.*, 1995). Cholesterol in particular has been proposed to mediate alphavirus binding to target membranes (Nieva *et al.*, 1994). Sphingolipids, on the other hand, are considered as co-factors involved in the E1 conformational change (Nieva *et al.*, 1994; Samsonov *et al.*, 2002). However, under various physiological conditions, cholesterol and sphingolipid are not required for alphaviral infection (Brown and Hernandez, 2012; Kononchik *et al.*, 2011a). Using techniques such as thin sections and freeze-fracture, it has been shown that SINV gains entry into cells without endocytosis and acidification. Interestingly, at low temperatures alphaviruses can infect cells despite inhibition of endocytosis and membrane fusion (Wang *et al.*, 2007). The virus-liposome model of membrane fusion describes a non-leaky process of viral entry; however, alphavirus entry into the host cells is leaky as it requires pore-like structures. This raises the possibility that membrane fusion induced under artificial conditions may not be physiologically relevant to the infection process (Vancini *et al.*, 2013). Taken together, alphaviruses are capable of releasing their genome into cytoplasm through an alternative mechanism, which involves pore formation on the host cell membrane (Kononchik *et al.*, 2011a). More recently, it has been shown that alphaviruses can directly release their genome at the cell membrane, however this is a time- and temperature-dependent process (Kononchik *et al.*, 2011a).

Alphavirus capsid disassembly

For release of the viral genome, the nucleocapsid needs to be disassembled and this process remains largely hypothetical. Since the nucleocapsid of alphaviruses binds to ribosomes it has been proposed that the ribosome competes with viral RNA for capsid protein, which results in viral disassembly and release of the viral genome (Glanville and Ulmanen, 1976; Ulmanen *et al.*, 1976, 1979; Wilschut *et al.*, 1995). Studies using SINV have identified the region between residues 94 and 105 of capsid as being involved in this interaction (Wengler and Wurkner, 1992). Alternatively, it has been proposed that exposure of the viral particle to low pH in the endosome can trigger the formation of ion channels in the viral particle, resulting in proton entry that subsequently destabilizes and disassembles the nucleocapsid (Lanzrein *et al.*, 1993).

Replication of alphaviruses

The alphavirus genome consists of single-stranded, positive-sense RNA with a 5′ cap structure and 3′ polyadenosine tail. The genome is divided into two major regions or cistrons: the first, located in the 5′ two-thirds of the genome, encodes the viral replication machinery, termed non-structural proteins (nsPs), whereas the structural proteins that form virus particles are encoded in the second cistron (Strauss and Strauss, 1994). Soon after the viral genome is released into the host cell, a minus-strand RNA is synthesized. The minus-strand synthesis stops abruptly 4–6 hours post infection, after which only plus-strand RNAs are produced (Shin *et al.*, 2012; Shirako and Strauss, 1994; Vasiljeva *et al.*, 2003). The processing of the non-structural proteins and polyproteins, which form part of the viral polymerase component, is a very complex and highly regulated process during replication. The genome is translated from two open reading frames (ORFs) to generate the non-structural (P1234) and structural polyproteins. Immunoprecipitation studies have shown that in the early steps of replication, P1234 is cleaved between the nsP3 and nsP4 in a cis fashion to give rise to P123 and P4 (de Groot *et al.*, 1990; Takkinen *et al.*, 1991). These two proteins form a replication complex, which is unable to process negative-strand RNA (De *et al.*, 1996; Lemm *et al.*, 1994; Shirako and Strauss, 1994). When the P123 polyprotein reaches high concentrations, it is cleaved into nsP1 and nsP23. This cleavage is mediated by a proteinase in the nsP2. Early studies have shown that mutations inactivating the nsP2 proteinase, resulting in uncleaved P123, lead to accumulation of minus-strand RNAs and reduced accumulation of genomic and subgenomic plus-strand RNAs.

When the P123 polyprotein reaches significantly high concentrations it is cleaved into nsP1 and nsP23. Together with nsP4, these two proteins form a replication complex that can synthetize negative strand, as well as genomic RNA, but not subgenomic RNA. This first replication complex is often found in cytopathic vacuoles (CPVs), which contain late endosomal and lysosomal markers (De *et al.*, 1996; Froshauer *et al.*, 1988; Kujala *et al.*, 2001; LaStarza *et al.*, 1994; Salonen *et al.*, 2003). The cytosolic surface of CPVs consists of regular membrane invaginations or spherules, which are the sites of RNA synthesis. To understand how these vacuoles form, Salonen *et al.* (2003) designed experiments in which the individual SFV non-structural proteins nsP1 to nsP4 were expressed in different combinations as well as their precursor polyprotein P1234 and its cleavage intermediates. They found that nsP1 domain is responsible for the membrane association of the non-structural polyprotein, whereas the nsP1 domain, together with the nsP3 domain, targets it to the intracellular endosomal vesicles (Salonen *et al.*, 2003). When complete cleavage of structural polyproteins into nsP1, nsP2, nsP3 and nsP4 is achieved, positive-strand genomic and subgenomic synthesis commences (Lemm *et al.*, 1994; Shirako and Strauss, 1994). The functions of the individual nsPs as polymerase components have been studied with the use of temperature-sensitive mutants. The nsP1 is needed for the initiation of minus-strand synthesis, the nsP2 protein contains the non-structural protease activity and regulates 26S RNA synthesis, and the nsP4 protein is the viral RNA polymerase. The role of the nsP3 protein is not completely clear, but it has been shown to be important for virus RNA replication in mosquito cells.

Host proteins are also involved in RNA synthesis, although the mechanisms have not been described. The first evidence for this came from studies that reported that mutations in viral RNA promoter elements affect RNA replication differently in different cells (Niesters and Strauss, 1990). It was later found that the 3′ end of the viral minus strand binds mosquito cellular proteins and that this interaction is an important step in the initiation of RNA replication (Pardigon *et al.*, 1993; Pardigon and Strauss, 1992). Viral structural proteins are also believed to play a role in alphavirus synthesis. For example, the capsid protein of SINV binds to the viral RNA, and deletions that affect this binding result in impaired viral replication (Weaver *et al.*, 1992). There is also evidence that E2 protein can also regulate RNA synthesis, but this regulation is host cell dependent (Vrati *et al.*, 1986).

The structural domain is translated as a polyprotein from a subgenomic mRNA, the 26S mRNA. This initial structural polyprotein is processed to give rise to the envelope proteins, the capsid protein and the two small polypeptides E3 and 6K. Cleavage of structural proteins begins with the autoproteolytic cleavage of the capsid protein (Cp), at its carboxy terminus, from rest of the polyprotein (Aliperti and Schlesinger, 1978; Garoff *et al.*, 1978). Once Cp has been released into the cytoplasm, polyprotein synthesis continues and is associated with ER membranes (Garoff and Li, 1998; Kuhn *et al.*, 1996; Strauss and Strauss, 1994). After Cp cleavage, the exposed amino terminus (part of the E3 protein) contains a signal sequence that interacts with ER membranes and directs the glycoprotein precursor (E3–E2–6K–E1) into the lumen of the ER vesicles, where it then becomes associated with the ER membrane (Strauss and Strauss, 1994). Cellular proteases then cleave this precursor at both ends of the 6K protein, generating the products PE2 (E3–E2), 6K and E1. PE2 and E1 then associate with each other to form dimers that travel with 6K through the vesicular system to the plasma membrane. In addition, the PE2–E1 dimers oligomerize, with the result that the dimers adopt a trimeric structure (Andersson *et al.*, 1997; Barth *et al.*, 1995; Duffus *et al.*, 1995). Later the PE2 precursor is cleaved at its luminal domain to give rise to E2 and E3 proteins (de Curtis and Simons, 1988; Sariola *et al.*, 1995; Ziemiecki *et al.*, 1980). As a result, the E1–E2 interaction is weakened, rendering the fusion peptide ready for activation at the right pH conditions (Lobigs and Garoff, 1990). Newly synthesized full-length genomic RNA is recognized by and associates with the Cp protein to form nucleocapsid-like particles (Owen and Kuhn, 1996; Weiss *et al.*, 1989). Cp recognizes a specific sequence towards the 5′ end of the genome, limiting the packaging to full-length genomic RNA only (Owen and Kuhn, 1996; Weiss *et al.*, 1989). The assembled nucleocapsids subsequently interact with the carboxy domain of E2. This interaction provokes the wrapping of the capsid by the lipid envelope, concomitant with the budding of virus particles (Metsikko and Garoff, 1990; Vaux *et al.*, 1988).

Alphavirus assembly and egress

Nucleocapsid processing and assembly

Alphaviral nucleocapsids, which assemble in the cytoplasm of infected cells, can be observed by electron microscopy and isolated from infected cells. Nucleocapsids have an icosahedral structure composed on 240 copies of capsid protein in a $T=4$ symmetry and contain a single copy of viral RNA (Jose *et al.*, 2009; Strauss and Strauss, 1994). Although the majority of alphaviruses produce nucleocapsid cores that have a $T=4$ symmetry, capsids with a $T=1$, $T=3$ and $T=7$ symmetry have also been described (Strauss and Strauss, 1994). Efficient assembly of nucleocapsid cores and packaging of genomic RNA is dependent on both capsid–capsid interactions and capsid–RNA interactions. It has been found that the C terminal two-thirds of the capsid protein contains the information required for these interactions. Capsid–RNA interactions are mediated by a packaging signal on the 5′ two-thirds of viral genomic RNA and the presence of positively charged amino acids in the capsid protein. Packaging signals are specific in that they direct encapsidation of only viral genomic RNA despite the abundance of other forms of RNA such as sub-genomic and cellular RNA in the cytoplasm and, to date, capsids devoid of RNA have not been reported (Strauss and Strauss, 1994) One exception to this is Aura virus where sub-genomic 26s mRNA is also packaged into either $T=4$ or $T=3$ capsids (Strauss and Strauss, 1994). Studies using SINV identified a 132-nt region in nsp1 (nt 945–1076) as the location of the RNA packaging signal (PS) (Frolova *et al.*, 1997; Weiss *et al.*, 1989, 1994). Further studies, using VEEV, identified the structure of the PS as a 4 to 6 stem–loop containing conserved GGG sequences at the base of the loop (Kim *et al.*, 2011) and that the presence of more than one stem–loop structure containing GGG motifs was required for encapsidation of the RNA. Conversely, mutation of this structure or the GGG sequences led to inefficient packaging and packaging of sub-genomic RNA. Organization of the PS is conserved amongst alphaviruses. The location of the packaging signal for the majority of alphaviruses is in nsp-1 and the structural and functional elements of the PS are conserved amongst alphaviruses. However, for viruses of the SFV clade the PS is located in nsp-2. Interestingly, the capsid protein of viruses from the SFV clade are still able to use the nsp-1 PS of other alphaviruses for encapsidation of RNA (Kim *et al.*, 2011).

The viral capsid gene contains three regions, which have been delineated based on function. The region of the capsid involved in binding of the viral RNA PS for SINV is amino acids 81–112, corresponding to region II; however, other regions of the capsid protein are also involved in nucleocapsid assembly (Weiss *et al.*, 1989, 1994). Region I (amino acids 1–80) plays a role in non-specific binding and charge neutralization with viral genomic RNA. This region also contains a series of 18 amino acids, which are highly conserved among alphaviruses, and which form a helix (helix I) that plays a role in nucleocapsid core assembly. Region III, the C-terminal domain (amino acids 114–264), is a serine protease domain that forms the capsomeres on the surface of the nucleocapsid core (Hong *et al.*, 2006).

Capsid–capsid interactions involving regions I and II of the capsid protein are required to generate a complete icosahedral core structure. Following the nucleation event whereby a capsid monomer binds to the RNA PS, a conformation change in the PS exposes a second binding site where an additional Cp molecule binds and Cp dimerization events occur involving region II (Linger *et al.*, 2004; Warrier *et al.*, 2008). The Cp–Cp interactions are mediated by the helix I region, which stabilizes the Cp dimers through coiled-coil interactions (Hong *et al.*, 2006). The exact details of the subsequent steps in the formation of the complete nucleocapsid core, oligomerization and formation of the $T=4$ icosahedral shell remain elusive.

Structural protein processing and assembly

Glycoprotein spike assembly requires the interaction of all the structural proteins including E3, E2, 6K, TF and E1. As described previously, the structural proteins are translated as a single polyprotein in the cytoplasm. Auto-cleavage of capsid exposes the 5′ end of E3, which acts as a signal to translocate the remaining polyprotein (pE2–6K–E1) to the endoplasmic reticulum (ER). Once in the ER the polyprotein is further cleaved by signalase to generate pE2 (E2 + E3), 6K and E1. Within the ER, pE2 and E1 form heterodimers through pairing interactions between the helices of their transmembrane (TM) regions. The heterodimers then oligomerize to form a trimer, the immature non-fusogenic spike complex, which is transported to the plasma membrane via the host secretory pathway, where it is further modified (Tang *et al.*, 2011). Importantly, E2 is cleaved from E3 by a cellular furin to yield a fusogenic spike, an essential step in the production of infectious virions. Although many groups have shown the necessity of E3 for particle assembly (Liljestrom and Garoff, 1991; Lobigs *et al.*, 1990) recent studies using chimeric viruses containing E2 and E3 genes from different virus species identified specific interactions between E2 and E3 that are crucial for folding of E2 in glycoprotein spike formation and assembly suggesting a new role for E3 (Snyder and Mukhopadhyay, 2012). At the plasma membrane the trimeric spikes arrange in hexagonal array structures through lateral spike–spike interactions.

Palmitoylation of the envelope glycoproteins occurs at cysteine residues in both E1 and pE2 proteins and at residues in the transmembrane and cytoplasmic tail regions. In vertebrate cells palmitoylation occurs after egress from the ER but prior to entry into the Golgi and post oligomerization of the E1 and E2 proteins. In contrast, in mosquito cells palmitoylation occurs on E2 at an earlier stage as pE2 is cleaved earlier in these cells. A single palmitoylation site in E1 has been identified for both SFV (C-433) and SINV (C-430), located in the transmembrane region of E1. Other alphaviruses such as EEE and VEE lack cysteine residues in the E1 region and are presumably not palmitoylated. For alphaviruses, pE2 can contain four to six palmitoylation sites. Using mutagenesis studies, four to six palmitoylation sites have been identified for SINV pE2 – three in the cytoplasmic domain, which are conserved amongst all alphaviruses examined to date, and one in the transmembrane region. Similarly, although identified by mass spectrometry studies, SFV pE2 contains four palmitoylation sites in the same regions as for SINV (Kordyukova *et al.*, 2010; Schmidt *et al.*, 1988; Strauss and Strauss, 1994). The molecules added during fatty acid acylation of the envelope proteins are either palmitic acid or stearic acid, and it is thought that these molecules are added to cytoplasmic and transmembrane cysteine residues, respectively (Kordyukova *et al.*, 2010). However, the type of molecule added to envelope glycoproteins varies between alphaviruses. The 6K protein is palmitoylated at four sites. Mutagenesis of these sites in SINV resulted in defective assembly and budding of new virions, but transport of the glycoproteins to the cell membrane was not disrupted, demonstrating that the 6K protein and its state of acylation are critical for assembly and budding (Gaedigk-Nitschko and Schlesinger, 1990).

Glycosylation of the envelope glycoproteins takes place in the lumen of the ER where high-mannose carbohydrates or complex oligosaccharides are added to the proteins. Modification of the carbohydrate chains takes place as the proteins transit through the Golgi apparatus. Glycosylation is required for correct folding, solubility and transport of the proteins, as ablation of glycosylation results in failure of virus production (Leavitt *et al.*, 1977). The type of sugar added to glycoproteins is dependent on the cell type and the growth state of the cells. For invertebrate cells all sites contain high-mannose sugars, whereas for vertebrate cells the addition of complex and high mannose sugars is dependent on the accessibility of the glycosylation site to processing enzymes in the Golgi apparatus (Knight *et al.*, 2009). For all alphaviruses, both pE2 and E1 are glycosylated, such as SINV where each of the glycoproteins has two N-linked glycosylation sites, at position 139 and 245 for E1 and 196 and 318 for E2 (Knight *et al.*, 2009); however, the number, position and type of carbohydrate chain attached are not conserved among alphaviruses (Strauss and Strauss, 1994).

Transport of viral components

Although it is established that interactions between capsid and glycoproteins occur at the plasma membrane, transport of these two components to the plasma membrane remains unresolved. Recent evidence suggests that interactions between these two components may occur prior to their arrival at the plasma membrane. Studies using SFV demonstrated that cytopathic vacuoles type II (CPV-II), which are derived from the *trans*-Golgi network and are the predominant vacuole type at late stages of infection, contain viral glycoproteins, and when CPV-II are not present budding does not occur (Garoff *et al.*, 1994; Soonsawad *et al.*, 2010). Using electron tomography and immune-electron microscopy Soonsawad *et al.* (2010) demonstrated that E1/E2 glycoproteins are arranged in CPV-II in a helical tubular array akin to the arrangement of E1/E2 glycoprotein trimers on the virion. Further evidence for the role of these structures in transport of envelope glycoproteins to the plasma membrane and viral budding is their location adjacent to the site of budding for alphaviruses (Soonsawad *et al.*, 2010). These studies also showed the presence of dense particles, similar in size and appearance to nucleocapsids on the CPV-II, suggesting that nucleocapsids may also be transported to the plasma membrane on these structures (Snyder *et al.*, 2012; Soonsawad *et al.*, 2010). Further evidence supporting the association of capsid and E2 in the cytoplasm is provided by studies performed using fluorescently tagged SINV capsid and live imaging where capsid was observed co-localized with E2 as motile internal puncta in the cytoplasm (Zheng and Kielian, 2013). These studies also demonstrated cytoplasmic association of capsid with G3BP and viral nsp3, proteins known to be involved in the regulation of stress granule formation and the presence of capsid at the plasma membrane, also associated with E2 (Zheng and Kielian, 2013). However, further studies are required to fully define the transport of both structural components of the virion to the site of budding.

Budding

Budding is the final process in the alphaviral life cycle whereby infectious virions are released from the plasma membrane of an infected cell. In this process virus nucleocapsid cores and glycoproteins come together and the virus egresses from the cell, obtaining a cell-derived lipid membrane studded with 240 copies of the E1 and E2 glycoproteins and substoichiometric amounts of 6K as it departs. The envelope glycoproteins are arranged as 80 trimers in an icosahedral $T=4$ symmetry constituting the outer shell of the virion. The mature virion also contains an inner shell, the nucleocapsid containing 240 copies of capsid protein and a single copy of viral genomic RNA, also arranged in a $T=4$ icosahedral symmetry. Clustering of nucleocapsid cores at the sites of budding on the plasma membrane and interaction between the cores and glycoprotein spikes is required for budding.

Two models of alphavirus budding have been proposed. The first, termed the 'membrane wrapping' model, entails capsid protein and genomic RNA forming nucleocapsid cores in the cytoplasm, which move to the plasma membrane where they interact with the glycoprotein spike, specifically E2, which drives budding of virions from the plasma membrane. The second model describes the loose association of the capsid protein and genomic RNA at the plasma membrane, whereby nucleocapsid core formation and budding are orchestrated by lateral interaction of the envelope glycoproteins in the plasma membrane (Garoff *et al.*, 2004)

Interactions between capsid and glycoproteins

Specific interactions between the nucleocapsid core and the envelope glycoproteins, in particular the cytoplasmic domain of E2 (cdE2), are essential for budding, and this interaction is thought to provide the energy required for budding. It is suggested that the glycoprotein spikes of E1–E2 heterodimers arrive at the plasma membrane and form a crystalline array structure and associate with capsid in a 1:1 ratio.

The capsid protein has a chymotrypsin-like fold containing a hydrophobic pocket in the carboxyl-terminal domain, which is the site of interactions with the cdE2 glycoprotein. cdE2 contains a helix–loop–helix structure and the loop of this structure fits into the hydrophobic pocket on the capsid protein and binds Cp, which stabilizes the virion. Initiation of the interaction between cdE2 and Cp is facilitated by charged residues on cdE2-RI and subsequent orientation of the cdE2-RII loop towards the Cp pocket (Tang *et al.*, 2011). The cdE2 loop region contains a YXL (Tyr-X-Leu) motif (residues 400–402) which has been identified as its binding site to the capsid pocket. The tyrosine (Y400) residue is conserved across all alphaviruses and has been identified as the most important residue in these interactions (Garoff *et al.*, 2004; Jose *et al.*, 2009, 2012). Binding of cdE2 in the Cp pocket is assisted by palmitoylated cysteine residues in cdE2. Recently, studies on Aura virus also identified P405, a residue in the loop region of cdE2 as a residue that also makes intermolecular hydrophobic contacts with the capsid (Aggarwal *et al.*, 2012). These studies also identified the residues Cys397, Cys418 and Tyr 401 as important as they stabilize the structure of E2 endodomain (Aggarwal *et al.*, 2012). From these data it has been suggested that the cdE2 region contains two determinants for budding, the first being the Y400 residue and the second palmitoylation of cysteine residues (Jose *et al.*, 2012). Interactions between the N terminal region of cdE2 and Cp have also been identified and shown to play a role in assembly and budding (Snyder and Mukhopadhyay, 2012). The role of E1 in budding is not yet resolved; however, interactions between Cp and cdE1 may contribute to stabilizing the structure (Tang *et al.*, 2011).

Cholesterol requirement in budding

Efficient budding of alphaviruses from the plasma membrane requires the presence of cholesterol. Cholesterol depletion affects both viral stability and infectivity. Initial studies using SFV and SINV demonstrated that budding efficiency was greatly reduced in cholesterol depleted cells (Lu *et al.*, 1999; Marquardt *et al.*, 1993). Further mutagenesis studies identified a point mutation in the E1 spike glycoprotein (P226S) of SFV that governed the requirement for cholesterol for efficient viral budding (Vashishtha *et al.*, 1998). Similar studies using SINV also identified that SINV required cholesterol for cellular egress and identified the E1 226 region as being responsible for cholesterol dependence (Lu *et al.*, 1999). Interestingly, budding of virus particles from mosquito cells is not affected by depletion of cholesterol.

Role of 6K and TF proteins in budding

Two additional structural proteins, 6K and the recently described TF, which is produced as a consequence of a frameshift during 6K translation (Firth *et al.*, 2008; Snyder *et al.*, 2013), may play accessory roles in assembly and budding. 6K protein (described previously) is produced in equivalent amounts to the envelope glycoprotein and travels with these glycoproteins to the plasma membrane where it is incorporated into virions at vastly reduced levels in comparison to E1 and E2. 6K contains two transmembrane domains, the first of which has a role in ion-channel activity and the second of which acts as a signal sequence for E1, and a role for 6K in the stabilization of the E1–E2 heterodimer has been suggested (McInerney *et al.*, 2004). Ablation of 6K decreased budding efficiency of SFV, but not infectivity, suggesting a role for 6K in the packaging of glycoproteins and in interactions with the lipid membrane (Loewy *et al.*, 1995). TF protein contains the same N-terminal region as 6K, including the first transmembrane domain, but has a unique C terminus containing basic amino acid residues. Studies aimed at delineating the roles of 6K and TF in assembly and budding demonstrated that deletion of both 6K and TF in SINV reduced particle release and the reduction was more pronounced than deletion of TF alone, suggesting that the role of TF is associated with assembly or release and that each protein has a unique role in this process (Snyder *et al.*, 2013).

References

Aggarwal, M., Tapas, S., Preeti, Siwach, A., Kumar, P., Kuhn, R.J., and Tomar, S. (2012). Crystal structure of aura virus capsid protease and its complex with dioxane: new insights into capsid-glycoprotein molecular contacts. PLOS ONE 7, e51288.

Akhrymuk, I., Kulemzin, S.V., and Frolova, E.I. (2012). Evasion of innate immune response: the Old World alphavirus nsP2 protein induces rapid degradation of Rpb1, a catalytic subunit of RNA polymerase II. J. Virol. *86*, 7180–7891.

Aliperti, G., and Schlesinger, M.J. (1978). Evidence for an autoprotease activity of sindbis virus capsid protein. Virology *90*, 366–369.

Andersson, H., Barth, B.U., Ekstrom, M., and Garoff, H. (1997). Oligomerization-dependent folding of the membrane fusion protein of Semliki Forest virus. J. Virol. *71*, 9654–9663.

Barth, B.U., Wahlberg, J.M., and Garoff, H. (1995). The oligomerization reaction of the Semliki Forest virus membrane protein subunits. J. Cell Biol. *128*, 283–291.

Bernard, E., Solignat, M., Gay, B., Chazal, N., Higgs, S., Devaux, C., and Briant, L. (2010). Endocytosis of chikungunya virus into mammalian cells: role of clathrin and early endosomal compartments. PLOS ONE *5*, e11479.

Boege, U., Wengler, G., and Wittmann-Liebold, B. (1981). Primary structures of the core proteins of the alphaviruses Semliki Forest virus and Sindbis virus. Virology *113*, 293–303.

Boggs, W.M., Hahn, C.S., Strauss, E.G., Strauss, J.H., and Griffin, D.E. (1989). Low pH-dependent Sindbis virus-induced fusion of BHK cells: differences between strains correlate with amino acid changes in the E1 glycoprotein. Virology *169*, 485–488.

Bonatti, S., Migliaccio, G., Blobel, G., and Walter, P. (1984). Role of signal recognition particle in the membrane assembly of Sindbis viral glycoproteins. FEBS *140*, 499–502.

Breakwell, L., Dosenovic, P., Karlsson Hedestam, G.B., D'Amato, M., Liljestrom, P., Fazakerley, J., and McInerney, G.M. (2007). Semliki Forest virus nonstructural protein 2 is involved in suppression of the type I interferon response. J. Virol. *81*, 8677–8684.

Bron, R., Wahlberg, J.M., Garoff, H., and Wilschut, J. (1993). Membrane fusion of Semliki Forest virus in a model system: correlation between fusion kinetics and structural changes in the envelope glycoprotein. EMBO J. *12*, 693–701.

Brown, D.T., and Hernandez, R. (2012). Infection of cells by alphaviruses. Adv. Exp. Med. Biol. *726*, 181–199.

Cao, S., and Zhang, W. (2013). Characterization of an early-stage fusion intermediate of Sindbis virus using cryoelectron microscopy. Proc. Natl. Acad. Sci. U.S.A. *110*, 13362–13367.

Cassell, S., Edwards, J., and Brown, D.T. (1984). Effects of lysosomotropic weak bases on infection of BHK-21 cells by Sindbis virus. J. Virol. *52*, 857–864.

Cheng, R.H., Kuhn, R.J., Olson, N.H., Rossmann, M.G., Choi, H.K., Smith, T.J., and Baker, T.S. (1995). Nucleocapsid and glycoprotein organization in an enveloped virus. Cell *80*, 621–630.

Choi, H.K., Tong, L., Minor, W., Dumas, P., Boege, U., Rossmann, M.G., and Wengler, G. (1991). Structure of Sindbis virus core protein reveals a chymotrypsin-like serine proteinase and the organization of the virion. Nature *354*, 37–43.

Coombs, K., Mann, E., Edwards, J., and Brown, D.T. (1981). Effects of chloroquine and cytochalasin B on the infection of cells by Sindbis virus and vesicular stomatitis virus. J. Virol. *37*, 1060–1065.

Corver, J., Moesby, L., Erukulla, R.K., Reddy, K.C., Bittman, R., and Wilschut, J. (1995). Sphingolipid-dependent fusion of Semliki Forest virus with cholesterol-containing liposomes requires both the 3-hydroxyl group and the double bond of the sphingolipid backbone. J. Virol. *69*, 3220–3223.

de Curtis, I., and Simons, K. (1988). Dissection of Semliki Forest virus glycoprotein delivery from the trans-Golgi network to the cell surface in permeabilized BHK cells. Proc. Natl. Acad. Sci. U.S.A. *85*, 8052–8056.

Dalrymple, J.M., Schlesinger, S., and Russell, P.K. (1976). Antigenic characterization of two sindbis envelope glycoproteins separated by isoelectric focusing. Virology *69*, 93–103.

De, I., Sawicki, S.G., and Sawicki, D.L. (1996). Sindbis virus RNA-negative mutants that fail to convert from minus-strand to plus-strand synthesis: role of the nsP2 protein. J. Virol. *70*, 2706–2719.

De, I., Fata-Hartley, C., Sawicki, S.G., and Sawicki, D.L. (2003). Functional analysis of nsP3 phosphoprotein mutants of Sindbis virus. J. Virol. *77*, 13106–13116.

DeTulleo, L., and Kirchhausen, T. (1998). The clathrin endocytic pathway in viral infection. EMBO J. *17*, 4585–4593.

Dubuisson, J., and Rice, C.M. (1993). Sindbis virus attachment: isolation and characterization of mutants with impaired binding to vertebrate cells. J. Virol. *67*, 3363–3374.

Duffus, W.A., Levy-Mintz, P., Klimjack, M.R., and Kielian, M. (1995). Mutations in the putative fusion peptide of Semliki Forest virus affect spike protein oligomerization and virus assembly. J. Virol. *69*, 2471–2479.

Faragher, S.G., Meek, A.D., Rice, C.M., and Dalgarno, L. (1988). Genome sequences of a mouse-avirulent and a mouse-virulent strain of Ross River virus. Virology *163*, 509–526.

Fields, B.N., Knipe, D.M., and Howley, P.M. (2013). Fields virology, 6th edn (Wolters Kluwer/Lippincott Williams & Wilkins Health, Philadelphia, PA, USA).

Fields, W., and Kielian, M. (2013). A key interaction between the alphavirus envelope proteins responsible for initial dimer dissociation during fusion. J. Virol. *87*, 3774–3781.

Firth, A.E., Chung, B.Y., Fleeton, M.N., and Atkins, J.F. (2008). Discovery of frameshifting in Alphavirus 6K resolves a 20-year enigma. Virol. J. *5*, 108.

Foo, S.S., Chen, W., Herrero, L., Bettadapura, J., Narayan, J., Dar, L., Broor, S., and Mahalingam, S. (2011). The genetics of alphaviruses. Future Virol. *6*, 1407–1422.

Frolova, E., Frolov, I., and Schlesinger, S. (1997). Packaging signals in alphaviruses. J. Virol. *71*, 248–258.

Frolova, E.I., Fayzulin, R.Z., Cook, S.H., Griffin, D.E., Rice, C.M., and Frolov, I. (2002). Roles of nonstructural protein nsP2 and Alpha/Beta interferons in determining the outcome of Sindbis virus infection. J. Virol. *76*, 11254–11264.

Froshauer, S., Kartenbeck, J., and Helenius, A. (1988). Alphavirus RNA replicase is located on the cytoplasmic surface of endosomes and lysosomes. J. Cell Biol. *107*, 2075–2086.

Gaedigk-Nitschko, K., and Schlesinger, M.J. (1990). The sindbis virus 6K protein can be detected in virions and is acylated with fatty acids. Virology *175*, 274–281.

Gaedigk-Nitschko, K., and Schlesinger, M.J. (1991). Site-directed mutations in Sindbis virus E2 glycoprotein's cytoplasmic domain and the 6K protein lead to similar defects in virus assembly and budding. Virology *183*, 206–214.

Garmashova, N., Gorchakov, R., Frolova, E., and Frolov, I. (2006). Sindbis virus nonstructural protein nsP2 is cytotoxic and inhibits cellular transcription. J. Virol. *80*, 5686–5696.

Garoff, H., and Li, K.J. (1998). Recent advances in gene expression using alphavirus vectors. Curr. Opin. Biotechnol. *9*, 464–469.

Garoff, H., Simons, K., and Dobberstein, B. (1978). Assembly of the Semliki Forest virus membrane glycoproteins in the membrane of the endoplasmic reticulum in vitro. J. Mol. Biol. *124*, 587–600.

Garoff, H., Frischauf, A.M., Simons, K., Lehrach, H., and Delius, H. (1980). Nucleotide sequence of cdna coding for Semliki Forest virus membrane glycoproteins. Nature *288*, 236–241.

Garoff, H., Wilschut, J., Liljestrom, P., Wahlberg, J.M., Bron, R., Suomalainen, M., Smyth, J., Salminen, A., Barth, B.U., Zhao, H., *et al.* (1994). Assembly and entry mechanisms of Semliki Forest virus. Arch. Virol. Suppl. *9*, 329–338.

Garoff, H., Sjoberg, M., and Cheng, R.H. (2004). Budding of alphaviruses. Virus Res. *106*, 103–116.

Geigenmuller-Gnirke, U., Nitschko, H., and Schlesinger, S. (1993). Deletion analysis of the capsid protein of Sindbis virus: identification of the RNA binding region. J. Virol. *67*, 1620–1626.

Gibbons, D.L., Vaney, M.C., Roussel, A., Vigouroux, A., Reilly, B., Lepault, J., Kielian, M., and Rey, F.A. (2004). Conformational change and protein–protein interactions of the fusion protein of Semliki Forest virus. Nature *427*, 320–325.

Glanville, N., and Ulmanen, I. (1976). Biological activity of in vitro synthesized protein: binding of Semliki Forest virus capsid protein to the large ribosomal subunit. Biochem. Biophys. Res. Commun. *71*, 393–399.

Glomb-Reinmund, S., and Kielian, M. (1998). The role of low pH and disulfide shuffling in the entry and fusion of Semliki Forest virus and Sindbis virus. Virology *248*, 372–381.

Gomez de Cedron, M., Ehsani, N., Mikkola, M.L., Garcia, J.A., and Kaariainen, L. (1999). RNA helicase activity of Semliki Forest virus replicase protein NSP2. FEBS Lett. *448*, 19–22.

Gonzalez, M.E., and Carrasco, L. (2003). Viroporins. FEBS Lett. *552*, 28–34.

Gorbalenya, A.E., Koonin, E.V., Donchenko, A.P., and Blinov, V.M. (1988). A conserved NTP-motif in putative helicases. Nature *333*, 22.

Gorbalenya, A.E., Koonin, E.V., Donchenko, A.P., and Blinov, V.M. (1989). Two related superfamilies of putative helicases involved in replication, recombination, repair and expression of DNA and RNA genomes. Nucleic Acids Res. *17*, 4713–4730.

Gorbalenya, A.E., Koonin, E.V., and Wolf, Y.I. (1990). A new superfamily of putative NTP-binding domains encoded by genomes of small DNA and RNA viruses. FEBS Lett. *262*, 145–148.

de Groot, R.J., Hardy, W.R., Shirako, Y., and Strauss, J.H. (1990). Cleavage-site preferences of Sindbis virus polyproteins containing the non-structural proteinase. Evidence for temporal regulation of polyprotein processing in vivo. EMBO J. *9*, 2631–2638.

Hahn, C.S., and Strauss, J.H. (1990). Site-directed mutagenesis of the proposed catalytic amino acids of the Sindbis virus capsid protein autoprotease. J. Virol. *64*, 3069–3073.

Hahn, Y.S., Grakoui, A., Rice, C.M., Strauss, E.G., and Strauss, J.H. (1989). Mapping of RNA- temperature-sensitive mutants of Sindbis virus: complementation group F mutants have lesions in nsP4. J. Virol. *63*, 1194–1202.

Haywood, A.M. (2010). Membrane uncoating of intact enveloped viruses. J. Virol. *84*, 10946–10955.

Heil, M.L., Albee, A., Strauss, J.H., and Kuhn, R.J. (2001). An amino acid substitution in the coding region of the E2 glycoprotein adapts Ross River virus to utilize heparan sulfate as an attachment moiety. J. Virol. *75*, 6303–6309.

Helenius, A., Morein, B., Fries, E., Simons, K., Robinson, P., Schirrmacher, V., Terhorst, C., and Strominger, J.L. (1978). Human (HLA-A and HLA-B) and murine (H-2K and H-2D) histocompatibility antigens are cell surface receptors for Semliki Forest virus. Proc. Natl. Acad. Sci. U.S.A. *75*, 3846–3850.

Helenius, A., Kartenbeck, J., Simons, K., and Fries, E. (1980). On the entry of Semliki forest virus into BHK-21 cells. J. Cell Biol. *84*, 404–420.

Helenius, A., Marsh, M., and White, J. (1982). Inhibition of Semliki forest virus penetration by lysosomotropic weak bases. J. Gen. Virol. *58*(1), 47–61.

Hernandez, R., Luo, T., and Brown, D.T. (2001). Exposure to low pH is not required for penetration of mosquito cells by Sindbis virus. J. Virol. *75*, 2010–2013.

Hodgman, T.C. (1988). A new superfamily of replicative proteins. Nature *333*, 22–23.

Hong, E.M., Perera, R., and Kuhn, R.J. (2006). Alphavirus capsid protein helix I controls a checkpoint in nucleocapsid core assembly. J. Virol. *80*, 8848–8855.

Hunt, S.R., Hernandez, R., and Brown, D.T. (2011). Role of the vacuolar-ATPase in Sindbis virus infection. J. Virol. *85*, 1257–1266.

Irurzun, A., Nieva, J.L., and Carrasco, L. (1997). Entry of Semliki forest virus into cells: effects of concanamycin A and nigericin on viral membrane fusion and infection. Virology *227*, 488–492.

Jose, J., Snyder, J.E., and Kuhn, R.J. (2009). A structural and functional perspective of alphavirus replication and assembly. Future Microbiol. *4*, 837–856.

Jose, J., Przybyla, L., Edwards, T.J., Perera, R., Burgner, J.W., 2nd, and Kuhn, R.J. (2012). Interactions of the cytoplasmic domain of Sindbis virus E2 with nucleocapsid cores promote alphavirus budding. J. Virol. *86*, 2585–2599.

Kamer, G., and Argos, P. (1984). Primary structural comparison of RNA-dependent polymerases from plant, animal and bacterial viruses. Nucleic Acids Res. *12*, 7269–7282.

Kielian, M. (1995). Membrane fusion and the alphavirus life cycle. Adv. Virus Res. *45*, 113–151.

Kielian, M.C., and Helenius, A. (1984). Role of cholesterol in fusion of Semliki Forest virus with membranes. J. Virol. 52, 281–283.

Kielian, M., and Helenius, A. (1985). pH-induced alterations in the fusogenic spike protein of Semliki Forest virus. J. Cell Biol. *101*, 2284–2291.

Kielian, M.C., Marsh, M., and Helenius, A. (1986). Kinetics of endosome acidification detected by mutant and wild-type Semliki Forest virus. EMBO J. *5*, 3103–3109.

Kielian, M., Chanel-Vos, C., and Liao, M. (2010). Alphavirus entry and membrane fusion. Viruses *2*, 796–825.

Kim, D.Y., Firth, A.E., Atasheva, S., Frolova, E.I., and Frolov, I. (2011). Conservation of a packaging signal and the viral genome RNA packaging mechanism in alphavirus evolution. J. Virol. *85*, 8022–8036.

Klimstra, W.B., Ryman, K.D., and Johnston, R.E. (1998). Adaptation of Sindbis virus to BHK cells selects for use of heparan sulfate as an attachment receptor. J. Virol. *72*, 7357–7366.

Klimstra, W.B., Nangle, E.M., Smith, M.S., Yurochko, A.D., and Ryman, K.D. (2003). DC-SIGN and L-SIGN can act as attachment receptors for alphaviruses and distinguish between mosquito cell- and mammalian cell-derived viruses. J. Virol. *77*, 12022–12032.

Knight, R.L., Schultz, K.L., Kent, R.J., Venkatesan, M., and Griffin, D.E. (2009). Role of N-linked glycosylation for sindbis virus infection and replication in vertebrate and invertebrate systems. J. Virol. *83*, 5640–5647.

Kononchik, J.P., Jr., Hernandez, R., and Brown, D.T. (2011a). An alternative pathway for alphavirus entry. Virol. J. *8*, 304.

Kononchik, J.P., Vancini, R., and Brown, D.T. (2011b). Alphavirus adsorption to mosquito cells as viewed by freeze fracture immunolabeling. Virology *415*, 132–140.

Koonin, E.V., and Dolja, V.V. (1993). Evolution and taxonomy of positive-strand RNA viruses: implications of comparative analysis of amino acid sequences. Crit. Rev. Biochem. Mol. Biol. *28*, 375–430.

Kordyukova, L.V., Serebryakova, M.V., Baratova, L.A., and Veit, M. (2010). Site-specific attachment of palmitate or stearate to cytoplasmic versus transmembrane cysteines is a common feature of viral spike proteins. Virology *398*, 49–56.

Kuhn, R.J., Griffin, D.E., Owen, K.E., Niesters, H.G., and Strauss, J.H. (1996). Chimeric Sindbis-Ross River viruses to study interactions between alphavirus nonstructural and structural regions. J. Virol. *70*, 7900–7909.

Kujala, P., Ikaheimonen, A., Ehsani, N., Vihinen, H., Auvinen, P., and Kaariainen, L. (2001). Biogenesis of the Semliki Forest virus RNA replication complex. J. Virol. *75*, 3873–3884.

La Linn, M., Eble, J.A., Lubken, C., Slade, R.W., Heino, J., Davies, J., and Suhrbier, A. (2005). An arthritogenic alphavirus uses the alpha1beta1 integrin collagen receptor. Virology *336*, 229–239.

Lanzrein, M., Weingart, R., and Kempf, C. (1993). pH-dependent pore formation in Semliki forest virus-infected Aedes albopictus cells. Virology *193*, 296–302.

LaStarza, M.W., Lemm, J.A., and Rice, C.M. (1994). Genetic analysis of the nsP3 region of Sindbis virus: evidence for roles in minus-strand and subgenomic RNA synthesis. J. Virol. *68*, 5781–5791.

Leavitt, R., Schlesinger, S., and Kornfeld, S. (1977). Tunicamycin inhibits glycosylation and multiplication of Sindbis and vesicular stomatitis viruses. J. Virol. *21*, 375–385.

Lee, H., and Brown, D.T. (1994). Mutations in an exposed domain of Sindbis virus capsid protein result in the production of noninfectious virions and morphological variants. Virology *202*, 390–400.

Lemm, J.A., and Rice, C.M. (1993). Roles of nonstructural polyproteins and cleavage products in regulating Sindbis virus RNA replication and transcription. J. Virol. *67*, 1916–1926.

Lemm, J.A., Rumenapf, T., Strauss, E.G., Strauss, J.H., and Rice, C.M. (1994). Polypeptide requirements for assembly of functional Sindbis virus replication complexes: a model for the temporal regulation of minus- and plus-strand RNA synthesis. EMBO J. *13*, 2925–2934.

Li, G.P., La Starza, M.W., Hardy, W.R., Strauss, J.H., and Rice, C.M. (1990). Phosphorylation of Sindbis virus nsP3 in vivo and in vitro. Virology *179*, 416–427.

Li, L., Jose, J., Xiang, Y., Kuhn, R.J., and Rossmann, M.G. (2010). Structural changes of envelope proteins during alphavirus fusion. Nature *468*, 705–708.

Liljestrom, P., and Garoff, H. (1991). Internally located cleavable signal sequences direct the formation of Semliki Forest virus membrane proteins from a polyprotein precursor. J. Virol. *65*, 147–154.

Linger, B.R., Kunovska, L., Kuhn, R.J., and Golden, B.L. (2004). Sindbis virus nucleocapsid assembly: RNA folding promotes capsid protein dimerization. RNA *10*, 128–138.

Lobigs, M., and Garoff, H. (1990). Fusion function of the Semliki Forest virus spike is activated by proteolytic cleavage of the envelope glycoprotein precursor p62. J. Virol. *64*, 1233–1240.

Lobigs, M., Zhao, H.X., and Garoff, H. (1990). Function of Semliki Forest virus E3 peptide in virus assembly: replacement of E3 with an artificial signal peptide abolishes spike heterodimerization and surface expression of E1. J. Virol. *64*, 4346–4355.

Loewy, A., Smyth, J., von Bonsdorff, C.H., Liljestrom, P., and Schlesinger, M.J. (1995). The 6-kilodalton membrane protein of Semliki Forest virus is involved in the budding process. J. Virol. *69*, 469–475.

Lu, Y.E., Cassese, T., and Kielian, M. (1999). The cholesterol requirement for sindbis virus entry and exit and characterization of a spike protein region involved in cholesterol dependence. J. Virol. *73*, 4272–4278.

Lusa, S., Garoff, H., and Liljestrom, P. (1991). Fate of the 6K membrane protein of Semliki Forest virus during virus assembly. Virology *185*, 843–846.

McInerney, G.M., Smit, J.M., Liljeström, P., and Wilschut, J. (2004). Semliki Forest virus produced in the absence of the 6K protein has an altered spike structure as revealed by decreased membrane fusion capacity. Virology *325*, 200–206.

Marquardt, M.T., Phalen, T., and Kielian, M. (1993). Cholesterol is required in the exit pathway of Semliki Forest virus. J. Cell Biol. *123*, 57–65.

Marsh, M., and Helenius, A. (1980). Adsorptive endocytosis of Semliki Forest virus. J. Mol. Biol. *142*, 439–454.

Marsh, M., Wellsteed, J., Kern, H., Harms, E., and Helenius, A. (1982). Monensin inhibits Semliki Forest virus penetration into culture cells. Proc. Natl. Acad. Sci. U.S.A. *79*, 5297–5301.

Marsh, M., Bolzau, E., and Helenius, A. (1983). Penetration of Semliki Forest virus from acidic prelysosomal vacuoles. Cell 32, 931–940.

Mayuri, Geders, T.W., Smith, J.L., and Kuhn, R.J. (2008). Role for conserved residues of sindbis virus nonstructural protein 2 methyltransferase-like domain in regulation of minus-strand synthesis and development of cytopathic infection. J. Virol. *82*, 7284–7297.

Melancon, P., and Garoff, H. (1987). Processing of the Semliki Forest virus structural polyprotein: role of the capsid protease. J. Virol. *61*, 1301–1309.

Metsikko, K., and Garoff, H. (1990). Oligomers of the cytoplasmic domain of the p62/E2 membrane protein of Semliki Forest virus bind to the nucleocapsid in vitro. J. Virol. *64*, 4678–4683.

Mi, S., and Stollar, V. (1991). Expression of Sindbis virus nsP1 and methyltransferase activity in *Escherichia coli*. Virology *184*, 423–427.

Moesby, L., Corver, J., Erukulla, R.K., Bittman, R., and Wilschut, J. (1995). Sphingolipids activate membrane fusion of Semliki Forest virus in a stereospecific manner. Biochemistry *34*, 10319–10324.

Montgomery, S.A., and Johnston, R.E. (2007). Nuclear import and export of Venezuelan equine encephalitis virus nonstructural protein 2. J. Virol. *81*, 10268–10279.

Mukhopadhyay, S., Zhang, W., Gabler, S., Chipman, P.R., Strauss, E.G., Strauss, J.H., Baker, T.S., Kuhn, R.J., and Rossmann, M.G. (2006). Mapping the structure and function of the E1 and E2 glycoproteins in alphaviruses. Structure *14*, 63–73.

Mulvey, M., and Brown, D.T. (1996). Assembly of the Sindbis virus spike protein complex. Virology *219*, 125–132.

Niesters, H.G., and Strauss, J.H. (1990). Defined mutations in the 5′ nontranslated sequence of Sindbis virus RNA. J. Virol. *64*, 4162–4168.

Nieva, J.L., Bron, R., Corver, J., and Wilschut, J. (1994). Membrane fusion of Semliki Forest virus requires sphingolipids in the target membrane. EMBO J. *13*, 2797–2804.

Oberste, M.S., Parker, M.D., and Smith, J.F. (1996). Complete sequence of Venezuelan equine encephalitis virus subtype IE reveals conserved and hypervariable domains within the C terminus of nsP3. Virology *219*, 314–320.

Omar, A., and Koblet, H. (1988). Semliki Forest virus particles containing only the E1 envelope glycoprotein are infectious and can induce cell–cell fusion. Virology *166*, 17–23.

Owen, K.E., and Kuhn, R.J. (1996). Identification of a region in the Sindbis virus nucleocapsid protein that is involved in specificity of RNA encapsidation. J. Virol. *70*, 2757–2763.

Owen, K.E., and Kuhn, R.J. (1997). Alphavirus budding is dependent on the interaction between the nucleocapsid and hydrophobic amino acids on the cytoplasmic domain of the E2 envelope glycoprotein. Virology *230*, 187–196.

Pardigon, N., and Strauss, J.H. (1992). Cellular proteins bind to the 3′ end of Sindbis virus minus-strand RNA. J. Virol. *66*, 1007–1015.

Pardigon, N., Lenches, E., and Strauss, J.H. (1993). Multiple binding sites for cellular proteins in the 3′ end of Sindbis alphavirus minus-sense RNA. J. Virol. *67*, 5003–5011.

Paredes, A.M., Simon, M.N., and Brown, D.T. (1992). The mass of the Sindbis virus nucleocapsid suggests it has T = 4 icosahedral symmetry. Virology *187*, 329–332.

Paredes, A.M., Brown, D.T., Rothnagel, R., Chiu, W., Schoepp, R.J., Johnston, R.E., and Prasad, B.V. (1993). Three-dimensional structure of a membrane-containing virus. Proc. Natl. Acad. Sci. U.S.A. *90*, 9095–9099.

Parrott, M.M., Sitarski, S.A., Arnold, R.J., Picton, L.K., Hill, R.B., and Mukhopadhyay, S. (2009). Role of conserved cysteines in the alphavirus E3 protein. J. Virol. *83*, 2584–2591.

Peranen, J. (1991). Localization and phosphorylation of Semliki Forest virus non-structural protein nsP3 expressed in COS cells from a cloned cDNA. J. Gen. Virol. 72(1), 195–199.

Peranen, J., Rikkonen, M., Liljestrom, P., and Kaariainen, L. (1990). Nuclear localization of Semliki Forest virus-specific nonstructural protein nsP2. J. Virol. *64*, 1888–1896.

Phalen, T., and Kielian, M. (1991). Cholesterol is required for infection by Semliki Forest virus. J. Cell Biol. *112*, 615–623.

Powers, A.M., Brault, A.C., Shirako, Y., Strauss, E.G., Kang, W., Strauss, J.H., and Weaver, S.C. (2001). Evolutionary relationships and systematics of the alphaviruses. J. Virol. 75, 10118–10131.

Rikkonen, M., Peranen, J., and Kaariainen, L. (1994). ATPase and GTPase activities associated with Semliki Forest virus nonstructural protein nsP2. J. Virol. *68*, 5804–5810.

Roussel, A., Lescar, J., Vaney, M.C., Wengler, G., and Rey, F.A. (2006). Structure and interactions at the viral surface of the envelope protein E1 of Semliki Forest virus. Structure *14*, 75–86.

Russo, A.T., White, M.A., and Watowich, S.J. (2006). The crystal structure of the Venezuelan equine encephalitis alphavirus nsP2 protease. Structure *14*, 1449–1458.

Ryan, C., Stevens, T.H., and Schlesinger, M.J. (1992). Inhibitory effects of HSP70 chaperones on nascent polypeptides. Protein Sci. *1*, 980–985.

Salonen, A., Vasiljeva, L., Merits, A., Magden, J., Jokitalo, E., and Kaariainen, L. (2003). Properly folded nonstructural polyprotein directs the semliki forest virus replication complex to the endosomal compartment. J. Virol. 77, 1691–1702.

Samsonov, A.V., Chatterjee, P.K., Razinkov, V.I., Eng, C.H., Kielian, M., and Cohen, F.S. (2002). Effects of membrane potential and sphingolipid structures on fusion of Semliki Forest virus. J. Virol. *76*, 12691–12702.

Sanchez-San Martin, C., Liu, C.Y., and Kielian, M. (2009). Dealing with low pH: entry and exit of alphaviruses and flaviviruses. Trends Microbiol. *17*, 514–521.

Sariola, M., Saraste, J., and Kuismanen, E. (1995). Communication of post-Golgi elements with early endocytic pathway: regulation of endoproteolytic cleavage of Semliki Forest virus p62 precursor. J. Cell Sci. *108*(6), 2465–2475.

Sawicki, D.L., and Sawicki, S.G. (1985). Functional analysis of the A complementation group mutants of Sindbis HR virus. Virology *144*, 20–34.

Sawicki, D.L., and Sawicki, S.G. (1993). A second nonstructural protein functions in the regulation of alphavirus negative-strand RNA synthesis. J. Virol. *67*, 3605–3610.

Sawicki, D.L., Sawicki, S.G., Keranen, S., and Kaariainen, L. (1981). Specific Sindbis virus-coded function for minus-strand RNA synthesis. J. Virol. *39*, 348–358.

Sawicki, D.L., Perri, S., Polo, J.M., and Sawicki, S.G. (2006). Role for nsP2 proteins in the cessation of alphavirus minus-strand synthesis by host cells. J. Virol. *80*, 360–371.

Scheidel, L.M., Durbin, R.K., and Stollar, V. (1987). Sindbis virus mutants resistant to mycophenolic acid and ribavirin. Virology *158*, 1–7.

Schmidt, M., Schmidt, M.F., and Rott, R. (1988). Chemical identification of cysteine as palmitoylation site in a transmembrane protein (Semliki Forest virus E1). J. Biol. Chem. *263*, 18635–18639.

Shabman, R.S., Morrison, T.E., Moore, C., White, L., Suthar, M.S., Hueston, L., Rulli, N., Lidbury, B., Ting, J.P., Mahalingam, S., *et al.* (2007). Differential induction of type I interferon responses in myeloid dendritic cells by mosquito and mammalian-cell-derived alphaviruses. J. Virol. *81*, 237–247.

Shin, G., Yost, S.A., Miller, M.T., Elrod, E.J., Grakoui, A., and Marcotrigiano, J. (2012). Structural and functional insights into alphavirus polyprotein processing and pathogenesis. Proc. Natl. Acad. Sci. U.S.A. *109*, 16534–16539.

Shirako, Y., and Strauss, J.H. (1994). Regulation of Sindbis virus RNA replication: uncleaved P123 and nsP4 function in minus-strand RNA synthesis, whereas cleaved products from P123 are required for efficient plus-strand RNA synthesis. J. Virol. *68*, 1874–1885.

Sieczkarski, S.B., and Whittaker, G.R. (2002). Influenza virus can enter and infect cells in the absence of clathrin-mediated endocytosis. J. Virol. *76*, 10455–10464.

Sieczkarski, S.B., and Whittaker, G.R. (2003). Differential requirements of Rab5 and Rab7 for endocytosis of influenza and other enveloped viruses. Traffic *4*, 333–343.

Simizu, B., Yamamoto, K., Hashimoto, K., and Ogata, T. (1984). Structural proteins of Chikungunya virus. J. Virol. *51*, 254–258.

Simmons, D.T., and Strauss, J.H. (1972). Replication of Sindbis virus. I. Relative size and genetic content of 26 s and 49 s RNA. J. Mol. Biol. *71*, 599–613.

Smit, J.M., Bittman, R., and Wilschut, J. (1999). Low-pH-dependent fusion of Sindbis virus with receptor-free cholesterol- and sphingolipid-containing liposomes. J. Virol. 73, 8476–8484.

Snyder, A.J., and Mukhopadhyay, S. (2012). The alphavirus E3 glycoprotein functions in a clade-specific manner. J. Virol. *86*, 13609–13620.

Snyder, J.E., Berrios, C.J., Edwards, T.J., Jose, J., Perera, R., and Kuhn, R.J. (2012). Probing the early temporal and spatial interaction of the Sindbis virus capsid and E2 proteins with reverse genetics. J. Virol. *86*, 12372–12383.

Snyder, J.E., Kulcsar, K.A., Schultz, K.L., Riley, C.P., Neary, J.T., Marr, S., Jose, J., Griffin, D.E., and Kuhn, R.J. (2013). Functional Characterization of the Alphavirus TF Protein. J. Virol. *87*, 8511–8523.

Solignat, M., Gay, B., Higgs, S., Briant, L., and Devaux, C. (2009). Replication cycle of chikungunya: a re-emerging arbovirus. Virology *393*, 183–197.

Soonsawad, P., Xing, L., Milla, E., Espinoza, J.M., Kawano, M., Marko, M., Hsieh, C., Furukawa, H., Kawasaki, M., Weerachatyanukul, W., *et al.* (2010). Structural evidence of glycoprotein assembly in cellular membrane compartments prior to Alphavirus budding. J. Virol. *84*, 11145–11151.

Strauss, J.H., and Strauss, E.G. (1994). The alphaviruses: gene expression, replication, and evolution. Microbiol. Rev. *58*, 491–562.

Suopanki, J., Sawicki, D.L., Sawicki, S.G., and Kaariainen, L. (1998). Regulation of alphavirus 26S mRNA transcription by replicase component nsP2. J. Gen. Virol. 79(2), 309–319.

Takkinen, K., Peranen, J., and Kaariainen, L. (1991). Proteolytic processing of Semliki Forest virus-specific non-structural polyprotein. J. Gen. Virol. *72*(7), 1627–1633.

Tang, J., Jose, J., Chipman, P., Zhang, W., Kuhn, R.J., and Baker, T.S. (2011). Molecular links between the E2 envelope glycoprotein and nucleocapsid core in Sindbis virus. J. Mol. Biol. *414*, 442–459.

Ulmanen, I., Soderlund, H., and Kaariainen, L. (1976). Semliki Forest virus capsid protein associates with the 60S ribosomal subunit in infected cells. J. Virol. *20*, 203–210.

Ulmanen, I., Soderlund, H., and Kaariainen, L. (1979). Role of protein synthesis in the assembly of Semliki forest virus nucleocapsid. Virology *99*, 265–276.

Vancini, R., Wang, G., Ferreira, D., Hernandez, R., and Brown, D.T. (2013). Alphavirus genome delivery occurs directly at the plasma membrane in a time- and temperature-dependent process. J. Virol. *87*, 4352–4359.

Vashishtha, M., Phalen, T., Marquardt, M.T., Ryu, J.S., Ng, A.C., and Kielian, M. (1998). A single point mutation controls the cholesterol dependence of Semliki Forest virus entry and exit. J. Cell Biol. *140*, 91–99.

Vasiljeva, L., Merits, A., Auvinen, P., and Kaariainen, L. (2000). Identification of a novel function of the alphavirus capping apparatus. RNA 5′-triphosphatase activity of Nsp2. J. Biol. Chem. *275*, 17281–17287.

Vasiljeva, L., Merits, A., Golubtsov, A., Sizemskaja, V., Kaariainen, L., and Ahola, T. (2003). Regulation of the sequential processing of Semliki Forest virus replicase polyprotein. J. Biol. Chem. *278*, 41636–41645.

Vaux, D.J., Helenius, A., and Mellman, I. (1988). Spike–nucleocapsid interaction in Semliki Forest virus reconstructed using network antibodies. Nature *336*, 36–42.

Vihinen, H., and Saarinen, J. (2000). Phosphorylation site analysis of Semliki forest virus nonstructural protein 3. J. Biol. Chem. *275*, 27775–27783.

Vihinen, H., Ahola, T., Tuittila, M., Merits, A., and Kaariainen, L. (2001). Elimination of phosphorylation sites of Semliki Forest virus replicase protein nsP3. J. Biol. Chem. *276*, 5745–5752.

Voss, J.E., Vaney, M.C., Duquerroy, S., Vonrhein, C., Girard-Blanc, C., Crublet, E., Thompson, A., Bricogne, G., and Rey, F.A. (2010). Glycoprotein organization of Chikungunya virus particles revealed by X-ray crystallography. Nature *468*, 709–712.

Vrati, S., Faragher, S.G., Weir, R.C., and Dalgarno, L. (1986). Ross River virus mutant with a deletion in the E2 gene: properties of the virion, virus-specific macromolecule synthesis, and attenuation of virulence for mice. Virology *151*, 222–232.

Wahlberg, J.M., Boere, W.A., and Garoff, H. (1989). The heterodimeric association between the membrane proteins of Semliki Forest virus changes its sensitivity to low pH during virus maturation. J. Virol. *63*, 4991–4997.

Wahlberg, J.M., Bron, R., Wilschut, J., and Garoff, H. (1992). Membrane fusion of Semliki Forest virus involves homotrimers of the fusion protein. J. Virol. *66*, 7309–7318.

Wang, G., Hernandez, R., Weninger, K., and Brown, D.T. (2007). Infection of cells by Sindbis virus at low temperature. Virology *362*, 461–467.

Wang, K.S., Kuhn, R.J., Strauss, E.G., Ou, S., and Strauss, J.H. (1992). High-affinity laminin receptor is a receptor for Sindbis virus in mammalian cells. J. Virol. *66*, 4992–5001.

Wang, Y.F., Sawicki, S.G., and Sawicki, D.L. (1991). Sindbis virus nsP1 functions in negative-strand RNA synthesis. J. Virol. *65*, 985–988.

Wang, Y.F., Sawicki, S.G., and Sawicki, D.L. (1994). Alphavirus nsP3 functions to form replication complexes transcribing negative-strand RNA. J. Virol. *68*, 6466–6475.

Warrier, R., Linger, B.R., Golden, B.L., and Kuhn, R.J. (2008). Role of sindbis virus capsid protein region II in nucleocapsid core assembly and encapsidation of genomic RNA. J. Virol. *82*, 4461–4470.

Weaver, S.C., Hagenbaugh, A., Bellew, L.A., and Calisher, C.H. (1992). Genetic characterization of an antigenic subtype of eastern equine encephalomyelitis virus. Arch. Virol. *127*, 305–314.

Weiss, B., Nitschko, H., Ghattas, I., Wright, R., and Schlesinger, S. (1989). Evidence for specificity in the encapsidation of Sindbis virus RNAs. J. Virol. *63*, 5310–5318.

Weiss, B., Geigenmuller-Gnirke, U., and Schlesinger, S. (1994). Interactions between Sindbis virus RNAs and a 68 amino acid derivative of the viral capsid protein further defines the capsid binding site. Nucleic Acids Res. *22*, 780–786.

Wengler, G., and Wurkner, D. (1992). Identification of a sequence element in the alphavirus core protein which mediates interaction of cores with ribosomes and the disassembly of cores. Virology *191*, 880–888.

Wengler, G., Koschinski, A., and Dreyer, F. (2003). Entry of alphaviruses at the plasma membrane converts the viral surface proteins into an ion-permeable pore that can be detected by electrophysiological analyses of whole-cell membrane currents. J. Gen. Virol. *84*, 173–181.

White, J., and Helenius, A. (1980). pH-dependent fusion between the Semliki Forest virus membrane and liposomes. Proc. Natl. Acad. Sci. U.S.A. *77*, 3273–3277.

Wilschut, J., Corver, J., Nieva, J.L., Bron, R., Moesby, L., Reddy, K.C., and Bittman, R. (1995). Fusion of Semliki Forest virus with cholesterol-containing liposomes at low pH: a specific requirement for sphingolipids. Mol. Membr. Biol. *12*, 143–149.

Zhang, W., Mukhopadhyay, S., Pletnev, S.V., Baker, T.S., Kuhn, R.J., and Rossmann, M.G. (2002). Placement of the structural proteins in Sindbis virus. J. Virol. *76*, 11645–11658.

Zheng, Y., and Kielian, M. (2013). Imaging of the alphavirus capsid protein during virus replication. J. Virol. *87*, 9579–9589.

Ziemiecki, A., Garoff, H., and Simons, K. (1980). Formation of the Semliki Forest virus membrane glycoprotein complexes in the infected cell. J. Gen. Virol. *50*, 111–123.

Zusinaite, E., Tints, K., Kiiver, K., Spuul, P., Karo-Astover, L., Merits, A., and Sarand, I. (2007). Mutations at the palmitoylation site of non-structural protein nsP1 of Semliki Forest virus attenuate virus replication and cause accumulation of compensatory mutations. J. Gen. Virol. *88*, 1977–1985.

Alphavirus Evolution

2

Lark L. Coffey, Scott C. Weaver and Naomi L. Forrester

Abstract

Alphaviruses exhibit a near-global distribution and use many vertebrates and arthropods as hosts. This chapter will discuss biological properties and phylogenetic relationships of the alphaviruses and patterns of alphavirus evolution marked by genetic changes that influence infectivity and promote emergence in epizootics and epidemics.

Phylogenetic relationships of alphaviruses

Overview

The evolutionary history of the alphaviruses is as diverse as the genus itself. The alphaviruses exist in every continent except Antarctica and use birds, mammals, fish, and potentially amphibians as amplifying and reservoir hosts (Bingham *et al.*, 2012; Forrester *et al.*, 2012b; Graham *et al.*, 2012; Strauss and Strauss, 1994; White *et al.*, 2011a). The arthropod vectors of the alphaviruses are also diverse, and include numerous mosquito species as well as potentially mites. Aquatic alphaviruses have been isolated from lice, but evidence that they transmit these viruses is lacking (Beaty *et al.*, 1989; La Linn *et al.*, 2001). Alphaviruses are classified phenotypically and phylogenetically into two groups: the arthralgic or Old World viruses and the encephalitic or New World viruses (Forrester *et al.*, 2012b; Powers *et al.*, 2001) (Fig. 2.1). Notable exceptions to this pattern exist: Sindbis virus (SINV) occurs in the Old World but is genetically more closely related to the New World encephalitides, and Mayaro and Una viruses (MAYV and UNAV) occur in South America but group most closely genetically and antigenically with the arthralgic Old World viruses.

Analysis of the alphavirus genus has given rise to two theories based on the most parsimonious phylogeographic reconstructions that explain the origins of alphaviruses; two possible patterns of virus spread are illustrated in Fig. 2.2. The first analysis (Fig. 2.2A) using alphavirus structural protein gene sequences (Powers *et al.*, 2001) suggests a New World origin in South or Central America and subsequent spread across the globe. There are a few issues that make this scenario unlikely. Firstly, a South American origin does not take account of the nearly global distribution of the aquatic alphaviruses, which widens the geographic area over which the original land-based introduction could occur. Moreover, if the alphaviruses originated in South America, MAYV and UNAV would need to have been secondarily re-introduced there at a later date. A subsequent analysis based on full genome sequences (Forrester *et al.*, 2012b) suggests that alphaviruses have a marine origin based on the basal positions of the marine viruses southern elephant seal virus and salmon pancreatic disease virus (SPDV) in the alphavirus phylogeny (Fig. 2.1). This theory removes the need for a secondary introduction into South America (Fig. 2.2B). However, it requires that alphaviruses have origins in South America as well as Africa/Oceania. Although SPDV has been isolated from mites that are sometimes associated with diseased fish (Weston *et al.*, 1999, 2002), no confirmatory vector competence studies have shown that they transmit SPDV. The limited systematic surveillance, coupled with recent discovery of the mosquito-specific

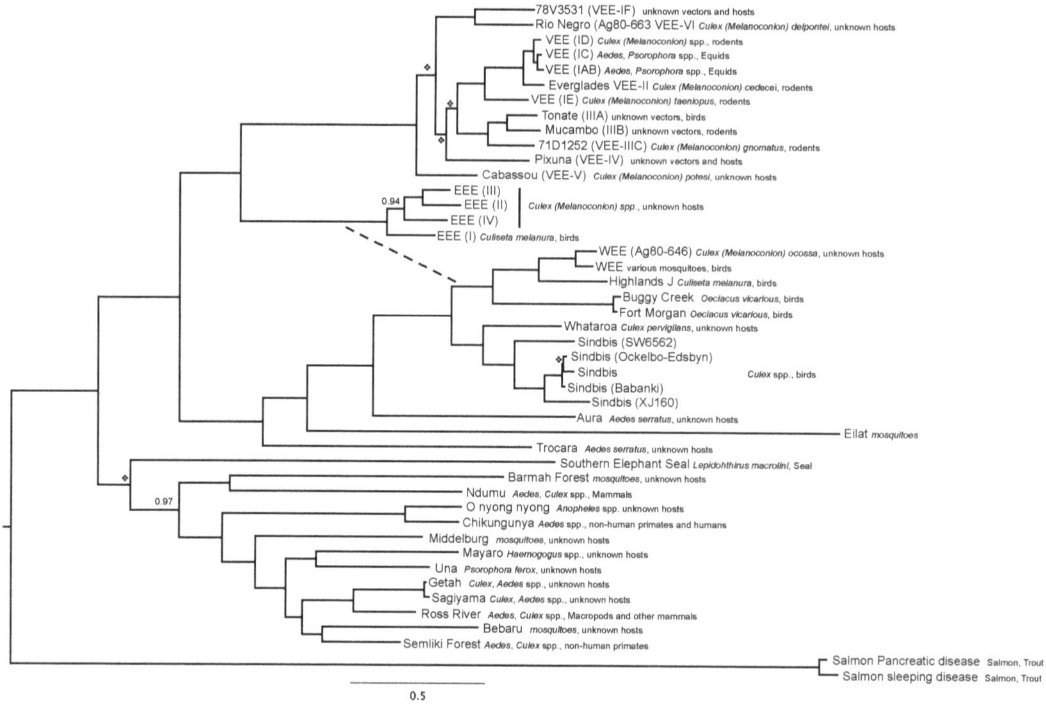

Figure 2.1 Phylogenetic tree of alphavirus species based on the structural protein (E2, 6K and E1) amino acid sequences made using Bayesian methods and midpoint rooting. The dashed line indicates the point at which the ancestor of SINV and EEEV recombined to form the recombinant WEEV group. All posterior probabilities were 1 unless shown. Nodes with a ❖ had posterior probabilities less than 0.9.

Eilat virus (EILV) (Nasar *et al.*, 2012), suggests that many alphaviruses remain to be discovered, probably including many more insect-specific members; thus, the addition of new alphavirus species to phylogenies will probably reveal new insights into their evolutionary histories. Regardless of where alphaviruses originated, they now exist on most continents and may spread via oceans. Given that alphaviruses have traversed oceans and continents, and the fact that many are pathogens of animals including humans, the potential for new emerging diseases from this genus of viruses should not be underestimated.

Arthralgic alphaviruses

The arthralgic alphaviruses exhibit a nearly global distribution, with Ross River (RRV) and Barmah Forest viruses (BFV) in Australia and South-East Asia, chikungunya (CHIKV) and O'nyong-nyong viruses (ONNV) in Africa and Asia (CHIKV only) and MAYV and UNAV in the Americas. Of these viruses, CHIKV is the best studied, partly due to its recent emergence in Oceania and South-East Asia, which has resulted in more than six million human cases since 2004 (Schwartz and Albert, 2010) (see Chapter 7 for clinical manifestations of arthritogenic Old World alphaviruses). Since 2013, an estimated 1.5M cases of CHIKV have occurred in the Caribbean and Latin America. Surprisingly, this epidemic resulted from the importation not of the *Aedes albopictus*-adapted Indian Ocean lineage strain, but of the Asian lineage strain that has been circulating in South East Asia since 1958 and is primarily transmitted by *A. aegyptii* (Weaver and Lecuit, 2015; Weaver and Forrester, 2015).

Chikungunya virus

CHIKV is classified into three major lineages – West African, Asian and the East, Central, and South African (ECSA) lineage (Poidinger *et al.*, 1997) – of which only the Asian lineage has been associated with urban disease (see below). CHIKV has historically circulated largely undetected in

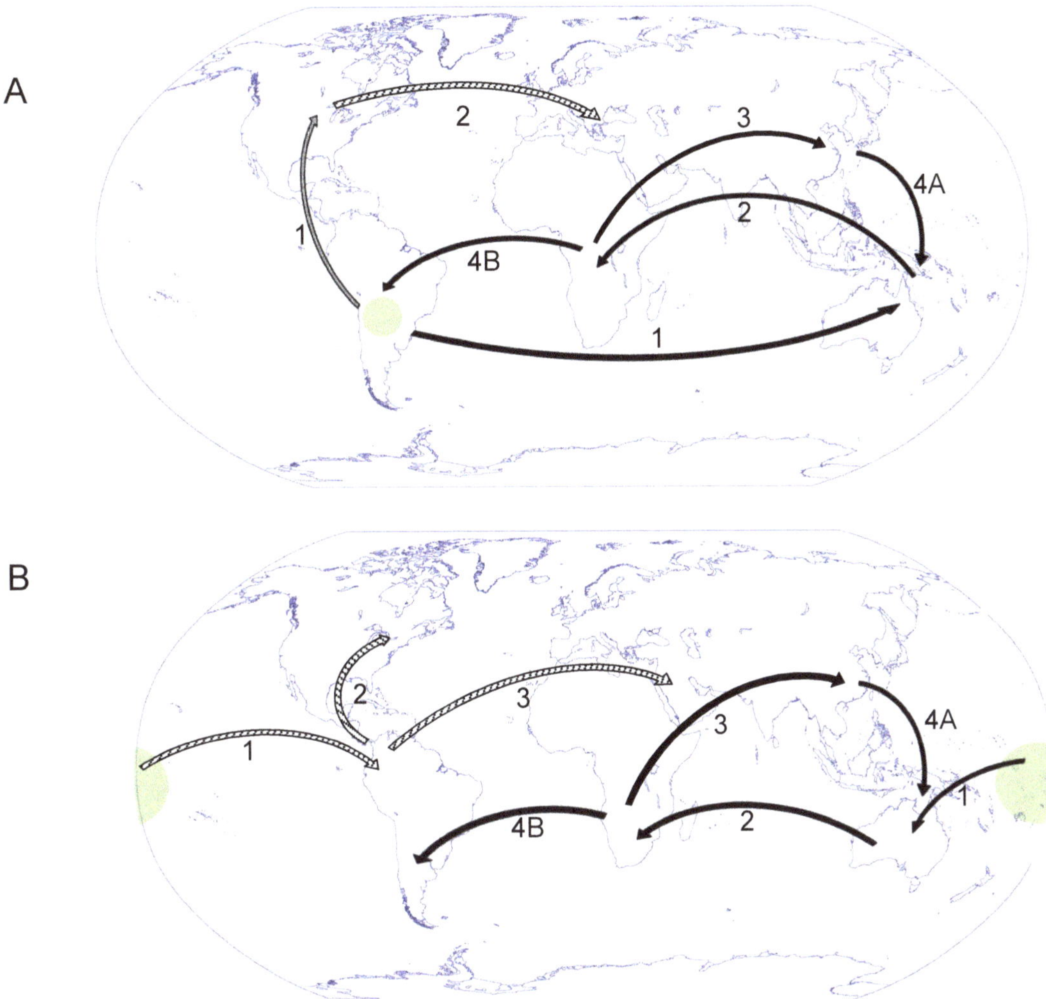

Figure 2.2 Two possible theories for the origin and geographic spread of alphaviruses. Spread of New World alphaviruses is indicated by hatched arrows; Old World alphavirus spread is shown in solid arrows. A: arrow 1, introduction of New World Alphaviruses into the Americas: and arrow 2, secondary introduction into the Old World. Old World alphaviruses are indicated by bold arrows: arrow 1, original introduction of old world alphaviruses into the Oceania region from the New World; arrow 2, secondary introduction into Southern Africa; arrow 3, tertiary introduction to Northern Africa and Eurasia; arrow 4A, secondary introduction of RRV to Australasia; and arrow 4B, secondary introduction of MAYV and UNAV to the New World. Diagram B. New World alphaviruses are indicated by hatched arrows: arrow 1, introduction from Oceania to the New World; and arrow 2, secondary introduction to the Old World. Old World viruses are indicated by black arrows: arrow 1, introduction from Oceania to Australasia; arrow 2, secondary introduction into Southern Africa; arrow 3, tertiary introduction to Northern Africa and Eurasia; arrow 4A, secondary introduction of RRV to Australasia; and arrow 4B, secondary introduction of MAYV and UNAV to the New World.

Africa and Asia, causing limited outbreaks. However, in 2005 a new Indian Ocean lineage (IOL), that arose out of the ECSA enzootic lineage, emerged (Parola *et al.*, 2006; Schuffenecker *et al.*, 2006), and has since spread nearly worldwide via infected travellers, although there have been only two introductions that resulted in autochthonous cycles, in Italy and France (Grandadam *et al.*, 2011; Watson, 2007). Epidemic human–mosquito–human transmission was established in India and Southern Asia causing extensive human disease. Prior to the emergence of the IOL, CHIKV had been recognized in African sylvatic cycles vectored by *Aedes furcifer* and *A. africanus* and forest-dwelling primates, and also cycled as an urban disease in people using *A. aegypti* as vectors (Weaver *et al.*, 2012). However, emergence of the CHIKV IOL has been associated with at least five mutations that increased infection of and dissemination within *A. albopictus* mosquitoes (Tsetsarkin *et al.*, 2007, 2011b, 2014). Further discussion of the effects of these mutations on the evolution and spread of CHIKV is presented below.

Barmah Forest and Ross River viruses

BFV and RRV co-circulate in Australia, and RRV also occurs in Indonesia and the Pacific Islands. BFV typically causes hundreds of cases of human disease with athralgia and rash annually in Australia, with kangaroos, wallabies and possibly birds serving as amplification hosts (Suhrbier *et al.*, 2012). In 2012, for example, BFV and RRV caused 7500 clinical cases of human disease in Australia (Knope *et al.*, 2013). Phylogenies of RRV show geographic clustering of lineages in Australia, which probably reflects the limited dispersal of rodent hosts, wallabies and kangaroos (Lindsay *et al.*, 1993; Sammels *et al.*, 1995). BFV, by contrast, exhibits distinct lineages with little geographic clustering, suggesting a more mobile vertebrate host such as birds (Poidinger *et al.*, 1997).

Mayaro virus

Although MAYV belongs to the Semliki Forest antigenic complex of Old World alphaviruses (Forrester *et al.*, 2012b), it occurs in South America. Given that MAYV causes clinical disease similar to the flavivirus dengue virus, human infection is likely widely under-reported. Vertebrate hosts of MAYV are not known but may include non-human primates. Strains of MAYV cluster into two distinct lineages, one of which comprises strains from geographically distant regions of South America that cluster temporally. This wide distribution suggests that the vertebrate host also has a wide geographic distribution, and may be a highly mobile animal such as a bird. The other lineage consists of strains isolated from geographically focal locations (Powers *et al.*, 2006). MAYV serosurveillance in areas where isolates cluster geographically, coupled with reservoir studies, suggests that the silvery marmoset (*Callithrix argentata*) may be an important host (LeDuc *et al.*, 1981). Marmosets have highly focal ranges and therefore may not serve as reservoirs for the widely distributed MAYV genotype. Regardless, MAYV occurs throughout northern South America and has the potential to produce urban cycles like CHIKV, possibly using *A. aegypti*, an efficient vector (Long *et al.*, 2011).

Other arthralgic alphaviruses

Semliki Forest virus (SFV) is often used as a model alphavirus. However, no recent studies have elucidated its distribution and evolution. The only other arthralgic alphavirus that causes epidemics is ONNV, the closest relative of CHIKV. Unlike other alphaviruses that use *Aedes* or *Culex* mosquitoes as vectors, the primary ONNV vectors are *Anopheles* mosquitoes. Because it is rarely isolated between epidemics, the only study of ONNV evolution was limited to four strains (Powers *et al.*, 2000). Not surprisingly, most alphavirus studies since 2004 have focused almost exclusively on CHIKV because of its importance to public health.

The remaining Old World alphavirus is Ndumu virus. Ndumu virus has been isolated from mosquitoes (Kokernot *et al.*, 1961; Ochieng *et al.*, 2013) and detected in pigs (Masembe *et al.*, 2012) but is not associated with disease.

Encephalitic alphaviruses

The encephalitic alphaviruses, excluding SINV, occur in the New World. Viruses in this group fall within the Venezuelan (VEEV), eastern (EEEV), and western equine encephalitis virus (WEEV) complexes. See Chapter 8 for more detail on encephalitic alphaviruses.

Eastern equine encephalitis virus

The most comprehensive EEEV analysis, by Arrigo *et al.* (2010b), used structural protein sequences of representatives from all four lineages. These analyses showed that topologies of North American (NA) lineage 1 EEEV are influenced primarily by temporal rather than geographical relationships, consistent with other studies (Armstrong *et al.*, 2008; Molaei *et al.*, 2013; White *et al.*, 2011b). Unlike NA EEEV, South American lineages II and III show geographically based evolutionary relationships. Given that NA lineage I EEEV use passerine birds as amplification/reservoir hosts, the observed temporal evolutionary distribution may reflect the highly mobile vertebrate host, resulting in a wide distribution of minimally diverse strains across the eastern seaboard of the United States. In contrast, EEEV lineages II and III circulate principally among small rodents such as cotton rats (*Sigmodon hispidus*) (Arrigo *et al.*, 2010a); the limited movement of these hosts may restrict the dispersal of EEEV and result in geographic clustering and more highly partitioned local evolution. Lineages II-IV are now classified as a new species, Madariaga virus (http://talk.ictvonline.org/files/ictv_official_taxonomy_updates_since_the_8th_report/m/vertebrate-official/4488.aspx).

Venezuelan equine encephalitis virus

The VEE complex consists of seven subtypes/varieties of various species; of these, subtype I, which includes all the species of VEEV, has been primarily associated with disease in humans and equids. For subtypes II–VI few isolates have been made and inter-subtype evolutionary studies have been restricted to IIIA (Auguste *et al.*, 2009). Despite limited strain sampling, geographic clustering within subtypes suggests limited movement, consistent with the rodent-borne nature of the VEE complex (Aguilar *et al.*, 2009; Kinney *et al.*, 1998; Quiroz *et al.*, 2009; Weaver *et al.*, 2004b). The best-studied VEEV subtype is I, which is further subdivided into four varieties: IAB, IC, ID and IE. Of these, IAB and IC have been associated with major epizootics, of which one, from 1969–1971, spread from Guatemala and El Salvador through Central America to Texas (Anischenko *et al.*, 2006; Brault *et al.*, 2002, 2004; Greene *et al.*, 2005; Rivas *et al.*, 1997; Sudia and Newhouse, 1975; Sudia *et al.*, 1975; Weaver *et al.*, 1996). The evolutionary origin of the epizootic VEEV subtypes is discussed further below. Subtypes ID and IE are primarily enzootic strains that circulate in rodent-mosquito-rodent forest cycles in Central and South America. Subtype II occurs in Florida and uses cotton rats (Coffey *et al.*, 2004) as reservoirs and *Culex* (*Melanoconion*) mosquitoes (Weaver *et al.*, 1986) as vectors. The evolutionary histories of these subtypes may be similar to those of South American EEEV, showing geographical clustering of strains (Adams *et al.*, 2012; Aguilar *et al.*, 2009; Kinney *et al.*, 1998; Quiroz *et al.*, 2009; Weaver *et al.*, 2004b). Thus the evolution of VEEV appears to be confined by the dispersal of rodent reservoirs.

Western equine encephalitis virus complex

Three members of the WEE antigenic complex, including WEEV, Highlands J virus (HJV) and Buggy Creek virus (BCRV), have a unique evolutionary history among the alphaviruses. These viruses are descendants of an ancient recombination event between ancestors of EEEV and SINV (Hahn *et al.*, 1988). WEE complex viruses have non-structural proteins from an EEEV-like ancestor and structural proteins (E2, 6K, E1) from a SINV-like ancestor. Although the recombination event that gave rise to these viruses is well characterized, sequencing of strains of WEEV, HJV and BCRV is limited, rendering their recent evolutionary histories poorly understood. Of the alphavirus complexes, the WEEV complex is probably the most diverse in terms of host use. BCRV and HJV occur in North America and circulate between mites (*Oeciacus vicarious*) and birds. WEEV is similar to VEEV and EEEV in that it uses mosquitoes as vectors, including *Culex tarsalis*. The incidence of WEEV in the western USA has declined over the past five decades for unknown reasons (Reisen *et al.*, 2008). WEEV is more widely distributed than either EEEV or VEEV: although EEEV is found across the Americas, lineage I is confined to North America, whereas lineages II–IV are found in South America. By contrast, WEEV is classified as a single lineage and is found throughout South and North America (Bergren *et al.*, 2014). This diffuse distribution may reflect that WEEV is transmitted by migratory birds between North and South America, or different ecological niches occupied by WEEV compared to EEEV.

Sindbis virus

SINV is the only other Old World clade alphavirus where genetic relationships have been delineated. Although it causes less severe disease than the clades described above, SINV exhibits a wide distribution across Eurasia, Africa and Oceania (Jost *et al.*, 2010; Norder *et al.*, 1996; Saleh *et al.*, 2003; Sammels *et al.*, 1999). This is consistent with the fact that the virus has an avian host, and therefore strains tend to cluster temporally rather than geographically, reflecting efficient dispersal. Interestingly, Whataroa virus (WHATV), whose vertebrate host is unknown, clusters geographically (Saleh *et al.*, 2004), suggesting that it uses a less mobile host than its close relative SINV.

Eilat virus

In 2012, EILV, an alphavirus that replicates only in insect cells, was identified and characterized (Nasar *et al.*, 2012). This is the first insect-specific alphavirus, in contrast to the flaviviruses, for which numerous insect-specific viruses have been identified (Cook *et al.*, 2006; Crabtree *et al.*, 2003; Hobson-Peters *et al.*, 2013; Hoshino *et al.*, 2007, 2009). EILV groups most closely with the WEE complex, sitting equidistant from Aura virus (AURAV) and Trocara virus (TROCV). Given its position in the alphavirus phylogeny and most parsimonious reconstructions of host range superimposed on the trees, EILV may have secondarily lost the ability to infect vertebrate cells. This theory concurs with recent studies that suggest alphaviruses were insect-only viruses that subsequently acquired the ability to infect vertebrate cells (Ventoso, 2012).

Microscale evolution

Introduction and overview of RNA virus mutant swarms

Origin and nature of heterogeneity

Alphaviruses exist as heterogeneous populations of viral RNAs called mutant swarms (Fig. 2.3). Heterogeneity arises from frequent nucleotide misincorporation during replication due to the inability of the viral RNA-dependent RNA polymerase to correct errors, coupled with rapid replication and exponential growth of viral populations. Given a 12 kb alphavirus genome and a mutation frequency of $\approx 10^{-4}$, each nascent RNA genome on average possesses at least one mutation. By comparison, this mutation frequency is several $\log_{10}$ higher than for double-stranded DNA viruses (Holmes, 2009). Most mutant genomes are detrimental and thus removed from the swarm (purifying selection) to prevent reductions in infectivity that could lead to extinction. A dynamic mutation-selection balance, therefore, determines the size and genetic diversity of a mutant swarm (Fig. 2.3). Mutant swarms are relatively understudied due to difficulties inherent to characterizing massive and ever-changing populations in sufficient depth. Despite this, an increasing body of evidence, mostly from non-vector-borne RNA viruses including HIV, picornaviruses and rhabdoviruses (reviewed comprehensively in Domingo *et al.*, 2012), shows that viral genetic diversity is critical to virus evolution. Furthermore, these studies show that the size and composition of a mutant swarm fundamentally affects the ability of an RNA virus to change and adapt to new environments.

By virtue of co-existing in a shared environment, the genomes that comprise mutant swarms are selected as a population based on their replicative success that determine tropism and host range, responses to immune pressures, and anatomical bottlenecks. This means that, strictly speaking, the whole swarm should be considered as one replicative unit in an infected cell (Del Portillo *et al.*, 2011). However, this concept is quickly complicated when heterogeneous progeny from one infected cell invade neighbouring cells, and a second level of competition among viral particles occurs between RNAs that have originated from different progenitors. Regardless, the term 'mutant swarm' is commonly used *in vivo* because it establishes the concept of ensembles of closely related but non-identical genomes that are generated via mutation-selection processes (Domingo *et al.*, 2012; Eigen, 1996; Lauring and Andino, 2010).

Fitness and adaptability

The size and diversity of an RNA virus mutant swarm influence fitness and adaptability. Positive selection as a result of a fit phenotype results in increased abundance of a genotype. In contrast, negative selection eliminates disadvantaged genomes from a population. A genetically diverse mutant swarm renders a population less prone to the consequences of negative pressures that target

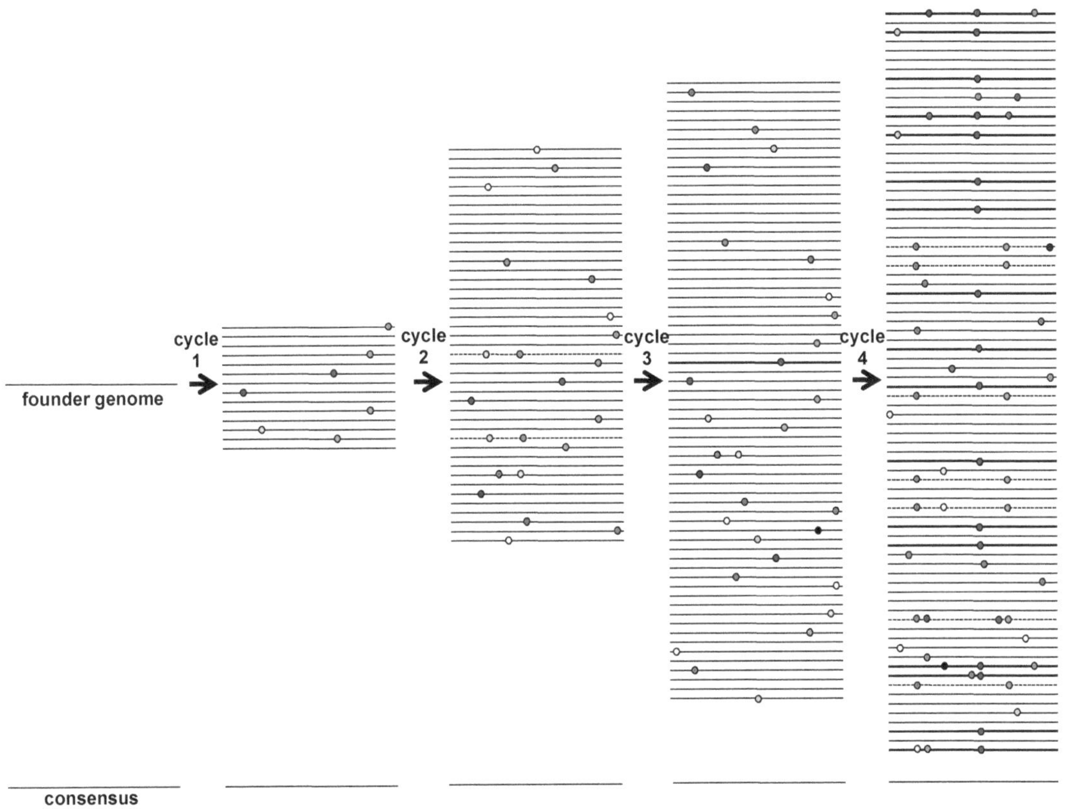

Figure 2.3 Schematic representation of the evolution of a viral mutant swarm beginning with a founding genome. Genomes are represented as lines and mutations are shown as coloured dots. Dotted lines indicate genomes that have acquired deleterious mutations that are negatively selected and do not survive to serve as templates for successive replication cycles, represented as arrows. Mutations that increase in frequency (blue dot in cycle 4) confer a phenotypic advantage (represented as a heavy line) and are positively selected. The bottom lines show the consensus (average) sequence. Despite constant evolution of the mutant swarm that leads to dynamic changes in the population, the consensus often remains unchanged. The number of genomes in a population in a cell can reach several thousand, providing a highly complex swarm of mutants. The genomes that circumvent anatomical barriers, to infect the mosquito midgut, e.g. serve as founders that act as templates for subsequent replication cycles of alphaviruses.

certain genotypes and more likely to contain variants with potential phenotypic advantages; together these features augment plasticity and adaptability. A high fidelity CHIKV variant (Coffey *et al.*, 2011) that produces fewer mutants than wild-type virus is less fit in vectors and vertebrates; if specific genotypes are removed from the wild-type population through negative selection or population bottlenecks, there is still a higher likelihood than for high fidelity populations that some remaining genomes can resist subsequent evolutionary pressures. However, genetic diversity is not the only factor required to maintain fitness. The composition and abundance of variants in a population also affect fitness. Furthermore, dynamic interactions between variants in a swarm may also determine the dominant phenotype. None of these interactions are well understood, but could occur via genome recombination, which necessitates co-infection of the same cell by two virus genotypes, or via complementation or suppression, processes that have been documented for flaviviruses (Aaskov *et al.*, 2006), picornaviruses (de la Torre and Holland, 1990; Gonzalez-Lopez *et al.*, 2004) and rhabdoviruses (Novella *et al.*, 2004), but not for alphaviruses. Although recombination is likely a rare event, the WEEV complex is a recombinant that was derived from EEEV and SINV-like ancestors (Hahn *et al.*, 1988) (discussed in detail above).

Dual host cycling and its impact on alphavirus evolution

Alphaviruses that obligately cycle between arthropods and vertebrates are subject to unique selective pressures compared to single-host RNA viruses that need only replicate in the same host species or in closely related hosts. This cycling pattern partly explains why strains of arboviruses isolated from nature show less variation than predicted by their high mutation rates (Davis *et al.*, 2005; Ebel *et al.*, 2004; Parameswaran *et al.*, 2012), and why genetic studies reveal that they undergo strong purifying selection (Jenkins and Holmes, 2003). Many experimental arbovirus evolution studies, including with EEEV, SINV, and VEEV (reviewed in Coffey *et al.*, 2013), show that disparate hosts present conflicting challenges to replication that limit adaptation to either host alone by imposing fitness costs where adaptations are antagonistic (Turner *et al.*, 2010). These trade-offs that arboviruses face in their obligate alternating host transmission cycles are evidenced by increased adaptability when one host is artificially removed via serial passage. Experiments with serially passaged CHIKV show that augmented adaptability, measured as higher fitness in novel cell types and enhanced neutralization and antiviral compound resistance, is accompanied by increased genetic diversity. In contrast, alternation between cell types restricts fitness and diversity increases, suggesting that only mutations beneficial or neutral in both hosts are maintained and that these variants retain fitness in alternating cycling (Coffey and Vignuzzi, 2011). Most trade-off studies have been based in cell monocultures or mutant swarms were not sequenced after serial or alternating *in vivo* passaging; future studies employing deep sequencing technologies will better characterize dynamic changes in mutant swarms after dual-host cycling.

Genetic bottlenecks and their impact on evolution

Intrahost bottlenecks

The ability of alphaviruses to survive bottlenecks within and between hosts in dynamic environments impacts their evolution and may be important for predicting host range shifts and expansions that cause human and veterinary disease. Bottlenecks that reduce arbovirus population size can modify the mutant swarm and affect viral fitness by restricting phenotypic plasticity that stems from genetic diversity. Muller's ratchet, an evolutionary theory, proposes that asexual organisms with high mutation rates and small population sizes irreversibly accumulate deleterious mutations unless compensatory mutations, e.g. delivered via recombination, restore mutation-free genomes to the population (Muller, 1964). Model studies support this idea (Lynch and Gabriel, 1990) and experimental data with EEEV (Weaver *et al.*, 1999a) and the sandfly-borne rhabdovirus vesicular stomatitis virus (VSV) validate this concept; fitness decreases after serial bottleneck passages can be rescued by subsequent large population passages, albeit with much replication needed to overcome the ratchet (Duarte *et al.*, 1993).

Despite poor characterization of anatomical barriers in vertebrate hosts, obstacles to productive alphavirus transmission by mosquito vectors are relatively well defined. As a first step, arboviruses in a bloodmeal ingested by a mosquito must infect the midgut epithelium. Studies with fluorescent protein-expressing epidemic VEEV replicon particles that are capable of only a single round of replication show that few midgut cells are infected, even at large ingested doses. Furthermore, when red and green fluorescent-expressing replicons are mixed in the same bloodmeal, the two colours co-localized in more cells than expected by independent chance (the product of the frequency of each individual cell infection rate), suggesting that only certain 'portal' midgut cells are susceptible (Smith *et al.*, 2008). By contrast, similar experiments with an enzootic VEEV subtype and its enzootic mosquito vector showed uniform susceptibility of the midgut epithelium to infection, suggesting a long-term virus–vector association that may promote higher midgut infectivity (Kenney *et al.*, 2012). Some arboviruses encounter secondary impediments to dissemination, where they are unable to escape from the midgut epithelium, infect salivary glands, and escape from salivary gland cells into saliva for transmission to vertebrates. Experiments tracking barcoded VEEV variants show that, even for virus species that successfully traverse these barriers, the number of variants is reduced at midgut escape and salivary gland infection, and that smaller initial bloodmeal populations are more prone to reductions in variant diversity (Forrester *et al.*, 2012a). Together these results suggest strong genetic bottlenecks that reduce diversity in parallel with population size changes. These observed changes in swarm genetic diversity in mosquitoes are somewhat contradictory to observations from alphaviruses (Weaver, 2006) and other mosquito-borne arboviruses isolated from nature (Brackney *et al.*, 2011), as well as experimental *in vivo* passaging studies that show the maintenance of genetic and phenotypic stability of the consensus (average sequence) (Coffey *et al.*, 2008). Genetic bottlenecks in vectors may be circumvented by re-establishment of diversity via subsequent replication. Studies with CHIKV support this idea; although population diversity in the midgut and salivary glands is reduced compared to the bloodmeal input or midgut population, respectively, it recovers downstream of each barrier and the consensus sequence remains unchanged (Stapleford *et al.*, 2014).

Interhost bottlenecks

Given strong intrahost genetic bottlenecks in mosquitoes, inefficient transmission could interrupt alphavirus cycling in nature (i.e. interhost bottlenecks). The degree of natural extinction of alphaviruses in vectors has not been assessed directly. However, the rate of extinction within individual mosquito–vertebrate–mosquito lineages is undoubtedly very high because most mosquitoes do not survive long enough in nature to feed more than once. The maintenance of consensus genetic stability in nature may be explained by highly intense circulation such that extinction of some or even most, but not all, individual lineages via vector bottlenecks in individual mosquitoes does not lead to overall regional extinction.

No alphavirus studies have examined mutant swarms in naturally infected mosquitoes, humans, or other vertebrate hosts, to study host-specific swarm patterns or to identify common minor variants shared across individuals and species. Barcoded VEEV variants in mosquitoes that transmitted to laboratory mice were also observed in the brains of animals, (Forrester *et al.*, 2012a), suggesting that mosquito–vertebrate transmission of alphaviruses does not present a major bottleneck [although few ($n=3$) infected mice were sampled]. Virus doses transmitted by mosquitoes vary greatly, and are probably overestimated by standard laboratory salivation assays where infected mosquitoes expectorate more virus into tubes than *in vivo* (Smith *et al.*, 2006). Variance in transmitted doses, especially when very few particles are transmitted, likely impacts the outcome of vertebrate infection and, by extension, maintenance of cycling in natural foci.

Interruptions in seasonal transmission may represent another significant bottleneck imposed on arboviruses. Phylogenetic studies show that most arbovirus maintenance is geographically focal and in some cases may involve persistent infection of vectors that survive winter. Infected swallow bugs, vectors of BCRV, can overwinter in bird nests in the Great Plains of the United States (Brown *et al.*, 2010), and RRV survives Australian winter in adult *Cx. annulirostris* (Kay, 1982) and in desiccation-resistant eggs of *Ae. vigilax* (Lindsay *et al.*, 1993). Some arboviruses also persistently infect vertebrate hosts (Kuno, 2001), but the likelihood that these animals maintain viraemias sufficient to infect feeding vectors is very low.

Alphavirus microevolution

Understanding dynamic changes in mutant swarms, so-called 'microevolution', is important for understanding evolutionary trajectories of all RNA viruses including alphaviruses. For CHIKV and VEEV, it is unclear whether adaptive mutations that undergo selective sweeps during epidemics (see below) are present as minority variants in populations and positively selected in the appropriate ecological settings, or whether they arise *de novo* after random mutation. In the case of the former, standard laboratory infections of vectors and vertebrate models, coupled with deep sequencing and reverse genetics to express candidate minority variants as dominant mutations, could be used to identify variants that could promote future epidemics.

Microevolution of mutant swarms may also influence pathogenesis in vertebrates. While strains of EEEV from North America cause encephalitis with 50–70% human case fatality, South American strains are not often associated with human disease (Zacks and Paessler, 2010). Although mutant swarms of EEEV have received little attention, this disparity in pathogenesis may be partially explained by contrasting selective pressures on swarms that result from different patterns of host use, transmission and maintenance. South American EEEV cycle year-round in discrete foci, probably using birds, rodents, marsupials, reptiles and *Culex* (*Melanoconion*) spp. as vectors (Weaver *et al.*, 1999b). In contrast, North American EEEV use ornithophilic *Culiseta melanura* mosquitoes and passerine birds in temperate hardwood swamps (Scott and Weaver, 1989). EEEV mutant swarms in both regions are probably subjected to focal bottlenecks arising from geographically restricted transmission (South America) and overwintering (North America) (Arrigo *et al.*, 2010b). However, the extent and mechanisms by which bottlenecks affect circulation and how these dynamics relate to differential EEEV pathogenesis remain largely unknown.

Consequences of alphavirus evolution

Alphavirus emergence and vector usage

Like many arboviruses, alphaviruses use a variety of vectors, but almost all are mosquitoes in the family Culicidae. By way of comparison, many flaviviruses also use ticks, while the only non-culicid vectors reported for alphaviruses are avian bugs in the family Cimicidae (Hanley and Weaver, 2008; Weaver *et al.*, 2004a). On the other hand, flaviviruses appear to evolve in concert with certain genera of mosquitoes, while alphaviruses show little correlation between their relationships and mosquito vector taxa. This apparent promiscuity in vector usage by alphaviruses has also led to their exploitation of new vector niches to enhance disease emergence, as described below (Fig. 2.4).

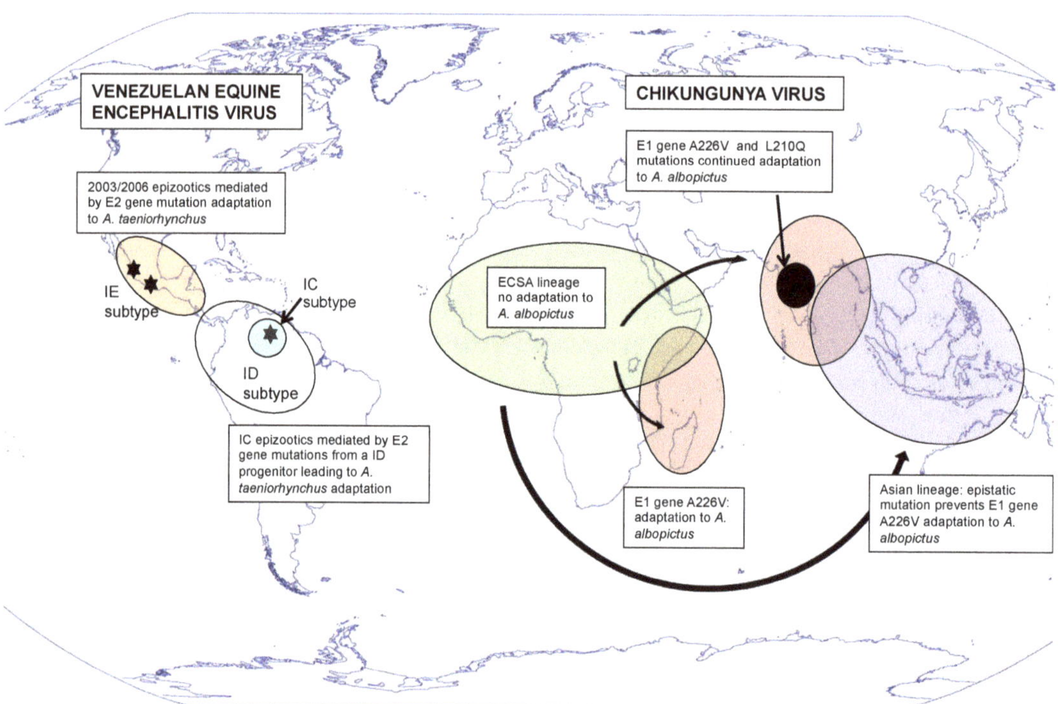

Figure 2.4 The emergence of epidemic CHIKV and VEEV viruses showing mutations associated with various lineages that promoted viral spread. Locations are approximate and not comprehensive.

Case study: CHIKV envelope gene mutations in Indian Ocean and Asian outbreaks and adaptation to new vectors

CHIKV in its ancestral state circulates in sylvatic habitats of Africa, where it is transmitted by several different arboreal *Aedes* spp. mosquitoes among non-human primates and possibly other vertebrate hosts (Coffey *et al.*, 2013; Diallo *et al.*, 1999; Weaver *et al.*, 2012). This sylvatic cycle can lead to spillover infections by mosquito vectors that contact humans in the forest or in nearby villages, especially by *Aedes* (*Diceromyia*) *furcifer* in eastern Senegal (Diallo *et al.*, 1999, Diallo *et al.*, 2012). However, most, if not all, major chikungunya fever (CHIK) epidemics occur when the virus reaches urban habitats occupied by urban mosquitoes in the subgenus *Aedes* (*Stegomyia*), namely *A. aegypti* and/or *A. albopictus*. These mosquitoes, especially *A. aegypti*, live close to people, and their behavioural characteristics and high susceptibility to CHIKV infection are critical to their role in transmission among humans, who become viraemic and thus serve as amplification hosts. There is no evidence that CHIKV adaptation is needed to initiate the urban epidemic cycle when *A. aegypti* is the principal vector, and indeed urban transmission of the ECSA CHIKV lineage introduced from Africa to Asia during or before the 1950s has been accompanied by no detectable adaptation to its principal vector. However, the recent urban emergence of CHIKV (and establishment of the IOL genotype) from the east coast of Kenya into islands of the Indian Ocean and into India was accompanied by at least five adaptive mutations that enhanced transmission by *A. albopictus* while having little or no effect on infection of *A. aegypti* (Fig. 2.4). Because *A. albopictus* has rapidly invaded many tropical and temperate regions of the world beyond its native Asian distribution since 1985, CHIKV now poses a threat not only to the tropics but also to temperate, industrialized countries like Italy (Rezza *et al.*, 2007) and France (Gould *et al.*, 2010; Grandadam *et al.*, 2011). The many documented importations of CHIKV-infected persons to the Americas (Lanciotti *et al.*, 2007) also underscore the potential for this virus to spread and become endemic nearly worldwide. The first adaptive mutation implicated in the recent outbreaks, a A226V substitution in the E1 envelope glycoprotein, increased by ca. 100-fold the infectivity of the ECSA CHIKV genotype for *A. albopictus* and probably played a major role in the explosive nature of the epidemics (Tsetsarkin *et al.*, 2007). This mutation, which was selected convergently at least four times (de Lamballerie *et al.*, 2008; Volk *et al.*, 2010), was followed in India by a second-step adaptive substitution in the E2 glycoprotein, L210Q. Like E1-A226V, this mutation also has no detectable phenotype for infection of *A. aegypti*. More recently, an additional three *A. albopictus*-adaptive E2 mutations were characterized, and combinations of these, not yet seen in nature, will probably result in further adaptation (Tsetsarkin *et al.*, 2014). A surprising outcome of these genetic studies is that, although the recently emerging IOL strains have undergone at least two *A. albopictus*-adaptive mutations, the Asian CHIKV lineage circulating in South-East Asia since the 1950s shows no evidence of adaptation to this vector in its native range. A single epistatic mutation in these Asian strains, probably the result of a founder effect when it was introduced there from East Africa, is responsible for its inability to adapt to *A. albopictus* for over 50 years (Tsetsarkin *et al.*, 2011a). This epistatic constraint has become even more important with the recent importation of an Asian lineage strain to the Caribbean in late 2013, followed by spread to South, Central and North America, where local transmission has been detected (Weaver, 2014). The lack of evidence of a major role for *A. albopictus* in transmission of Asian lineage strains since the 1950s, combined with the epistatic constraint on adaptation of the Asian lineage to this vector, may limit CHIKV spread into temperate American areas where *A. aegypti* cannot survive cold winters.

Case study: VEEV envelope gene mutations and epizootics/epidemics via adaptation to increased infectivity and virulence

The other alphavirus for which there is strong evidence that adaptive mutations change its host range to mediate emergence is VEEV. This virus occurs in four different subtypes: IAB, IC, ID and IE. The first two are considered epizootic/epidemic strains because they cause major outbreaks in equids and people when one or more of several large-mammalophilic mosquito vectors such as *Aedes* and *Psorophora* spp. transmit in an equid–mosquito–equid cycle that can rapidly spread and infect up to hundreds of thousands of people via spillover (Weaver *et al.*, 2004b). However, these epizootic/epidemic strains are short lived and disappear when the equine amplification hosts are killed or become immune during outbreaks. Subtypes ID and IE are enzootic VEEV strains that circulate continuously in sylvatic *Culex* (*Melanoconion*) spp.-borne mosquito–rodent–mosquito cycles nearly throughout the neotropics. Like the epidemic strains, these subtypes also infect people via spillover to cause a highly debilitating acute febrile illness that can be fatal. Although most subtype ID and many IE strains are avirulent for equids, the burden of human disease that results from enzootic spillover has been estimated at the tens of thousands of cases annually (Aguilar *et al.*, 2011).

The origins of the epizootic/epidemic IAB and IC strains remained enigmatic for many years. However, phylogenetic (Powers *et al.*, 1997; Weaver *et al.*, 1992) and reverse genetic (Anischenko *et al.*, 2006; Brault *et al.*, 2002) studies eventually revealed that enzootic, subtype ID VEEV strains can evolve to epizootic, subtype IC strains via single point mutations in the E2 envelope glycoprotein gene. These mutations increase equine virulence and viraemia and also mediate an antigenic transformation that changes their serotype. Amplification efficiency can also be augmented by mosquito vector-adaptive mutations, as was implicated in equine VEE outbreaks in southern Mexico on the Pacific coast of Chiapas and Oaxaca states. There, an *A.* (*Ochlerotatus*) *taeniorhynchus*-adaptive E2 mutation increased VEEV infectivity for this vector of local subtype IE strains (Brault *et al.*, 2004), which had also become equine virulent although they retained a low amplification competence (viraemia) phenotype (Gonzalez-Salazar *et al.*, 2003).

Conclusion

Evolution of alphavirus species is influenced by patterns of host and vector use that affect transmission and dispersal, and may be subjected to temporal interruptions in cycling that influence genetic structure. On a microscale, alphaviruses that persist must survive dynamic mutation-selection processes to traverse anatomical barriers in individual hosts. Increasing evidence from in-depth viral population studies indicate that the size, complexity and composition of minority variants in the mutant swarm influence the phenotype. Point mutations in structural genes of alphaviruses have produced adaptive CHIKV and VEEV variants with augmented infectiousness for vertebrate and vector hosts in epizootic and epidemic settings. Future studies will increase understanding of how different evolutionary processes give rise to this important group of pathogens and how alphaviruses evolve and adapt to cause human and veterinary disease.

References

Aaskov, J., Buzacott, K., Thu, H.M., Lowry, K., and Holmes, E.C. (2006). Long-term transmission of defective RNA viruses in humans and *Aedes* mosquitoes. Science *311*, 236–238.

Adams, A.P., Navarro-Lopez, R., Ramirez-Aguilar, F.J., Lopez-Gonzalez, I., Leal, G., Flores-Mayorga, J.M., Travassos da Rosa, A.P., Saxton-Shaw, K.D., Singh, A.J., Borland, E.M., *et al.* (2012). Venezuelan equine encephalitis virus activity in the Gulf Coast region of Mexico, 2003–2010. PLoS Negl. Trop. Dis. *6*, e1875.

Aguilar, P.V., Adams, A.P., Suarez, V., Beingolea, L., Vargas, J., Mancock, S., Freire, J., Espinoza, W.R., Felices, V., Diaz, A., *et al.* (2009). Genetic characterization of Venezuelan equine encephalitis virus from Bolivia, Ecuador and Peru: identification of a new subtype ID lineage. PLoS Negl. Trop. Dis. *15*, e514.

Aguilar, P.V., Estrada-Franco, J.G., Navarro-Lopez, R., Ferro, C., Haddow, A.D., and Weaver, S.C. (2011). Endemic Venezuelan equine encephalitis in the Americas: hidden under the dengue umbrella. Future Virol. *6*, 721–740.

Anischenko, M., Bowen, R.A., Paessler, S., Austgen, L., Greene, I.P., and Weaver, S.C. (2006). Venezuelan encephalitis emergence mediated by a phylogenetically predicted viral mutation. Proc. Natl. Acad. Sci. U.S.A. *103*, 4994–4999.

Armstrong, P.M., Andreadis, T.G., Anderson, J.F., Stull, J.W., and Mores, C.N. (2008). Tracking eastern equine encephalitis virus perpetuation in the northeastern United States by phylogenetic analysis. Am. J. Trop. Med. Hyg. *79*, 291–296.

Arrigo, N.C., Adams, A.P., Watts, D.M., Newman, P.C., and Weaver, S.C. (2010a). Cotton rats and house sparrows as hosts for North and South American strains of eastern equine encephalitis virus. Emerg. Infect. Dis. *16*, 1373–1380.

Arrigo, N.C., Adams, A.P., and Weaver, S.C. (2010b). Evolutionary patterns of eastern equine encephalitis

virus in North versus South America suggest ecological differences and taxonomic revision. J. Virol. *84*, 1014–1025.
Auguste, A.J., Volk, S.M., Arrigo, N.C., Martinez, R., Ramkissoon, V., Adams, A.P., Thompson, N.N., Adesiyun, A.A., Chadee, D.D., Foster, J.E., *et al.* (2009). Isolation and phylogenetic analysis of Mucambo virus (Venezuelan equine encephalitis complex subtype IIIA) in Trinidad. Virology *392*, 123–130.
Beaty, B.J., Calisher, C.H., and Shope, R.E. (1989). Arboviruses. In Diagnostic procedures for viral, rickettsial and chlamydial infections, N.J. Schmidt, and R.W. Emmons, eds. (American Public Health Association, Washington, DC, USA), pp. 797–855.
Bergren, N.A., Auguste, A.J., Forrester, N.L., Negi, S.S., Braun, W.A., and Weaver, S.C. (2014). Western equine encephalitis virus: evolutionary analysis of a declining aphavirus based on complete genome sequences. J. Virol. *88*, 9260–9267.
Bingham, A.M., Graham, S.P., Burkett-Cadena, N.D., White, G.S., Hassan, H.K., and Unnasch, T.R. (2012). Detection of eastern equine encephalomyelitis virus RNA in North American snakes. Am. J. Trop. Med. Hyg. *87*, 1140–1144.
Brackney, D.E., Pesko, K.N., Brown, I.K., Deardorff, E.R., Kawatachi, J., and Ebel, G.D. (2011). West Nile virus genetic diversity is maintained during transmission by *Culex pipiens quinquefasciatus* mosquitoes. PLOS ONE *6*, e24466.
Brault, A.C., Powers, A.M., Holmes, E.C., Woelk, C.H., and Weaver, S.C. (2002). Positively charged amino acid substitutions in the E2 envelope glycoprotein are associated with the emergence of Venezuelan equine encephalitis virus. J. Virol. *76*, 1718–1730.
Brault, A.C., Powers, A.M., Ortiz, D., Estrada-Franco, J.G., Navarro-Lopez, R., and Weaver, S.C. (2004). Venezuelan equine encephalitis emergence: Enhanced vector infection from a single amino acid substitution in the envelope glycoprotein. Proc. Natl. Acad. Sci. U.S.A. *101*, 11344–11349.
Brown, C.R., Strickler, S.A., Moore, A.T., Knutie, S.A., Padhi, A., Brown, M.B., Young, G.R., O'Brien, V.A., Foster, J.E., and Komar, N. (2010). Winter ecology of Buggy Creek virus (*Togaviridae*, Alphavirus) in the Central Great Plains. Vector Borne Zoonotic Dis. *10*, 355–363.
Coffey, L.L., and Vignuzzi, M. (2011). Host alternation of chikungunya virus increases fitness while restricting population diversity and adaptability to novel selective pressures. J. Virol. *85*, 1025–1035.
Coffey, L.L., Carrara, A.S., Paessler, S., Haynie, M.L., Bradley, R.D., Tesh, R.B., and Weaver, S.C. (2004). Experimental Everglades virus infection of cotton rats (*Sigmodon hispidus*). Emerg. Infect. Dis. *10*, 2182–2188.
Coffey, L.L., Vasilakis, N., Brault, A.C., Powers, A.M., Tripet, F., and Weaver, S.C. (2008). Arbovirus evolution *in vivo* is constrained by host alternation. Proc. Natl. Acad. Sci. U.S.A. *105*, 6970–6975.
Coffey, L.L., Beeharry, Y., Borderia, A.V., Blanc, H., and Vignuzzi, M. (2011). Arbovirus high fidelity variant loses fitness in mosquitoes and mice. Proc. Natl. Acad. Sci. U.S.A. *108*, 16038–16043.
Coffey, L.L., Forrester, N., Tsetsarkin, K., Vasilakis, N., and Weaver, S.C. (2013). Factors shaping the adaptive landscape for arboviruses: implications for the emergence of disease. Future Microbiol. *8*, 155–176.
Cook, S., Bennett, S.N., Holmes, E.C., De Chesse, R., Moureau, G., and de Lamballerie, X. (2006). Isolation of a new strain of the flavivirus cell fusing agent virus in a natural mosquito population from Puerto Rico. J. Gen. Virol. *87*, 735–748.
Crabtree, M.B., Sang, R.C., Stollar, V., Dunster, L.M., and Miller, B.R. (2003). Genetic and phenotypic characterization of the newly described insect flavivirus, Kamiti River virus. Arch. Virol. *148*, 1095–1118.
Davis, C.T., Ebel, G.D., Lanciotti, R.S., Brault, A.C., Guzman, H., Siirin, M., Lambert, A., Parsons, R.E., Beasley, D.W., Novak, R.J., *et al.* (2005). Phylogenetic analysis of North American West Nile virus isolates, 2001–2004: evidence for the emergence of a dominant genotype. Virology *342*, 252–265.
Del Portillo, A., Tripodi, J., Najfeld, V., Wodarz, D., Levy, D.N., and Chen, B.K. (2011). Multiploid inheritance of HIV-1 during cell-to-cell infection. J. Virol. *85*, 7169–7176.
Diallo, D., Sall, A.A., Buenemann, M., Chen, R., Faye, O., Diagne, C.T., Ba, Y., Dia, I., Watts, D., Weaver, S.C., *et al.* (2012). Landscape ecology of sylvatic chikungunya virus and mosquito vectors in southeastern Senegal. PLoS Negl. Trop. Dis. *6*, e1649.
Diallo, M., Thonnon, J., Traore-Lamizana, M., and Fontenille, D. (1999). Vectors of Chikungunya virus in Senegal: current data and transmission cycles. Am. J. Trop. Med. Hyg. *60*, 281–286.
Domingo, E., Sheldon, J., and Perales, C. (2012). Viral quasispecies evolution. Microbiol. Mol. Biol. Rev. *76*, 159–216.
Duarte, E.A., Clarke, D.K., Moya, A., Elena, S.F., Domingo, E., and Holland, J. (1993). Many-trillionfold amplification of single RNA virus particles fails to overcome the Muller's ratchet effect. J. Virol. *67*, 3620–3623.
Ebel, G.D., Carricaburu, J., Young, D., Bernard, K.A., and Kramer, L.D. (2004). Genetic and phenotypic variation of West Nile virus in New York, 2000–2003. Am. J. Trop. Med. Hyg. *71*, 493–500.
Eigen, M. (1996). On the nature of virus quasispecies. Trends Microbiol. *4*, 216–218.
Forrester, N.L., Guerbois, M., Seymour, R.L., Spratt, H., and Weaver, S.C. (2012a). Vector-borne transmission imposes a severe bottleneck on an RNA virus population. PLoS Pathog. *8*, e1002897.
Forrester, N.L., Palacios, G., Tesh, R.B., Savji, N., Guzman, H., Sherman, M., Weaver, S.C., and Lipkin, W.I. (2012b). Genome scale phylogeny of the Alphavirus genus suggests a marine origin. J. Virol. *85*, 8709–8717.
Gonzalez-Lopez, C., Arias, A., Pariente, N., Gomez-Mariano, G., and Domingo, E. (2004). Preextinction viral RNA can interfere with infectivity. J. Virol. *78*, 3319–3324.
Gonzalez-Salazar, D., Estrada-Franco, J.G., Carrara, A.S., Aronson, J.F., and Weaver, S.C. (2003). Equine

amplification and virulence of subtype IE Venezuelan equine encephalitis viruses isolated during the 1993 and 1996 Mexican epizootics. Emerg. Infect. Dis. *9*, 161–168.

Gould, E.A., Gallian, P., De Lamballerie, X., and Charrel, R.N. (2010). First cases of autochthonous dengue fever and chikungunya fever in France: from bad dream to reality! Clin. Microbiol. Infect. *16*, 1702–1704.

Graham, S.P., Hassan, H.K., Chapman, T., White, G., Guyer, C., and Unnasch, T.R. (2012). Serosurveillance of eastern equine encephalitis virus in amphibians and reptiles from Alabama, USA. Am. J. Trop. Med. Hyg. *86*, 540–544.

Grandadam, M., Caro, V., Plumet, S., Thiberge, J.-M., Souares, Y., Failloux, A.-B., Tolou, H.J., Budelot, M., Cosserat, D., Leparc-Goffart, I., *et al.* (2011). Chikungunya virus, Southeastern France. Emerg. Infect. Dis. *17*, 910–913.

Greene, I.P., Paessler, S., Austgen, L., Anischenko, M., Brault, A.C., Bowen, R.A., and Weaver, S.C. (2005). Envelope glycoprotein mutations mediate equine amplificiation and virulence of epizootic Venezuelan equine encephalitis virus. J. Virol. *79*, 9128–9133.

Hahn, C.S., Lustig, S., Strauss, E.G., and Strauss, J.H. (1988). Western equine encephalitis virus is a recombinant virus. Proc. Natl. Acad. Sci. U.S.A. *85*, 5997–6001.

Hanley, K.A., and Weaver, S.C. (2008). Arbovirus Evolution. In Origin and evolution of viruses, E. Domingo, C.R. Parrish, and J.J. Holland, eds. (Academic Press), pp. 351–391.

Hobson-Peters, J., Yam, A.W., Lu, J.W., Setoh, Y.X., May, F.J., Kurucz, N., Walsh, S., Prow, N.A., Davis, S.S., Weir, R., *et al.* (2013). A new insect-specific flavivirus from northern Australia suppresses replication of West Nile virus and Murray Valley encephalitis virus in co-infected mosquito cells. PLOS ONE *8*, e56534.

Holmes, E.C. (2009). The evolutionary genetics of emerging viruses. Annu. Rev. Ecol. Evol. Syst. *40*, 353–372.

Hoshino, K., Isawa, H., Tsuda, Y., Yano, K., Sasaki, T., Yuda, M., Takasaki, T., Kobayashi, M., and Sawabe, K. (2007). Genetic characterization of a new insect flavivirus isolated from *Culex pipiens* mosquito in Japan. Virology *359*, 405–414.

Hoshino, K., Isawa, H., Tsuda, Y., Sawabe, K., and Kobayashi, M. (2009). Isolation and characterization of a new insect flavivirus from Aedes albopictus and Aedes flavopictus mosquitoes in Japan. Virology *391*, 119–129.

Jenkins, G.M., and Holmes, E.C. (2003). The extent of codon usage bias in human RNA viruses and its evolutionary origin. Virus Res. *92*, 1–7.

Jost, H., Bialonski, A., Storch, V., Gunther, S., Becker, N., and Schmidt-Chanasit, J. (2010). Isolation and phylogenetic analysis of Sindbis viruses from mosquitoes in Germany. J. Clin. Microbiol. *48*, 1900–1903.

Kay, B.H. (1982). Three modes of transmission of Ross River virus by *Aedes vigilax* (Skuse). Aust. J. Exp. Biol. Med. *60*, 339–344.

Kenney, J.L., Adams, A.P., Gorchakov, R., Leal, G., and Weaver, S.C. (2012). Genetic and anatomic determinants of enzootic Venezuelan equine encephalitis virus infection of *Culex* (*Melanoconion*) *taeniopus*. PLoS Negl. Trop. Dis. *6*, e1606.

Kinney, R.M., Pfeffer, M., Tsuchiya, K.R., Chang, G.J., and Roehrig, J.T. (1998). Nucleotide sequences of the 26S mRNA's of the viruses defining the Venezuelan equine encephalitis antigenic complex. Am. J. Trop. Med. Hyg. *59*, 952–964.

Knope, K., Whelan, P., Smith, D., Nicholson, J., Moran, R., Doggett, S., Sly, A., Hobby, M., and Wright, P. (2013). Arboviral diseases and malaria in Australia, 2010–11: Annual report of the National Arbovirus and Malaria Advisory Committee. Commun. Dis. Intell. Q Rep *37*, E1–E20.

Kokernot, R.H., McIntosh, B.M., and Worth, C.B. (1961). Ndumu virus, a hitherto unknown agent, isolated from culicine mosouitoes collected in northern Natal. Union of South Africa. Am. J. Trop. Med. Hyg. *10*, 383–386.

Kuno, G. (2001). Persistence of arboviruses and antiviral antibodies in vertebrate hosts: its occurrence and impacts. Rev. Med. Virol. *11*, 165–190.

La Linn, M., Gardner, J., Warrilow, D., Darnell, G.A., McMahon, C.R., Field, I., Hyatt, A.D., Slade, R.W., and Suhrbier, A. (2001). Arbovirus of marine mammals: a new alphavirus isolated form the elephant seal louse, *Lepidohthirus macrorhini*. J. Virol. *75*, 4103–4109.

de Lamballerie, X., Leroy, E., Charrel, R.N., Ttsetsarkin, K., Higgs, S., and Gould, E.A. (2008). Chikungunya virus adapts to tiger mosquito via evolutionary convergence: a sign of things to come? Virol. J. *5*, 33.

Lanciotti, R.S., Kosoy, O.L., Laven, J.J., Panella, A.J., Velez, J.O., Lambert, A.J., and Campbell, G.L. (2007). Chikungunya virus in US travelers returning from India, 2006. Emerg. Inf. Dis. *13*, 764–767.

Lauring, A.S., and Andino, R. (2010). Quasispecies theory and the behavior of RNA viruses. PLoS Pathog. *6*, e1001005.

LeDuc, J., Pinheiro, F., and Travassos da Rosa, A. (1981). An outbreak of Mayaro virus disease in Belterra, Brazil. II. Epidemiology. Am. J. Trop. Med. Hyg. *30*, 682–688.

Lindsay, M.D., Broom, A.K., Wright, A.E., Johansen, C.A., and Mackenzie, J.S. (1993). Ross River virus isolations from mosquitoes in arid regions of Western Australia: implication of vertical transmission as a means of persistence of the virus. Am. J. Trop. Med. Hyg. *49*, 686–696.

Long, K.C., Ziegler, S.A., Thangamani, S., Hausser, N.L., Kochel, T.J., Higgs, S., and Tesh, R.B. (2011). Experimental transmission of Mayaro virus by *Aedes aegypti*. Am. J. Trop. Med. Hyg. *85*, 750–757.

Lynch, M., and Gabriel, W. (1990). Mutation load and the survival of small populations. Evolution *44*, 1725–1737.

Masembe, C., Michuki, G., Onyango, M., Rumberia, C., Norling, M., Bishop, R.P., Djikeng, A., Kemp, S.J., Orth, A., Skilton, R.A., *et al.* (2012). Viral metagenomics demonstrates that domestic pigs are a potential reservoir for Ndumu virus. Virol. J. *9*, 218.

Molaei, G., Andreadis, T.G., Armstrong, P.M., Thomas, M.C., Deschamps, T., Cuebas-Incle, E., Montgomery,

W., Osborne, M., Smole, S., Matton, P., *et al.* (2013). Vector–host interactions and epizootiology of eastern equine encephalitis virus in massachusetts. Vector Borne Zoonotic Dis. *13*, 312–323.

Muller, H.J. (1964). The relation of recombination to mutational advance. Mutat. Res. *106*, 2–9.

Nasar, F., Palacios, G., Gorchakov, R.V., Guzman, H., Travassos da Rosa, A.P., Savji, N., Popov, V.L., Sherman, M.B., Lipkin, W.I., Tesh, R.B., *et al.* (2012). Eilat virus, a unique alphavirus with host range restricted to insects by RNA replication. Proc. Natl. Acad. Sci. U.S.A. *109*, 14622–14627.

Norder, H., Lundström, J.O., Kozuch, O., and Magnius, L.O. (1996). Genetic relatedness of Sindbis virus strains from Europe, Middle East and Africa. Virology *222*, 440–445.

Novella, I.S., Reissig, D.D., and Wilke, C.O. (2004). Density-dependent selection in vesicular stomatitis virus. J. Virol. *78*, 5799–5804.

Ochieng, C., Lutomiah, J., Makio, A., Koka, H., Chepkorir, E., Yalwala, S., Mutisya, J., Musila, L., Khamadi, S., Richardson, J., *et al.* (2013). Mosquito-borne arbovirus surveillance at selected sites in diverse ecological zones of Kenya; 2007–2012. Virol. J. *10*, 140.

Parameswaran, P., Charlebois, P., Tellez, Y., Nunez, A., Ryan, E.M., Malboeuf, C.M., Levin, J.Z., Lennon, N.J., Balmaseda, A., Harris, E., *et al.* (2012). Genome-wide patterns of intrahuman dengue virus diversity reveal associations with viral phylogenetic clade and interhost diversity. J. Virol. *86*, 8546–8558.

Parola, P., de Lamballerie, X., Jourdan, J., Rovery, C., Vaillant, V., Minodier, P., Brouqui, P., Flahault, A., Raoult, D., and Charrel, R.N. (2006). Novel chikungunya virus variant in travelers returning from Indian Ocean islands. Emerg. Infect. Dis. *12*, 1493–1499.

Poidinger, M., Roy, S., Hall, R.A., Turley, P.J., Scherret, J.H., Lindsay, M.D., Broom, A.K., and Mackenzie, J.S. (1997). Genetic stability among temporally and geographically diverse isolates of Barmah Forest virus. Am. J. Trop. Med. Hyg. *57*, 230–234.

Powers, A.M., Oberste, M.S., Brault, A.C., Rico-Hesse, R., Schmura, S.M., Smith, J.F., Kang, W., Sweeney, W.P., and Weaver, S.C. (1997). Repeated emergence of epidemic/epizootic Venezuelan equine encephalitis from a single genotype of enzootic subtype ID virus. J. Virol. *71*, 6697–6705.

Powers, A.M., Brault, A.C., Tesh, R.B., and Weaver, S.C. (2000). Re-emergence of Chikungunya and O'nyong nyong viruses: evidence for distinct geographical lineages and distant evolutionary relationships. J. Gen. Virol. *81*, 471–479.

Powers, A.M., Brault, A.C., Shirako, Y., Strauss, E.G., Kang, W., Strauss, J.H., and Weaver, S.C. (2001). Evolutionary relationship and systematics of the Alphaviruses. J. Virol. 75, 10118–10131.

Powers, A.M., Aguilar, P.V., Chandler, L.J., Brault, A.C., Meakins, T.A., Watts, D., Russell, K.L., Olson, J., Vasconcelos, P.F.C., Travassos da Rosa, A., *et al.* (2006). Genetic relationships among Mayaro and Una viruses suggest distinct patterns of transmission. Am. J. Trop. Med. Hyg. 75, 461–469.

Quiroz, E., Aguilar, P.V., Cisneros, J., Tesh, R.B., and Weaver, S.C. (2009). Venezuelan equine encephalitis in Panama: fatal endemic disease and genetic diversity of etiologic viral strains. PLoS Negl. Trop. Dis. *3*, e472.

Reisen, W.K., Fang, Y., and Brault, A.C. (2008). Limited interdecadal variation in mosquito (Diptera: Culicidae) and avian host competence for Western Equine Encephalomyelitis virus (*Togaviridae: Alphavirus*). Am. J. Trop. Med. Hyg. *78*, 681–686.

Rezza, G., Nicoletti, L., Angelini, R., Romi, R., Finarelli, A.C., Panning, M., Cordioli, P., Fortuna, C., Boros, S., Magurano, F., *et al.* (2007). Infection with chikungunya virus in Italy: an outbreak in a temperate region. Lancet *370*, 1840–1846.

Rivas, F., Diaz, L.A., Cardenas, V.M., Daza, E., Bruzon, L., Alcala, A., De la Hoz, O., Caceres, F.M., Aristizabal, G., Martinez, J.W., *et al.* (1997). Epidemic Venezuelan equine encephalitis in La Guajira, Colombia, 1995. J. Infect. Dis. *175*, 828–832.

Saleh, S.M., Poidinger, M., Mackenzie, J.S., Broom, A.K., Lindsay, M.D., and Hall, R.A. (2003). Complete genomic sequence of the Australian south-west genotype of Sindbis virus: comparisons with other Sindbis strains and identification of a unique deletion in the 3′-untranslated region. Virus Genes *26*, 317–327.

Saleh, S.M., Poidinger, M., Mackenzie, J.S., Broom, A.K., Lindsay, M.D.A., Russell, R.C., Cloonan, M.J., Calisher, C.H., Burgess, G.W., and Hall, R.A. (2004). Antigenic and genetic typing of Whataroa viruses in Australia. Am. J. Trop. Med. Hyg. *71*, 262–267.

Sammels, L.M., Coelen, R.J., Lindsay, M.D., and Mackenzie, J.S. (1995). Geographic distribution and evolution of Ross River virus in Australia and the Pacific Islands. Virology *212*, 20–29.

Sammels, L.M., Lindsay, M.D., Poidinger, M., Coelen, R.J., and Mackenzie, J.S. (1999). Geographic distribution and evolution of Sindbis virus in Australia. J. Gen. Virol. *80*, 739–748.

Schuffenecker, I., Iteman, I., Michault, A., Murri, S., Frangeul, L., Vaney, M.C., Lavenir, R., Pardigon, N., Reynes, J.M., Pettinelli, F., *et al.* (2006). Genome microevolution of chikungunya viruses causing the Indian ocean outbreak. PLoS Med. 3, e263.

Schwartz, O., and Albert, M.L. (2010). Biology and pathogenesis of chikungunya virus. Nat. Rev. Micobiol. *8*, 491–500.

Scott, T.W., and Weaver, S.C. (1989). Eastern equine encephalomyelitis virus: epidemiology and evolution of mosquito transmission. Adv. Virus Res. *37*, 277–328.

Smith, D.R., Aguilar, P.V., Coffey, L.L., Gromowski, G.D., Wang, E., and Weaver, S.C. (2006). Venezuelan equine encephalitis virus transmission and effect on pathogenesis. Emerg. Infect. Dis. *12*, 1190–1196.

Smith, D.R., Adams, A.P., Kenney, J.L., Wang, E., and Weaver, S.C. (2008). Venezuelan equine encephalitis virus in the mosquito vector *Aedes taeniorhynchus*: infection initiated by a small number of susceptible epithelial cells and a population bottleneck. Virology *372*, 176–186.

Stapleford, K.A., Coffey, L.L., Lay, S., Borderia, A.V., Duong, V., Isakov, O., Rozen-Gagnon, K., Arias-Goeta, C., Blanc, H., Beaucourt, S., *et al.* (2014).

Emergence and transmission of arbovirus evolutionary intermediates with epidemic potential. Cell Host Microbe *15*, 706–716.

Strauss, J.H., and Strauss, E.G. (1994). The alphaviruses: gene expression, replication, and evolution. Microbiol. Rev. *58*, 491–562.

Sudia, W.D., and Newhouse, V.F. (1975). Epidemic Venezuelan equine encephalitis in North America: a summary of virus–vector–host relationships. Am. J. Epidemiol. *101*, 1–13.

Sudia, W.D., Newhouse, V.F., Beadle, L.D., Miller, D.L., Johnston Jr, J.G., Young, R., Calisher, C.H., and Maness, K. (1975). Epidemic Venezuelan equine encephalitis in North America in 1971: Vector studies. Am. J. Epidemiol. *101*, 17–35.

Suhrbier, A., Jaffar-Bandjee, M.C., and Gasque, P. (2012). Arthritogenic alphaviruses--an overview. Nat. Rev. Rheumatol. *8*, 420–429.

de la Torre, J.C., and Holland, J.J. (1990). RNA virus quasispecies populations can suppress vastly superior mutant progeny. J. Virol. *64*, 6278–6281.

Tsetsarkin, K.A., Vanlandingham, D.L., McGee, C.E., and Higgs, S. (2007). A single mutation in chikungunya virus affects vector specificity and epidemic control. PLoS Pathog. 3, e201.

Tsetsarkin, K.A., Chen, R., Leal, G., Forrester, N., Higgs, S., Huang, J., and Weaver, S.C. (2011a). Chikungunya virus emergence is constrained in Asia by lineage-specific adaptive landscapes. Proc. Natl. Acad. Sci. U.S.A. *108*, 7872–7877.

Tsetsarkin, K.A., Chen, R., Sherman, M.B., and Weaver, S.C. (2011b). Chikungunya virus: evolution and genetic determinants of emergence. Curr. Opin. Virol. *1*, 310–317.

Tsetsarkin, K.A., Chen, R., Yun, R., Rossi, S.L., Plante, K.S., Guerbois, M., Forrester, N., Perng, G.C., Sreekumar, E., Leal, G., *et al.* (2014). Multi-peaked adaptive landscape for chikungunya virus evolution predicts continued fitness optimization in Aedes albopictus mosquitoes. Nat. Commun. 5, 4084.

Turner, P.E., Morales, N.M., Alto, B.W., and Remold, S.K. (2010). Role of evolved host breadth in the initial emergence of an RNA virus. Evolution *64*, 3273–3286.

Ventoso, I. (2012). Adaptive changes in Alphavirus mRNA translation allowed colonization of vertebrate hosts. J. Virol. *86*, 9484–9494.

Volk, S.M., Chen, R., Tsetsarkin, K.A., Adams, A.P., Garcia, T.I., Sall, A.A., Nasar, F., Schuh, A.J., Holmes, E.C., Higgs, S., *et al.* (2010). Genome-scale phylogenetic analyses of chikungunya virus reveal independent emergences of recent epidemics and various evolutionary rates. J. Virol. *84*, 6497–6504.

Watson, R. (2007). Chikungunya fever is transmitted locally in Europe for first time. BMJ *335*, 532–533.

Weaver, S.C. (2006). Evolutionary influences in arboviral disease. Curr. Top. Microbiol. Immunol. *299*, 285–314.

Weaver, S.C. (2014). Arrival of chikungunya virus in the new world: prospects for spread and impact on public health. PLoS Negl. Trop. Dis. *8*, e2921.

Weaver, S.C., Scherer, W.F., Taylor, C.A., Castello, D.A., and Cupp, E.W. (1986). Laboratory vector competence of *Culex* (*Melanoconion*) *cedecei* for sympatric and allopatric Venezuelan equine encephalomyelitis viruses. Am. J. Trop. Med. Hyg. *35*, 619–623.

Weaver, S.C., Bellew, L.A., and Rico-Hesse, R. (1992). Phylogenetic analysis of alphaviruses in the Venezuelan equine encephalitis complex and identification of the source of epizootic viruses. Virology *191*, 282–290.

Weaver, S.C., Salas, R., Rico-Hesse, R., Ludwig, G., Oberste, M.S., Boshell, J., and Tesh, R.B. (1996). Re-emergence of epidemic Venezuelan equine encephalomyelitis in South America. Lancet *348*, 436–440.

Weaver, S.C., Brault, A.C., Kang, W., and Holland, J.J. (1999a). Genetic and fitness changes accompanying adaptation of an arbovirus to vertebrate and invertebrate cells. J. Virol. *73*, 4316–4326.

Weaver, S.C., Powers, A.M., Brault, A.C., and Barrett, A.D. (1999b). Molecular epidemiological studies of veterinary arboviral encephalitides. Vet. J. *157*, 123–138.

Weaver, S.C., Coffey, L.L., Nussenzveig, R., Ortiz, D., and Smith, D.R. (2004a). Vector competence. In Microbe-Vector Interactions in Vector-Borne Diseases, S.H. Gillespie, G.L. Smith, and A. Osbourn, eds. (Cambridge University Press, Cambridge, UK), pp. 139–180.

Weaver, S.C., Ferro, C., Barrera, R., Boshell, J., and Navarro, J.C. (2004b). Venezuelan equine encephalitis. Ann. Rev. Entomol. *49*, 141–174.

Weaver, S.C., Winegar, R., Manger, I.D., and Forrester, N.L. (2012). Alphaviruses: population genetics and determinants of emergence. Antiviral Res. *94*, 242–257.

Weaver, S.C., and Forrester, N.L. (2015). Chikungunya: evolutionary history and recent epidemic spread. Viruses. In press.

Weaver, S.C., and Lecuit, M. (2015). Chikungunya virus and the global spread of a mosquito-borne disease. N. Engl. J. Med. *372*, 1231–1239.

Weston, J.H., Welsh, M.D., McLoughlin, M.F., and Todd, D. (1999). Salmon pancreas disease virus, an alphavirus infecting farmed Atlantic salmon, *Salmo salar* L. Virology *256*, 188–195.

Weston, J., Villoing, S., Bremont, M., Castric, J., Pfeffer, M., Jewhurst, V., McLoughlin, M., Rodseth, O., Christie, K.E., Koumans, J., *et al.* (2002). Comparison of two aquatic alphaviruses, Salmon pancreas disease virus and Sleeping disease virus, by using genome sequence analysis, monoclonal reactivity, and cross-infection. J. Virol. *76*, 6155–6163.

White, G., Ottendorfer, C., Graham, S.P., and Unnasch, T.R. (2011a). Competency of reptiles and amphibians for eastern equine encephalitis virus. Am. J. Trop. Med. Hyg. *85*, 421–425.

White, G.S., Pickett, B.E., Lefkowitz, E.J., Johnson, A.G., Ottendorfer, C., Stark, L.M., and Unnasch, T.R. (2011b). Phylogenetic analysis of eastern equine encephalitis virus isolates from Florida. Am. J. Trop. Med. Hyg. *84*, 709–717.

Zacks, M.A., and Paessler, S. (2010). Encephalitic alphaviruses. Vet. Microbiol. *140*, 281–286.

Laboratory Diagnosis and Detection of Alphaviruses

3

Linda Hueston

Abstract

Alphaviruses have been the cause of significant morbidity and mortality in both humans and animals. While infection is suspected clinically it cannot be confirmed without the use of laboratory testing. Laboratory diagnostic techniques have changed dramatically in recent years. The advent of nucleic acid testing has increased our understanding of these viruses and provided a means of rapid diagnosis in the viraemic phase of illness. The development of various ELISA techniques has significantly improved the speed and accuracy with which antibody can be detected, providing an inexpensive means of diagnosing recent infection in both humans and animals. However, the older techniques for both antibody and antigen detection still have value and should not be discarded or relegated to history. This chapter discusses the use of traditional and modern techniques suitable for detecting and diagnosing alphavirus infections.

Introduction

The alphaviruses derive their name from the first letter of the Greek alphabet and have a global distribution including Antarctica. Some have been responsible for large epidemics and significant morbidity and mortality in mammals, birds and even fish. Importantly, several alphaviruses have the potential to be used as weapons.

Currently there are 36 recognized alphaviruses (Calisher and Karabatsos, 1988; Chang *et al.*, 2006; Griffin, 2007; Ha *et al.*, 1995; Nasar *et al.*, 2012; Powers *et al.*, 2001; Strauss and Strauss, 1994). They have been grouped into Old and New World viruses depending on their geographic distribution. The Old World alphaviruses are associated with rheumatic disease in humans and include Ross River (RRV), Barmah Forest (BFV), chikungunya (CHIKV), O'nyong-nyong (ONNV), and Ockelbo viruses (OCKV) (Peters and Dalrymple, 1990). The New World alphaviruses are pathogenic for humans, horses, and some bird species and include Venezuelan, Eastern and Western Equine Encephalitis viruses (VEEV, EEEV, and WEEV) (Olsen *et al.*, 1997; Peters and Dalrymple, 1990). Two aquatic alphaviruses, Sleeping Disease virus (SDV) and Salmon Pancreas Disease virus (SPDV), are pathogenic to salmon and trout. The majority of alphaviruses are arboviruses, but in recent years non-arboviral alphaviruses such as Eilat and Negev viruses have been recognized (Nasar *et al.*, 2012).

As a result of climate changes, increased frequency of travel to endemic areas, reservoir and vector spread and their potential use as weapons there is a growing need for accurate and timely recognition, diagnosis and detection of alphaviruses.

Detection and diagnosis

Alphavirus infection is suspected clinically but cannot be confirmed without laboratory testing. Laboratory confirmation of a clinical diagnosis involves three test systems: virus isolation, direct detection of antigen, and serology. Within each of these systems there are choices of test methodology, and that choice will depend on the type of information being sought, the speed with which the answer is required, the technical constraints of the laboratory, and cost.

The diagnosis of human and animal alphavirus infection is not the only role for laboratory testing, the methods discussed in this chapter are also used for surveillance in sentinel animals and in vectors.

Sample types

The most common sample type submitted for human and animal diagnosis of alphaviral infection is serum. However, synovial fluid, skin biopsies, CSF and brain tissue obtained at autopsy are also submitted on occasion. Serum, plasma and whole blood (often collected onto filter-type papers) from both humans and animals are also used for serosurveillance. Finally pools of arthropods can be homogenized and processed for virus isolation and/or direct antigen detection by immunofluorescence (IFA), enzyme-linked immunosorbent assay (ELISA) and polymerase chain reaction (PCR).

Virus isolation and antigen detection

Virus isolation and antigen detection from clinical samples or vectors is often a difficult and frequently unsuccessful undertaking. Alphaviral viraemia particularly in humans is short lived (2– 3 days post onset of symptoms) and has often ceased by the time a patient seeks medical attention. However, Chikungunya viraemia is high titred in humans and often detectable for 4 to 6 days post onset using culture techniques and up to 12 days by PCR (Lanciotti *et al.*, 2007; Staples *et al.*, 2009).

Virus isolation

Virus isolation depends on growing an unknown virus in suckling animals, cell culture or insects. It is the *sine qua non* of diagnostic testing and while it requires the support of dedicated culture laboratories and experienced staff this technique provides invaluable information not available from other test systems. Isolating an alphavirus not only provides a definitive diagnosis, it increases our understanding of disease pathogenesis through animal models, facilitates evolutionary research, may provide suitable vaccine candidates and allows the production of diagnostic materials.

There is no single, widely accepted, universal system for alphavirus isolation, and success is dependent on numerous factors including sample type, time of collection, maintenance of the cold chain, transportation, storage of samples prior to transportation and inoculation and processing of samples. The presence of antibody and interferon can also inhibit virus isolation from serum samples.

Historically virus isolation involved intracranial inoculation of suckling mice, guinea pigs or hamsters. The advent of vertebrate cell culture provided an alternative isolation system. However, with the use of direct inoculation of mosquitoes in the 1930s and the development of mosquito cell lines in the 1960s it became apparent that many viruses and some virus strains had been previously undetected because of their inability to replicate in the usual animal models or in vertebrate cell culture.

Laboratory animals

The use of animals for virus isolation is not as widespread now as it was in the past. Today diagnostic and research institutes must follow strict animal ethics procedures to ensure the animals are treated humanely, respectfully and ethically. Indeed laboratory staff are actively encouraged to seek other means of virus isolation whenever possible. This combined with the cost of maintaining animal colonies has seen this work reduced to a few reference centres, research laboratories and production facilities. However, to date the largest number of arboviral isolates have been generated by the use of suckling mice. While this technique was superseded by the introduction of cell culture it is important to understand how it was used. It is beyond the scope of this chapter to provide details of this isolation method; the reader will find more detailed instructions in the literature (Beaty *et al.*, 1989).

Mosquito inoculation

Direct inoculation of mosquitoes for virus isolation of arboviruses has been undertaken for more than 70 years. While the majority of work cited in the literature was undertaken with flaviviruses and bunyaviruses, these methods are equally valid for use with other arboviruses including the alphaviruses (Merrill and Tenbroeck, 1935; Mourya *et al.*, 2007; Philip Samuel and Tyagi, 2006; Rosen and Gubler, 1974). Intrathoracic and intracerebral inoculation of mosquitoes and mosquito larvae have proven to be the most sensitive techniques for the isolation of several arboviruses (Gubler and Rosen, 1976; Mourya *et al.*, 2001; Rosen *et al.*, 1985). Four mosquito species have been successfully used for this purpose, *Aedes aegypti, Aedes albopictus, Toxorhynchites amboinensis and Toxorhynchites splendens.* Male and female mosquitoes are equally susceptible. When using Aedes species, males are preferred as they do not ingest blood and therefore cannot infect humans or animals. *Toxorhynchites* species offer a number of advantages for virus isolation. Since neither sex ingests blood, they do not present a safety risk and both sexes can be used. Members of this species are larger in physical size than *Aedes* species, making them easier to inoculate, and they tolerate a larger inoculum than any other species.

This technique is labour intensive, requiring an insectary to produce mosquitoes in sufficient quantity for diagnostic use. It also requires expertise, which may not be routinely available in all laboratories. There are safety issues if using female *Aedes* species, which needs careful consideration. Despite these disadvantages the technique remains the most sensitive isolation method for a variety of arboviruses with isolates being obtained within a few days. As this method propagates virus in a natural host, the virus it yields avoids potential modification of the viral properties, which can be attributed to adaptation to cell culture or mice. Unfortunately, this method is often not available in many countries with high arboviral burdens.

Detailed instructions for mosquito inoculation can be found in the literature, particularly the work of Rosen and Gubler (1974) and Mourya *et al.* (2007). Arbovirus infections do not cause morbidity or mortality in mosquitoes; therefore it is necessary to demonstrate the presence of virus by other means. The easiest and most common way to achieve this is by direct IFA using the heads of the mosquitoes, which are squashed onto glass slides, fixed and then examined by direct IFA. The remaining body parts can be used to prepare virus stocks. Alternatively, the mosquitoes can be triturated in a suitable buffer and tested using ELISA or PCR.

Mosquito cell culture

The first continuous mosquito cell line was established by Grace (1966) in 1966 from *Aedes aegypti* (L) mosquitoes. This was followed a year later by the establishment of three cell lines derived from *Aedes albopictus* (Skuse) and two from *Aedes aegypti* (L) by Singh (1967). The C6/36 clone of *Aedes albopictus* developed by Singh is the most commonly used, although AP-61 (*Aedes pseudoscutellaris*), TR-248 (*Toxorhynchites amboinensis*) and *Aedes krombeini* cells are also successfully used for isolation.

Mosquito cell lines are more sensitive than mice or mammalian cell lines for isolating a broader range of arboviruses. They are stable, have optimal growth at lower temperatures than mammalian cell lines and can be taken into the field without the need for incubators. Their main disadvantage is they do not demonstrate cytopathic effect (CPE), so an additional step must be employed to demonstrate the presence of virus such as passage into a mammalian cell line, which will show CPE or direct detection serologically or by molecular techniques.

Vertebrate cell culture

Numerous vertebrate cell lines have been used to isolate and propagate alphaviruses, including continuous cell lines: Vero, Vero E-6, BHK-21, LLC-MK_2, CER and PSEK and primary cells derived from hamster kidney, chick embryo and duck embryo. The advantages of vertebrate cell line cultures are that they are readily available and commonly used in many laboratories. In addition they have the advantage of displaying CPE when infected. One limitation of these cell lines is the observation that not all alphaviruses can replicate in vertebrate cell culture. Eliat virus (Nasar *et al.*, 2012), a recently recognized insect specific alphavirus, cannot infect vertebrate cell lines and the fish alphaviruses only infect fish cell lines and have a narrow temperature sensitivity (Graham *et al.*, 2008). Some strains of alphaviruses show variation in their ability to infect some vertebrate cell lines. For example one strain of Barmah Forest virus isolated from coastal NSW will not grow in BHK-21 cells but will grow in Vero, PSEK and, of course, C6/36 cells.

Cell culture procedures

Cell cultures are used not only for the isolation and identification of alphaviruses but also for the production of serological antigens. Most laboratories now obtain their cell lines and culture media from commercial sources and the techniques used for culturing alphaviruses are no different to those used to grow any other virus, although the use of invertebrate cells is more commonly associated with arbovirology. Detailed protocols for the production and maintenance of cell cultures have been published (Schmidt, 1989), and it is beyond the scope of this chapter to discuss the general procedure of cell culture. However, some modifications have been shown to be particularly useful in the isolation of alphaviruses, and these will be discussed. A testing algorithm for alphavirus isolation in cell culture is given in Fig. 3.1.

Isolation of virus from clinical samples or field-collected specimens is a difficult task. Viraemia in humans is short-lived and low titred for most alphaviruses, and avoiding toxicity, bacterial and fungal contamination of vector pools can be challenging. In addition alphaviruses are quite labile so it is important to preserve virus infectivity by transporting samples for virus isolation as soon as possible after collection and storing samples appropriately.

Centrifugation of samples assists in virus isolation by causing minor trauma to the cell membrane, which appears to facilitate viral uptake (Engler and Selepak, 1994).

Direct detection of antigen

Direct detection of alphaviruses in clinical samples can provide a rapid diagnosis, which

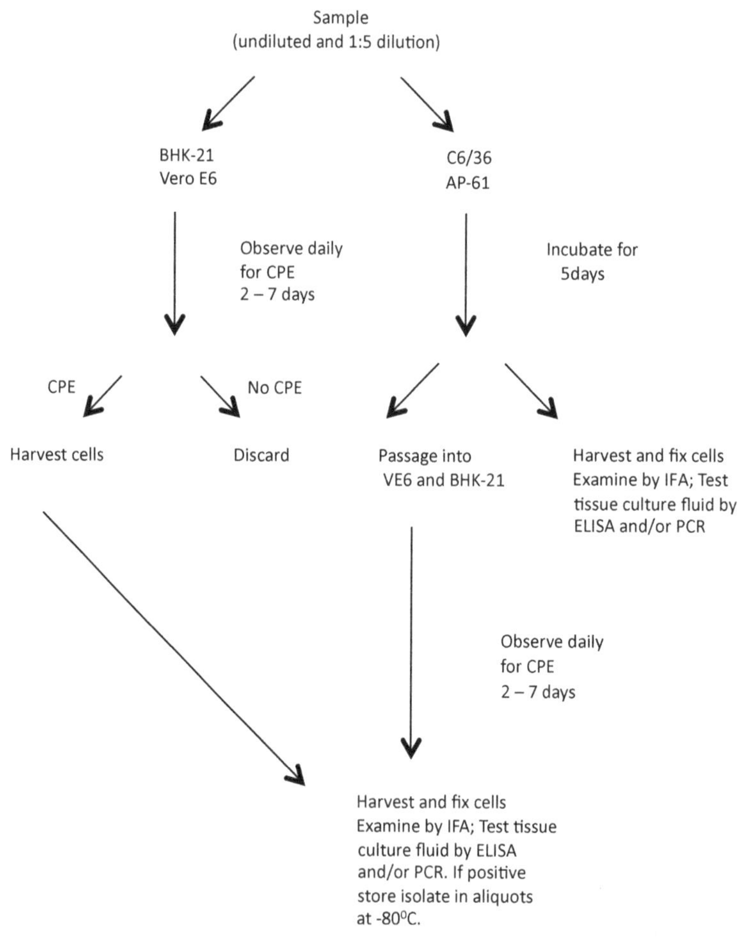

Figure 3.1 Testing algorithm for isolation of alphaviruses in cell culture.

facilitates treatment options. This, in conjunction with detection of virus in vectors, can help direct public health responses to outbreaks. Three techniques are currently used for direct antigen detection: IFA, ELISA and PCR. IFA and ELISA are more widely used in arbovirology than PCR, largely because they can be applied in the field or a clinical setting, are rapid and inexpensive. This is particularly important in those countries with high arboviral burdens and poor public health infrastructures. Despite being the most expensive of the antigen detection techniques, PCR has been developed to identify a number of alphaviruses in both clinical samples and vectors. This technique is evolving rapidly and, with improvements in technology, costs are likely to reduce significantly in the future (Kuno, 1988).

Immunofluorescence

IFA is a simple and reliable microscope based technique used clinically to detect both alphavirus antigen and antibody in diagnostic samples. It allows the visualization of an antigen in cells or tissue sections by binding a specific antibody, which has been labelled with a fluorescent dye such as fluorescein isothiocyanate (FITC)

Antigen detection by IFA is rapid and inexpensive to perform. While it lacks the sensitivity of ELISA and PCR, it has been and continues to be used in clinical settings (Monath *et al.*, 1981). Frozen sections, tissue smears, mosquito head squashes and infected cell cultures are all suitable samples. Formalin fixed tissue can also be used but requires pre-treatment with trypsin to expose antigenic sites. Two types of assay are available: direct immunofluorescence (DIF) and indirect immunofluorescence (IIF). DIF uses fluorescein-labelled antibodies to bind directly to the target antigen in tissues, cell cultures and vectors. While this technique is simpler and more rapid than the IIF, it is less sensitive.

IIF is a two-step technique in which a secondary antibody labelled with a fluorochrome is used to recognize a primary (antiviral detector) antibody. This test is not only sensitive but, with the availability of a large range of monoclonal antibodies, samples can be tested for a broad range of alphaviruses quickly and easily.

Results from IIF are scored semi quantitatively according to the intensity of fluorescence. The technique requires skill in interpreting the intensity of fluorescence and is somewhat subjective. Other limitations include the loss of fluorescence or bleaching over time. Non-specific staining can be minimized by prior absorption of the detector antibody and conjugate with normal tissue suspension or cell lines. The application of IFA to serological diagnosis will be described later in the chapter.

ELISA

ELISA is the most commonly and widely used technique in the laboratory diagnosis of alphaviral infections. It can be used for both antigen and antibody detection. It is rapid, cost-effective, easy to perform, scalable, easily automated, sensitive and, with the use of monoclonal antibodies, it can be both virus specific and group specific. There are numerous reports in the literature describing antigen detection ELISA for alphaviruses (Brown *et al.*, 2001; Hildreth and Beaty, 1984; Hildreth *et al.*, 1984; Hundekar *et al.*, 2002; Konishi and Takahashi, 1985; Kumar *et al.*, 2012; Scott and Olson, 1986). The most commonly used technique is the double sandwich ELISA The capture antibody should have a high avidity for the virus and can be monoclonal or polyclonal. However, given the large array of virus specific and alphavirus group reactive monoclonal antibodies now available, the use of monoclonal antibodies is preferable. There is a limit to the amount of protein, which can bind to the wells of an ELISA plate and whatever protein binds to the well is available to be 'seen' by all the reagents, which follow it in the test protocol. This can limit the sensitivity of the test and lead to high background problems. To remove these problems the capture antibody should be purified.

PCR

Nucleic acid testing has revolutionized antigen detection with many groups of viruses. However, its use in arbovirology has not been as widespread. There are several reasons for this, the viraemia in humans is quite short for the majority of alphaviruses and patients often present too late in the viral illness for antigen to be detected. In addition these tests are expensive to perform and require the use of specialized equipment and laboratories.

For countries with high arboviral burdens, the expense of these tests is often prohibitive and finally a lack of standardization and appropriate evaluation and validation can negatively impact on the use of these tests. It is for these reasons that many countries still rely on the classic techniques of virus isolation, serological typing of isolates and direct antigen detection.

Each PCR must be optimized for use: RNA extraction techniques must be chosen with the sample type in mind, care must be taken in the design of primers and individual reaction components (time and temperature parameters and reagent concentrations) must be adjusted to ensure efficient amplification of specific targets. It is beyond the scope of this chapter to provide details of PCR design and optimization which can be found in the textbooks of Sambrook and colleagues (Sambrook *et al.*, 1989)

One limitation of PCR is that one must know which virus to look for to design primers that will bind to the target, unlike virus isolation where any virus may be grown, including previously unrecognized viruses. However, in recent years alphavirus group reactive PCRs (Bronzoni *et al.*, 2004; Eshoo *et al.*, 2007; Grywna *et al.*, 2010; Wang *et al.*, 2006) have been developed which improves our ability to detect new viruses. However, it is necessary to regularly review primer choice to ensure strain variants are not being missed.

A search of the literature reveals increasing reports of the use of PCR in the detection of specific alphaviruses, particularly those of public health concern (Hueston *et al.*, 2013; Lambert *et al.*, 2003; Pfeffer *et al.*, 2002; Studdert *et al.*, 2003). Interestingly the technique is being applied more widely in the diagnosis of infection in mammals, birds and fish. It has proven to be most successful in surveillance by detection of virus in vectors, and it has revolutionized the study of the evolutionary relationships of the alphaviruses (Powers *et al.*, 2001).

Detection of alphavirus antibody

Serological techniques are used to diagnose infection, to determine prevalence, vaccine efficacy and to identify virus isolates. There are five major serological test systems: neutralization (NT), haemagglutination inhibition (HI), complement fixation (CF), immunofluorescence (IF) and ELISA. The first three measure total antibody; they cannot determine antibody class, so to definitively diagnose an infection with these techniques two samples are required to demonstrate seroconversion or a 4-fold or greater rise in titre. However, a presumptive diagnosis can be made on the basis of a single high titre. Each of these techniques measures a different type of antibody, with different sensitivities and specificities which appear at different times in the course of infection, and understanding these differences is essential in obtaining a diagnosis.

The use of immunofluorescence and later ELISA allowed antibody class to be determined. For the first time, IgM was routinely used to indicate recent infection. But it was the flexibility of ELISA that facilitated more widespread diagnostic capability for alphavirus infections. Neither ELISA nor immunofluorescence completely replaced the other techniques, but they are increasingly used as first line screening tests and can facilitate rapid diagnosis by the detection of IgM antibody.

Regardless of the technique used to diagnose infection the interpretation of arboviral serology results presents challenges often not found with other virus groups. Arboviruses exhibit extensive serological cross-reactivity with closely related viruses. In addition infection with multiple arboviruses and/or vaccinations complicates interpretation of serology results (Beaty *et al.*, 1989).

Neutralization tests

The neutralization test is the most sensitive and type-specific serological technique and its results are probably the easiest to interpret. Neutralizing antibodies are the first to appear, last longer than other antibodies and in many instances can persist for life. We equate this antibody with protection, which makes the technique useful for serosurveys to determine antibody prevalence and the efficacy of vaccines. The assay is based on the principle of interaction of virus and antibody, resulting in the inactivation of the virus so that it can no longer infect and replicate in cell culture. There are several types of neutralization tests available, e.g. plaque reduction (PRNT) (Beaty *et al.*, 1989), micro-neutralization (MNT) (Cloonan *et al.*, 1982) and fluorescent focus inhibition test (FFIT) (Canakoglu *et al.*, 2013).

The PRNT was developed in the 1950s and is the most technically demanding of the neutralization tests. Although this test is the oldest, most reliable and most used neutralization technique it takes longer to complete and is not amenable to large test numbers. Detailed instructions for its performance can be found in the literature (Beaty *et al.*, 1989). PRNT allows virus–antibody interaction to take place. Each virus that initiates an infection produces a localized area of infection or plaque. Plaques are counted manually and are compared to the starting concentration of virus to determine the percentage reduction in virus infectivity.

The micro-neutralization test is performed on a smaller scale in 96-well plates and, as all alphaviruses are highly lytic, the endpoint is easy to read. The 96-well plate format is amenable to larger test numbers than the PRNT, which facilitates serosurveys. (Cloonan *et al.*, 1982).

FFIT is particularly useful for viruses that do not readily plaque in mammalian cells and, as it does not rely on plaque formation or CPE, it provides results in the shortest time frame – usually 24 hours (Digoutte *et al.*, 1976; Hernandez *et al.*, 2010; Tignor and Digoutte, 1976).

All neutralization tests require the use of live virus, which limits their use to laboratories with culture facilities, and are also labour intensive, expensive and require significant technical expertise to perform. In addition, neutralization tests are biological assays and therefore display a degree of inter-laboratory variation.

Haemagglutination inhibition

Many viruses possess the ability to agglutinate the red blood cells (RBC) of mammalian or avian species, for example influenza, parainfluenza, rubella, picorna- and adenoviruses, in addition to the alpha-, flavi- and bunyaviruses.

Antibodies against the viral protein responsible for haemagglutination can prevent haemagglutination, and this is the principle behind the haemagglutination inhibition (HI) test. The specificity of the HI test varies with different viruses; for example, with influenza viruses the haemagglutination antigen is the same antigen responsible for virus adsorption and, therefore, the HI test is strain specific for influenza (Harmon and Kendal, 1989; Klenk *et al.*, 1983). However, for the alphaviruses it is not specific and the test exhibits cross-reactivity with other alphaviruses, which complicates interpretation. Alphavirus HI antibodies are more group specific and less type specific than NT antibodies; they appear after NT antibodies but before CF antibodies. They are detectable longer than CF antibodies but are not as long lasting as neutralizing antibodies. This test is more sensitive than the CF test but less sensitive than NT and ELISA. Detailed instructions for the performance of HI tests for alphaviruses can be found in the literature (Beaty *et al.*, 1989; Casals and Brown, 1954).

The advantages of HI tests for alphavirus diagnosis are that they are relatively easy and inexpensive to perform and require simple inexpensive equipment that is readily available. Tests can be performed in 96-well plate format allowing larger numbers to be tested than the PRNT, and results are available in 2 days.

There are however limitations in the performance of this test. There is no commercial source of antigen and therefore antigens must be made 'in-house'. The production of haemagglutinin (HA) antigens for some alphaviruses is difficult; for example, Barmah Forest virus does not produce haemagglutinin in mouse brain and carcass preparations (Dalgarno *et al.*, 1984; Marshall *et al.*, 1982). HA antigens are pH sensitive and, in the case of alphaviruses, the pH range over which the antigen will agglutinate is very narrow, so care must be taken in the production of reagents. The indicator in HI tests is red blood cells (RBCs). For alphaviruses, RBCs from adult white geese are the first choice, with RBCs from ganders being preferred. HA activity of RBCs from females can be quite variable due to fluctuations in hormonal levels and this can lead to difficulty with reading the test endpoint. Test sera contain non-specific inhibitors, which can lead to false positive results and natural agglutinins, which can lead to false-negative results, so pre-treatment of sera is necessary.

Complement fixation

The complement fixation (CF) test exploits the ability of antibody to activate the complement system. There are two separate reactions within the test: the first involving the reaction between antibody in the test sample, antigen and complement and the second reaction in which haemolysin sensitized sheep RBCs act as the indicator of the amount of complement remaining after the first reaction is complete. If complement has been 'fixed' by the primary antigen–antibody complex, haemolysis cannot occur and the red cells will settle into a button. The highest serum dilution that 'fixes' 75–100% of the complement represents the antibody titre. Detailed protocols for the preparation of reagents and performance of the test can be found in the literature (Beaty *et al.*, 1989).

CF antibody is produced later than either NT or HI antibody and is relatively short lived so it cannot be used in serosurveys or to determine immune status. Its specificity is intermediate between NT and HI. Once again, antigens are not commercially available and so must be produced in-house. Test sera frequently develop anticomplementary activity on prolonged or inappropriate storage. There is an element of subjectivity in reading endpoints and some individuals do not produce a CF antibody response. These limitations mean this test is not used alone for diagnosis of alphaviral infection; however, it has been widely used over many decades and in combination with other serological tests it can be helpful in determining recent from remote infection.

Immunofluorescence

The use of indirect IFA techniques for alphaviral antibody detection allowed antibody class to be determined rapidly, easily and inexpensively (Aaskov and Davies, 1978; El Mekki *et al.*, 1979). IFA for antibody detection is a two-step 'sandwich' technique and can be performed at a single dilution or it can be quantitated by testing serial dilutions to obtain a titre.

Immunofluorescent slides can be easily prepared in house. Other than the need to purchase an immunofluorescent microscope, the IFA technique does not require specialized equipment. Reading the immunofluorescent pattern is somewhat subjective but with appropriate training this can be minimized. Immunofluorescent slides are now commercially available for some alphaviruses (Litzba *et al.*, 2008). Indeed, some manufacturers are producing multiparametric immunofluorescence slides, which allow a number of viruses to be tested simultaneously (Sutherland *et al.*, 2011).

While IFA is less sensitive than ELISA, it is more rapid and allows antibody class to be determined, which NT, HI and CF are unable to achieve.

ELISA

Three types of ELISA assay have been applied to antibody detection of alphaviruses: indirect for IgG detection, antibody class capture for IgM detection and defined epitope blocking ELISA for the detection of total antibody.

Indirect IgG ELISA

In an indirect ELISA, antigen is coated onto the wells of a plastic 96-well microtitre plate. Unbound surfaces are blocked (often with protein such as bovine serum albumin or fetal calf serum) to reduce non-specific reactions and to stabilize the antigen. Patient serum is diluted and added to the appropriate wells of the plate, if antibody is present it will bind to the antigen. Unbound material is removed by washing. To determine if antibody bound to the antigen an anti-human IgG conjugate labelled with horseradish peroxidase (HRP) is added. Unbound conjugate is removed by washing and a suitable colorimetric substrate is added and allowed to react. If the conjugate bound to the antigen–antibody complex a colour will develop which is proportional to the amount of antibody present. After a predetermined time period the reaction is stopped and the optical density of the wells is read in a spectrophotometer. Purified antigen and anti-human IgG conjugate are titrated in chequerboard or box titrations using neutralizing antibody negative and positive samples to determine the optimal dilution of each reagent. Each test plate is run with known positive and negative controls.

This IgG assay provides results in just over 2 hours. Plates can be precoated and stored until required, and the test is automatable and can easily handle large test numbers. It is similar in sensitivity to NT, but it is not virus specific, and will cross-react with other closely related alphaviruses.

Antibody class capture IgM ELISA

In this assay anti-human IgM is coated onto the wells of a plastic 96-well microtitre plate. Only IgM antibody will bind to the plate removing the need for pre-treatment of the sample to remove IgG. Extraneous unbound material is removed by washing. Alphavirus specific antigen is added and if virus specific antibody is present it will bind to the antigen. This antibody–antigen complex is detected by the addition of a virus specific monoclonal antibody labelled with HRP. Unbound monoclonal conjugate is removed by washing and a colorimetric substrate is added and allowed to react. If the conjugate bound to the antigen–antibody complex, a colour will develop that is proportional to the amount of antibody present. After a predetermined time period the reaction is stopped and the optical density of the wells is read in a spectrophotometer. As for the IgG assay, all reagents must be pre-titrated to determine appropriate working concentrations and all runs must have suitable positive and negative controls.

This test can be as sensitive as NT and, by incorporating monoclonal antibodies, it is extremely specific. The test is automatable and capable of handling large sample numbers. Detection of IgM antibody in serum provides presumptive evidence of recent infection, although in some alphavirus infections IgM has been shown to persist for long periods (more than six months) post infection and this needs to be considered when determining the clinical significance of the results. Detection of IgM in cerebrospinal fluid (CSF) is extremely useful because IgM antibodies cannot cross the blood–brain barrier; therefore, finding IgM in CSF implies antibody synthesis in response to central nervous system infection.

Defined epitope blocking (DEB) ELISA

The DEB ELISA is a modification of an indirect ELISA in which test sera compete for binding sites with a virus specific monoclonal antibody HRP conjugate on an antigen-coated plate. In the first step, diluted test sera are allowed to react with the antigen coated on the ELISA plate. Without removing the test sera the monoclonal conjugate is added to all wells. If antibody was present in the test sample it will have bound to the plate, 'blocking' the ability of the monoclonal conjugate to bind to the antigen. If no antibody was present in the test sera the conjugate will bind. The plate is washed and a substrate is added. Unlike the indirect assay, negative sera develop colour as there was nothing to prevent the monoclonal conjugate from binding to the antigen. Positive sera do not develop colour as the antibody prevented the conjugate from binding. This test is virus specific and its specificity is determined by the monoclonal antibody conjugate. It is sensitive and, like NT, it measures total antibody so is invaluable for use in serosurveys. It also has the advantage of being able to detect antibody in any species, so it is useful in surveillance programmes.

Conclusions

There are no silver bullets in the laboratory diagnosis of alphavirus infections, no 'one test' that can be used above all others to achieve a definitive diagnosis. Rather, it requires a broad-spectrum approach, which encompasses numerous test systems. It would be easy to dismiss the use of virus isolation and the classic serological techniques as 'old fashioned' in favour of more recent technologies. But this would be short sighted. Each of the test systems discussed in this chapter has value and provides different insights. Often when dealing with alphaviruses it is necessary to use multiple serology assays, attempt virus isolation and employ nucleic acid techniques. The development of luminescent substrates in ELISA technology will continue to improve sensitivity and the employment of recombinant antigen could lead to improved specificity. The use of biochip technology in both ELISA and immunofluorescence will allow the testing of multiple antigens, which could reduce overall costs. Improvements in nucleic acid testing should see a reduction in reagent costs and the development of improved testing platforms will see a reduction in testing time and overall cost reduction of PCR.

However, in addition to these technologies it should not be forgotten that interpretation of results should not be attempted in the absence of clinical, travel and vaccination histories.

References

Aaskov, J.G., and Davies, C. (1978). An indirect immunofluorescence assay for human antibodies to Ross River virus. J. Immunol. Methods *25*, 37–41.

Beaty, B., Calisher, C., and Shope, R. (1989). Arboviruses. In Diagnostic Procedures for Viral, Rickettsial and Chlamydial Infections, N.J. Schmidt, and R.W. Emmons, eds. (American Public Health Association, Inc., Washington, DC, USA), pp. 797–855.

Bronzoni, R.V., Moreli, M.L., Cruz, A.C., and Figueiredo, L.T. (2004). Multiplex nested PCR for Brazilian Alphavirus diagnosis. Trans. R. Soc. Trop. Med. Hyg. *98*, 456–461.

Brown, T.M., Mitchell, C.J., Nasci, R.S., Smith, G.C., and Roehrig, J.T. (2001). Detection of eastern equine encephalitis virus in infected mosquitoes using a monoclonal antibody-based antigen-capture enzyme-linked immunosorbent assay. Am. J. Trop. Med. Hyg. *65*, 208–213.

Calisher, C., and Karabatsos, N. (1988). Arbovirus Serogroups: Definition and Geographic Distribution. In The Arboviruses:Epidemiology and Ecology, T. Monath, ed. (CRC Press, Boca Raton, FL, USA), pp. 19–57.

Canakoglu, N., Berber, E., Ertek, M., Yoruk, M.D., Tonbak, S., Bolat, Y., Aktas, M., Kalkan, A., and Ozdarendeli, A. (2013). Pseudo-plaque reduction neutralization test (PPRNT) for the measurement of neutralizing antibodies to Crimean-Congo hemorrhagic fever virus. Virol. J. *10*, 6.

Casals, J., and Brown, L.V. (1954). Hemagglutination with arthropod-borne viruses. J. Exp. Med. *99*, 429–449.

Chang, C.Y., Huang, C.C., Huang, T.S., Deng, M.C., Jong, M.H., and Wang, F.I. (2006). Isolation and characterization of a Sagiyama virus from domestic pigs. J. Vet. Diagn. Invest. *18*, 156–161.

Cloonan, M.J., O'Neill, B.J., Vale, T.G., Carter, I.W., and Williams, J.E. (1982). Ross River virus activity along the south coast of New South Wales. Aust. J. Exp. Biol. Med. Sci. *60*, 701–706.

Dalgarno, L., Short, N.J., Hardy, C.M., Bell, J.R., Strauss, J.H., and Marshall, I.D. (1984). Characterization of Barmah forest virus: an alphavirus with some unusual properties. Virology *133*, 416–426.

Digoutte, J.P., Tignor, G.H., Smith, A.L., and Knudson, D.L. (1976). [A rapid assay of Sindbis virus infectivity by counting immunofluorescence foci in "Aedes albopictus" cell culture (author's transl.)]. Ann. Microbiol. (Paris) *127B*, 573–576.

El Mekki, A.A., Van der Groen, G., and Pattyn, S.R. (1979). Evaluation of immunofluorescence and immunoperoxidase methods for antibody determination against Chikungunya, West Nile and yellow fever viruses. Ann. Soc. Belg. Med. Trop. *59*, 121–125.

Engler, H.D., and Selepak, S.T. (1994). Effect of centrifuging shell vials at 3,500 × *g* on detection of viruses in clinical specimens. J. Clin. Microbiol. *32*, 1580–1582.

Eshoo, M.W., Whitehouse, C.A., Zoll, S.T., Massire, C., Pennella, T.T., Blyn, L.B., Sampath, R., Hall, T.A., Ecker, J.A., Desai, A., *et al.* (2007). Direct broad-range detection of alphaviruses in mosquito extracts. Virology *368*, 286–295.

Grace, T.D. (1966). Establishment of a line of mosquito (*Aedes aegypti* L.) cells grown *in vitro*. Nature *211*, 366–367.

Graham, D.A., Wilson, C., Jewhurst, H., and Rowley, H. (2008). Cultural characteristics of salmonid alphaviruses--influence of cell line and temperature. J. Fish Dis. *31*, 859–868.

Griffin, D. (2007). Alphaviruses. In Fields Virology, D. Knipe, P. Howley, D. Griffin, R. Lamb, M. Martin, B. Roizman, and S. Straus, eds. (Lippincott Williams and Wilkins, Philadelphia, PA, USA), pp. 1023–1067.

Grywna, K., Kupfer, B., Panning, M., Drexler, J.F., Emmerich, P., Drosten, C., and Kummerer, B.M. (2010). Detection of all species of the genus Alphavirus by reverse transcription-PCR with diagnostic sensitivity. J. Clin. Microbiol. *48*, 3386–3387.

Gubler, D.J., and Rosen, L. (1976). A simple technique for demonstrating transmission of dengue virus by mosquitoes without the use of vertebrate hosts. Am. J. Trop. Med. Hyg. *25*, 146–150.

Ha, D.Q., Calisher, C.H., Tien, P.H., Karabatsos, N., and Gubler, D.J. (1995). Isolation of a newly recognized alphavirus from mosquitoes in Vietnam and evidence for human infection and disease. Am. J. Trop. Med. Hyg. *53*, 100–104.

Harmon, M., and Kendal, A. (1989). Influenza viruses. In Diagnostic Procedures for Viral, Rickettsial and Chlamydial Infections, N.J. Schmidt, and R.W. Emmons, eds. (American Public Health Association, Washington, DC, USA), pp. 631–668.

Hernandez, R., Sinodis, C., and Brown, D.T. (2010). Sindbis virus: propagation, quantification, and storage. Curr. Protoc. Microbiol. *Chapter 15*, Unit15B 11.

Hildreth, S.W., and Beaty, B.J. (1984). Detection of eastern equine encephalomyelitis virus and Highlands J virus antigens within mosquito pools by enzyme immunoassay (EIA). I. A laboratory study. Am. J. Trop. Med. Hyg. *33*, 965–972.

Hildreth, S.W., Beaty, B.J., Maxfield, H.K., Gilfillan, R.F., and Rosenau, B.J. (1984). Detection of eastern equine encephalomyelitis virus and Highlands J virus antigens within mosquito pools by enzyme immunoassay (EIA). II. Retrospective field test of the EIA. Am. J. Trop. Med. Hyg. *33*, 973–980.

Hueston, L., Toi, C.S., Jeoffreys, N., Sorrell, T., and Gilbert, G. (2013). Diagnosis of Barmah Forest virus infection by a nested real-time SYBR green RT-PCR assay. PLOS ONE *8*, e65197.

Hundekar, S.L., Thakare, J.P., Gokhale, M.D., Barde, P.V., Argade, S.V., and Mourya, D.T. (2002). Development of monoclonal antibody based antigen capture ELISA to detect Chikungunya virus in mosquitoes. Indian J. Med. Res. *115*, 144–148.

Klenk, H.D., Keil, W., Niemann, H., Geyer, R., and Schwarz, R.T. (1983). The characterization of influenza A viruses by carbohydrate analysis. Curr. Top. Microbiol. Immunol. *104*, 247–257.

Konishi, E., and Takahashi, J. (1985). Detection of chikungunya virus antigen in *Aedes albopictus* mosquitoes by enzyme-linked immunosorbent assay. J. Virol. Methods *12*, 279–285.

Kumar, J.S., Khan, M., Gupta, G., Bhoopati, M., Rao, P.V.L., and Parida, M. (2012). Production, characterization and application of monoclonal antibodies specific to recombinant (E2) structural protein in antigen-capture ELISA for clinical diagnosis of Chikungunya virus. Viral Immunol. *25*, 153–160.

Lambert, A.J., Martin, D.A., and Lanciotti, R.S. (2003). Detection of North American eastern and western equine encephalitis viruses by nucleic acid amplification assays. J. Clin. Microbiol. *41*, 379–385.

Litzba, N., Schuffenecker, I., Zeller, H., Drosten, C., Emmerich, P., Charrel, R., Kreher, P., and Niedrig, M. (2008). Evaluation of the first commercial chikungunya virus indirect immunofluorescence test. J. Virol. Methods *149*, 175–179.

Marshall, I.D., Woodroofe, G.M., and Hirsch, S. (1982). Viruses recovered from mosquitoes and wildlife serum collected in the Murray Valley of South-eastern Australia, February 1974, during an epidemic of encephalitis. Aust. J. Exp. Biol. Med. Sci. *60*(Pt 5), 457–470.

Merrill, M.H., and Tenbroeck, C. (1935). The transmission of equine encephalomyelitis virus by *Aedes aegypti*. J. Exp. Med. *62*, 687–695.

Monath, T.P., McLean, R.G., Cropp, C.B., Parham, G.L., Lazuick, J.S., and Calisher, C.H. (1981). Diagnosis of eastern equine encephalomyelitis by immunofluorescent staining of brain tissue. Am. J. Vet. Res. *42*, 1418–1421.

Mourya, D.T., Thakare, J.R., Gokhale, M.D., Powers, A.M., Hundekar, S.L., Jayakumar, P.C., Bondre, V.P., Shouche, Y.S., and Padbidri, V.S. (2001). Isolation of chikungunya virus from *Aedes aegypti* mosquitoes collected in the town of Yawat, Pune District, Maharashtra State, India. Acta Virol. *45*, 305–309.

Mourya, D.T., Gokhale, M.D., and Kumar, R. (2007). Xenodiagnosis: use of mosquitoes for the diagnosis of arboviral infections. J. Vector Borne Dis. *44*, 233–240.

Nasar, F., Palacios, G., Gorchakov, R.V., Guzman, H., Da Rosa, A.P., Savji, N., Popov, V.L., Sherman, M.B., Lipkin, W.I., Tesh, R.B., *et al.* (2012). Eilat virus, a unique alphavirus with host range restricted to insects by RNA replication. Proc. Natl. Acad. Sci. U.S.A. *109*, 14622–14627.

Olsen, G.H., Turell, M.J., and Pagac, B.B. (1997). Efficacy of eastern equine encephalitis immunization in whooping cranes. J. Wildl. Dis. 33, 312–315.

Peters, C., and Dalrymple, J. (1990). Alphaviruses. In Fields Virology, B. Fields, D. Knipe, and R. Chanok, eds. (Raven Press, New York, NY, USA), pp. 713–761.

Pfeffer, M., Linssen, B., Parke, M.D., and Kinney, R.M. (2002). Specific detection of chikungunya virus using a RT-PCR/nested PCR combination. J. Vet. Med. B Infect. Dis. Vet. Public Health *49*, 49–54.

Philip Samuel, P., and Tyagi, B.K. (2006). Diagnostic methods for detection & isolation of dengue viruses from vector mosquitoes. Indian J. Med. Res. *123*, 615–628.

Powers, A.M., Brault, A.C., Shirako, Y., Strauss, E.G., Kang, W., Strauss, J.H., and Weaver, S.C. (2001). Evolutionary relationships and systematics of the alphaviruses. J. Virol. *75*, 10118–10131.

Rosen, L., and Gubler, D. (1974). The use of mosquitoes to detect and propagate dengue viruses. Am. J. Trop. Med. Hyg. *23*, 1153–1160.

Rosen, L., Roseboom, L.E., Gubler, D.J., Lien, J.C., and Chaniotis, B.N. (1985). Comparative susceptibility of mosquito species and strains to oral and parenteral infection with dengue and Japanese encephalitis viruses. Am. J. Trop. Med. Hyg. *34*, 603–615.

Sambrook, J., Fritsch, E.F., and Maniatis, T. (1989). Molecular Cloning: A Laboratory Manual Vol 1–3 (Cold Spring Harbor University Press, Spring Harbor, NY, USA).

Schmidt, N.J. (1989). Cell Culture Procedures for Diagnostic Virology, 6th edn (American Public Health Association, Inc., Washington, DC, USA).

Scott, T.W., and Olson, J.G. (1986). Detection of eastern equine encephalomyelitis viral antigen in avian blood by enzyme immunoassay: a laboratory study. Am. J. Trop. Med. Hyg. 35, 611–618.

Singh, K.R.P. (1967). Cell cultures derived from larvae of *Aedes albopictus* (SKUSE) and *Aedes aegypti* (L). Curr. Sci. 36, 506–508.

Strauss, J.H., and Strauss, E.G. (1994). The alphaviruses: gene expression, replication, and evolution. Microbiol. Rev. 58, 491–562.

Studdert, M.J., Azuolas, J.K., Vasey, J.R., Hall, R.A., Ficorilli, N., and Huang, J.A. (2003). Polymerase chain reaction tests for the identification of Ross River, Kunjin and Murray Valley encephalitis virus infections in horses. Aust. Vet. J. *81*, 76–80.

Sutherland, L.J., Cash, A.A., Huang, Y.J., Sang, R.C., Malhotra, I., Moormann, A.M., King, C.L., Weaver, S.C., King, C.H., and LaBeaud, A.D. (2011). Serologic evidence of arboviral infections among humans in Kenya. Am. J. Trop. Med. Hyg. *85*, 158–161.

Tignor, G.H., and Digoutte, J.P. (1976). [Comparative studies of two alphavirus by macroplaque assays and by a fluorescent focus method in Vero cell lines (author's transl)]. Ann. Microbiol. *127B*, 439–444.

Wang, E., Paessler, S., Aguilar, P.V., Carrara, A.S., Ni, H., Greene, I.P., and Weaver, S.C. (2006). Reverse transcription-PCR-enzyme-linked immunosorbent assay for rapid detection and differentiation of alphavirus infections. J. Clin. Microbiol. *44*, 4000–4008.

The Interaction of Alphaviruses with the Interferon System

4

William B. Klimstra and Kate D. Ryman

Abstract

As reviewed elsewhere in this volume, the *Alphavirus* genus of the family Togaviridae includes a number of important human and zoonotic pathogens that cause a variety of diseases, ranging from asymptomatic through mild, self-limiting febrile disease to polyarthritis/arthralgia (see Chapter 7 for details of arthritogenic alphaviruses) or fatal encephalomyelitis (see Chapter 8 for details of encephalitic alphaviruses), depending upon the virus. In traditionally used cell lines of murine or primate origin, however, wild-type alphaviruses tend to replicate rapidly. While the synthesis of virus proteins by host translation machinery is highly efficient, and thousands of progeny virions are released from each infected cell within 18–24 hours post infection, host gene expression is almost completely arrested within a few hours via independent inhibition of transcription and translation. Consequently, *in vitro* the antiviral stress responses of the permissive host cell are very effectively suppressed by most, if not all, wild-type alphaviruses. In contrast, *in vivo* the relative ability of each alphavirus to resist or evade inhibition by the type I interferon (IFN)-mediated antiviral response is arguably the most important factor determining virus virulence, and this can be reproduced to some degree *in vitro* by using cell types relevant to the pathogenesis of each virus and maintaining the differentiated phenotype of such cells as completely as possible.

In this chapter, in order to understand the mechanisms involved in alphavirus virulence/attenuation, we will place in context recent findings about the interactions between the alphaviruses and the IFN-α/β system at the host, cellular and molecular levels. The current status of our knowledge in these areas is described in the following sections, focusing first from the host's perspective on IFN-α/β induction and effector mechanisms, and then on the mechanisms employed by the alphaviruses to inhibit, resist or avoid these responses.

The role of IFN-α/β in control of alphavirus infection

To counter the host's potent innate immune system, pathogenic viruses have developed diverse mechanisms to evade or antagonize the IFN-α/β response (reviewed in Garcia-Sastre and Biron, 2006; Randall and Goodbourn, 2008), either to directly combat antiviral activities and/or to subvert the transition from innate to adaptive immunity (Finlay and McFadden, 2006). Most, if not all, animal-virulent viruses encode viral gene products which: (i) avoid/suppress detection by the cell's pathogen sensing machinery; (ii) repress transcription and/or translation of the type I IFN genes, virus stress-induced genes (VSIGs) and/or IFN stimulated genes (ISGs); (iii) interfere with the IFN-α/β signalling cascade triggered by binding to the IFNAR receptor; and/or iv) directly inhibit/resist IFN-α/β-controlled antiviral gene products. The many observations suggesting that the relative sensitivity of alphaviruses to IFN-α/β-mediated antiviral activity is a major determinant of virulence and attenuation will be reviewed in this chapter. Alphaviruses with limited ability to evade or antagonize mammalian IFN-α/β tend to be attenuated in mice (e.g. Couderc *et al.*, 2008; Cruz *et al.*, 2010; Johnson *et al.*, 1972; Ryman *et al.*, 2000), whereas the virulence of the pathogenic alphaviruses for adult mice (and probably humans) is determined, at least in part, by antagonism and/or avoidance of the IFN-α/β response (e.g. Gardner *et al.*, 2008, 2012; Simmons *et al.*, 2009; White *et al.*, 2001; Yin *et al.*, 2009). In this section, we describe broadly the role of type I IFN *in vivo* in the best-characterized alphavirus infections of mice.

New World encephalitic alphaviruses

The three encephalitic New World alphaviruses have long been assumed to be fairly similar in their biology. However, in the last few years it has become apparent that their cell and tissue tropisms differ critically *in vivo* and *in vitro* (Gardner *et al.*, 2008; Vogel *et al.*, 2005), due at least in part to the evolution of different mechanisms to resist the type I IFN response (Gardner *et al.*, 2008; Ryman and Klimstra, 2008; Trobaugh *et al.*, 2013).

Venezuelan equine encephalitis virus

Best representing what one typically thinks of as the pathogenic sequence of an alphavirus, after deposition in the subcutaneous tissues of mice (mimicking mosquito bite), Venezuelan equine encephalitis virus (VEEV) infects patrolling dendritic cells (DCs; MacDonald and Johnston, 2000; Tonkin *et al.*, 2012) and possibly macrophages (unpublished observations) at the site of inoculation in the skin, and effectively commandeers these cells for transportation to regional draining lymph nodes (DLNs; Gardner *et al.*, 2008; MacDonald and Johnston, 2000). The extent to which free virus enters the bloodstream/lymphatics and disseminates to lymphoid and other tissues independently of these migratory cells remains unknown. In the DLN, the virus amplifies rapidly in the first few hours of infection (e.g. Gardner *et al.*, 2008; Grieder *et al.*, 1995), then disseminates systemically via serum viraemia, infecting other lymphoid tissues, muscle, and fibroblast connective tissues, distant from the site of inoculation. After neuroinvasion via infection of the olfactory neuroepithelium (Charles *et al.*, 2001), VEEV infects central nervous system (CNS) cells, primarily neurons (Jackson and Rossiter, 1997; Schoneboom *et al.*, 2000a).

Systemic levels of type I IFN in sera peak at approximately 10^5 international units (IU) per ml (Charles *et al.*, 2001; Gardner *et al.*, 2008; Grieder *et al.*, 1997), and yet VEEV remains highly virulent in spite of this extremely robust antiviral response, causing fatality of mice within 7–10 days of infection. From these observations, we can infer that VEEV is able to resist, evade and/or disable type I IFN response components in mice. Supporting this conjecture, wild-type VEEV is the most resistant alphavirus to type I IFN priming (i.e. pretreatment of cells with type I IFN several hours prior to infection) *in vitro* among those compared in head-to-head assay (Ryman and Klimstra, 2008). At least *in vivo*, antagonism of systemic type I IFN induction is not obvious but cannot be ruled out at the local level.

In spite of VEEV's resistance to the type I IFN-mediated antiviral effect, in mice deficient in type I IFN signalling due to the absence of the IFNAR1 receptor subunit (IFNAR1$^{-/-}$), the disease course is accelerated, and all animals reach euthanasia criteria within 1–2 days post infection (Grieder and Vogel, 1999; White *et al.*, 2001). Comparative pathogenesis studies in IFNAR$^{+/+}$ mice revealed that the IFN-α/β response mediates partial clearance of virus from the periphery, but fails to protect the CNS tissues from pathology or the host from death (Charles *et al.*, 1995; White *et al.*, 2001). This may reflect the fact that, at least in mice, wild-type VEEV induces the highest levels of serum IFN levels of all alphaviruses tested (reviewed by Ryman and Klimstra, 2008). VEEV mutants with increased sensitivity to IFN-α/β priming *in vitro* due to mutation in the 5′ untranslated region (UTR), including the live-attenuated TC83 vaccine strain (Spotts *et al.*), are highly attenuated in mice (Spotts *et al.*, 1998; White *et al.*, 2001). Important questions remain as to how wild-type VEEV overcomes the innate immune response, why the virulence of VEEV is so much greater for mice than for humans, and whether or not this could be due to more effective type I IFN antagonism in the natural (rodent) versus opportunistic (human) hosts, or a genetic deficiency in inbred mice.

Eastern equine encephalitis virus

In striking contrast to VEEV, infection of mice with North American (NA) wild-type strains of EEEV typically occurs with little/no detectable serum IFN-α/β circulation (Gardner *et al.*, 2008, 2009, 2011). Indeed the severity of EEEV encephalitic disease in mice is negatively correlated with serum IFN-α/β production (Gardner *et al.*, 2009, 2011; Trobaugh *et al.*, 2013) and this is consistent with analyses of human data showing that the magnitude of prodromal disease is positively associated with survival following EEEV infection (Silverman *et al.*, 2013). Naturally circulating NA EEEV strains are restricted by heparan sulfate binding from efficient dissemination to the DLN (Gardner *et al.*, 2011, 2013) and replicate poorly in human or murine myeloid-lineage cells (Gardner *et al.*, 2008; Levitt *et al.*, 1979; Vogel *et al.*, 2005), via a type I IFN signalling-independent inhibition of genome translation (Gardner *et al.*, 2008; Trobaugh *et al.*, 2013). Extrapolating from the observation that infection of myeloid-lineage cells such as DCs and macrophages by VEEV results in a wave of type I IFN release, we infer that avoidance of myeloid cell and lymphoid tissue replication by EEEV constitutes a unique antiviral escape mechanism (Gardner *et al.*, 2008, 2011; Trobaugh *et al.*, 2013).

Despite the lack of systemic type I IFN in NA EEEV-infected mice, the average survival of IFNAR1$^{-/-}$ mice is reduced by 1–2 days compared with IFNAR1$^{+/+}$ controls, and viral titres are increased, suggesting that local IFN-α/β-related responses are involved in a modest protection (unpublished observations). NA EEEV is believed to seed viraemia through replication in mesenchymal-lineage cells such as fibroblasts and osteoblasts (Gardner *et al.*, 2008; Vogel *et al.*, 2005) which are also capable of type I IFN production and response, presumably largely overwhelmed by the virus infection as these cells appear dead within 12–24 hours post infection *in vivo* (Vogel *et al.*, 2005). However, EEEV is significantly less resistant than VEEV to the antiviral effects of type I IFN priming *in vitro* (Ryman and Klimstra, 2008). In keeping with these observations, EEEV virulence is correlated with IFN-α/β sensitivity (Aguilar *et al.*, 2005); strains of EEEV that elicit serum IFN-α/β in mice (Gardner *et al.*, 2009, 2011) are attenuated *in vivo*, and experimental induction of an antiviral state *in vivo* during infection greatly attenuates disease (Aguilar *et al.*, 2005; Gardner *et al.*, 2009). These observations suggest that treatment of EEEV infection in the clinic with type I IFN or inducers thereof may be effective.

Western equine encephalitis virus

The much less studied western equine encephalitis virus (WEEV) infection is attenuated in mice, similar to human disease, unless the peripheral stages of disease are bypassed by intracerebral inoculation. In hamsters, however, severe neurotropic disease can be controlled by exogenous IFN-α administration (Julander *et al.*, 2007). As WEEV is a naturally occurring chimeric virus, resembling EEEV in the non-structural (nsP) and capsid protein and 3′ UTR sequences, but with Sindbis virus (SINV)-like glycoproteins (Hahn *et al.*, 1988; Weaver *et al.*, 1997), certain inferences may be made as to its biology and interaction with the type I IFN system; however, this remains to be tested.

Old World arthritogenic alphaviruses

Wild-type strains of Old World alphaviruses [including Sindbis (SINV), Semliki Forest (SFV), chikungunya (CHIKV) and O'nyong nyong (ONNV) viruses] cause severe, typically fatal disease in neonatal mice, but virulence is strongly dependent upon age/developmental status of the host, such that these infections are attenuated in mice by one to two weeks of age (Couderc *et al.*, 2008; Fazakerley, 2002; Griffin, 1976; Ryman *et al.*, 2000, 2007b). Although it is well known that these viruses are highly sensitive to IFN-α/β priming of cells *in vitro*, type I IFN system potentiation is frequently dismissed as a major determinant of disease attenuation (Griffin, 1976; Lenschow *et al.*, 2005) because high levels of systemically circulating serum IFN-α/β are positively correlated with higher virus titres (Klimstra *et al.*, 1999; Rabinovich and Liu, 1968; Trgovcich *et al.*, 1996, 1997; Vilcek, 1964). However, it has also been shown that, although the induction of systemic serum type I IFN is not typically robust in adult mice (Ryman and

Klimstra, 2008), IFNAR1$^{-/-}$ mice are extremely susceptible to infection by SINV (Ryman *et al.*, 2000, 2007c), SFV (Fragkoudis *et al.*, 2007; van den Broek *et al.*, 1995), CHIKV (Couderc *et al.*, 2008; Gardner *et al.*, 2012; Schilte *et al.*, 2010) and ONNV (Seymour *et al.*, 2013), succumbing to a systemic inflammatory response-like disease within a few days, despite the presence of an otherwise intact immune system.

Sindbis virus

Subcutaneous inoculation of adult mice with wild-type SINV is subclinical, with self-limiting virus replication and little tissue damage, and virus clearance begins 3–4 days post-infection (Ryman *et al.*, 2000). Although viral clearance roughly coincides with the appearance of virus-specific antibodies, virus control during primary infection does not require the presence of T or B lymphocytes or NK cells (Griffin, 1976; Wesselingh *et al.*, 1994), and early studies indicated a central role for the IFN-α/β response (Postic *et al.*, 1969).

The permissivity of myeloid-lineage cells to SINV infection is strictly controlled by the IFN-α/β-mediated antiviral response (Ryman *et al.*, 2000), so that in IFNAR1$^{-/-}$ mice, SINV replicates extensively in DCs and macrophages in the DLN, disseminates efficiently by high titre serum viraemia to similar cells in spleen, liver and other tissues, and induces a dysregulated, proinflammatory cytokine cascade (Ryman *et al.*, 2000, 2007c). Thus, the type I IFN response functions as a *de facto* tropism determinant (reviewed in McFadden *et al.*, 2009). In the absence of IFN-α/β signalling, a normally circumscribed SINV infection becomes rapidly fatal, with similarities to VEEV infection, highlighting the extreme sensitivity of this virus to mammalian IFN-α/β-mediated antiviral activity and the pivotal role of IFN-α/β in protection against infection in this model.

Chikungunya virus

Although CHIKV appears superficially similar to SINV in murine models, exhibiting strong dependence on age and type I IFN response for virulence (Couderc *et al.*, 2008), more detailed studies of CHIKV-induced musculoskeletal disease (MSD) in mice have revealed essential differences, reminiscent of CHIK fever with polyarthritis/arthralgia in humans (Gardner *et al.*, 2010; Morrison *et al.*, 2011). With some similarity to NA EEEV, CHIKV may partially avoid the type I IFN response by reduced tropism for myeloid-lineage cells, particularly macrophages, and lymphoid tissues (Couderc *et al.*, 2008; Gardner *et al.*, 2012; Kumar *et al.*, 2012) and, again, the type I IFN response has a relatively minor role in the refractory phenotype of these cells (Couderc *et al.*, 2008; Gardner *et al.*, 2012). Preliminary evidence suggests, however, that the factors controlling myeloid tropism differ between the two viruses (Trobaugh *et al.*, 2013; unpublished observations). CHIKV also appears to seed viraemia primarily from infected mesenchymal-lineage cells such as osteoblasts and fibroblasts (Couderc *et al.*, 2008; Schilte *et al.*, 2010). CHIKV can replicate in human monocyte/macrophage cells to some extent and elicits IFN-α/β (Her *et al.*, 2010; Hoarau *et al.*, 2010), but non-hematopoietic cells are pivotal in both type I IFN secretion and responsiveness (Schilte *et al.*, 2010, 2012). Interestingly, MSD is greatly exacerbated in IFNAR1$^{-/-}$ mice (Gardner *et al.*, 2012), or in mice with deficiency in IFN regulatory factor (IRF) 3 or IRF7 virus detection, type I IFN induction pathways (Rudd *et al.*, 2012), distinguishing this virus from SINV.

Type I IFN induction by alphaviruses

Upon virus infection of a host cell, a cascade of events occurs resulting in the induction of innate immune cytokines including the type I IFNs, IFN-α (multiple subtypes) and IFN-β (Fig. 4.1). Type II IFN (IFN-g) is released in the host as an indirect effect of alphavirus infection, by specialized cells such as T, NK and NKT cells, and some DCs. The most recently discovered type III IFN (IFN-l/IL-29/IL-28A/B) is secreted in response to some virus infections (reviewed in Witte *et al.*, 2010), but has not been studied with alphaviruses. As the type I IFN pathway is clearly the primary innate defence mechanism in alphavirus infections, we will focus in this section on the induction of type I IFNs from alphavirus-infected cells.

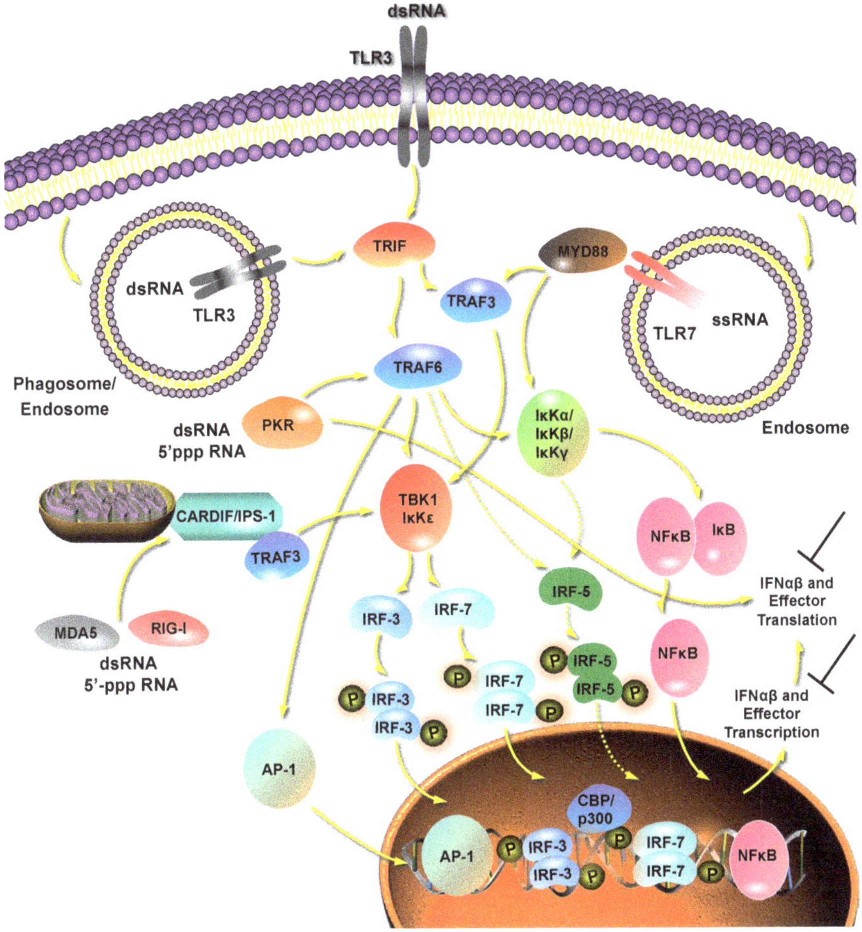

Figure 4.1 Diagram of the IFN-α/β induction pathways and the points at which they are antagonized by alphaviruses. Pathogen recognition receptors with potential relevance to alphaviruses include the surface and phago/endosomal toll-like receptor TLR-3 (dsRNA), endosomal TLR 7 (ligand ssRNA), the cytoplasmic dsRNA-dependent protein kinase, PKR (ligand dsRNA and 5′ triphosphate RNA), and the DEAD box RNA helicases MDA5 (ligand dsRNA) and RIG-I (ligands dsRNA and 5′ triphosphate RNA). For full expression of IFN genes, a complement of the AP-1 transcription factor, accessory proteins CBP/p300, either of the IFN-specific transcription factors IRF3 or IRF7 and the cytokine-specific transcription factor NFkB assemble on IFN-a or IFN-b gene promoters. Ratios of IRF3 versus IRF7 activity and homo- versus hetero-dimerization determine transcription of the different IFN subtypes in a cell-specific manner. IRF-5, originally thought to be unresponsive to viral infection has recently been shown to respond to TLR signalling in a virus- and cell-specific manner (Lazear *et al.*, 2013). In addition, transcription of some antiviral effector genes is facilitated through IRF3 or IRF7 transactivation (Grandvaux *et al.*, 2002). Steps in the process at which alphaviruses have been demonstrated to antagonize these responses are shown with a 'T' blockade symbol. Adapted from figures in Randall and Goodbourn (2008) using Protein Lounge software.

Cells bear sensory machinery that detects the presence of pathogen-associated molecular patterns (PAMPs) as non-self and belonging to invading microbes through pattern recognition receptors (PRRs). PRRs include extracellular and endosomal receptors such as the toll-like (TLRs) and C-type lectin scavenger receptors (Iwasaki and Medzhitov, 2004; McGreal *et al.*, 2004), as well as cytosolic receptors such as the RNA helicases, RIG-I and MDA-5 (Kang *et al.*, 2002; Yoneyama *et al.*, 2004), and the double-stranded (ds)RNA-dependent protein kinase (PKR; Jiang *et al.*, 2003). Potential alphaviral PAMPs would be speculated to include the capsid and envelope glycoproteins (C-type lectins; TLR2; PKR), and viral replicative intermediates including single-stranded (ss) RNAs (TLR7), double-stranded (ds) or structured RNAs (TLR3, PKR, RIG-I and MDA5), or ssRNA with an uncapped 5′ triphosphate (RIG-I) (reviewed in Kang *et al.*, 2002; Meylan *et al.*, 2006; Pollara *et al.*, 2005). The cytoplasmic PRRs are expressed by a majority of cell types (reviewed in Kang *et al.*, 2002; Meylan *et al.*, 2006; Pollara *et al.*, 2005), whereas the expression of functional cell-surface/endosomal PRRs is often specific to subsets of myeloid- or lymphoid-lineage cells [e.g. TLR7 is primarily expressed by plasmacytoid DCs (pDCs) and C-type lectin receptors are expressed on various myeloid cell subsets]. Therefore, viral induction of IFN-α/β can exhibit cell-type specificity, depending upon the pathway(s) triggered.

Several signalling pathways resulting in IFN induction have been elucidated downstream of the PRRs. Binding of the RIG-I or MDA5 RNA helicases to viral RNAs is followed by recruitment of the CARDIF/IPS-1 adapter molecule and activation of TANK binding kinase 1 (TBK1), or IKKε in lymphoid cells, which may directly phosphorylate constitutively expressed IRF3 (Paz *et al.*, 2006). IRF7 shares some activation characteristics with IRF3 but exhibits low baseline expression in most cells until it is up-regulated by IFN-α/β signalling (Paz *et al.*, 2006). IRF5 may also stimulate IFN-α/β gene transcription in a virus and cell-specific manner downstream of TLR7 (Barnes *et al.*, 2001; Cheng *et al.*, 2006; Lazear *et al.*, 2013). PKR, a dsRNA-binding kinase, appears to augment the induction of type I IFN by signalling through the nuclear factor (NF)-kB pathway and stabilization of IFN-β mRNA (Schulz *et al.*, 2010), as well as serving as a downstream modulator (Silva *et al.*, 2004) of the TLR3-triggered TRIF/TRAF6 pathway to activation of both NF-kB and TBK1-IRF3 (reviewed in Honda *et al.*, 2006). All other TLRs signal through a pathway that includes MyD88-IRAK1 adaptor proteins to activate NF-kB and induce the IFN-β promoter.

In designing and/or evaluating type I IFN induction experiments with relevance to *in vivo* alphavirus pathogenesis, it is critical to consider the cell types to be used, the virus strain and the natural tropism of the particular alphavirus in animals. It is known that type I IFN is induced *in vitro* by infection of a range of cell types (e.g. Atkins and Lancashire, 1976; Finter, 1964; Griffin, 2006; Marcus and Carver, 1967; Marcus and Fuller, 1979; Ryman and Klimstra, 2008) and, for myeloid-lineage cell-tropic alphaviruses, serum IFN-α/β is often correlated with systemic virus replication *in vivo* (Baron *et al.*, 1966; Bradish *et al.*, 1972; Klimstra *et al.*, 1999; Ryman *et al.*, 2000, 2002, 2005; Sherman and Griffin, 1990; White *et al.*, 2001). However, recent demonstrations of marked differences in *in vivo* tropisms for closely related viruses such as VEEV versus EEEV (Gardner *et al.*, 2008; Trobaugh *et al.*, 2013; Vogel *et al.*, 2005), or SINV versus CHIKV (Couderc *et al.*, 2008; Gardner *et al.*, 2012) suggest that the role of myeloid-lineage versus mesenchymal-lineage cells, IFN-α/β production and responses in the control of replication may be highly variable across the genus.

Type I IFN induction by alphaviruses in non-myeloid cells

Commonly used, highly tractable and readily available cells for studying the roles of particular proteins in pathogen sensing and type I IFN induction cascades are murine embryo fibroblasts (MEFs): (i) derived from genetically modified mice and often immortalized; (ii) engineered to overexpress specific proteins; or iii) treated with siRNAs to knock-down gene expression. However, there are a number of problems inherent in the use of MEFs for alphavirus studies, which can bias experimental results. First, even control 'protein-of-interest$^{+/+}$' MEF cultures can vary widely in their permissiveness and response to alphavirus infection (Burke *et al.*, 2009; Clavarino *et al.*, 2012), and are susceptible to clonal variation during passage and/or immortalization (e.g. Balachandran *et al.*, 2000; Wang *et al.*, 2009). In addition, when MEFs or other non-myeloid cultures relevant to pathogenesis (e.g. neurons or osteoblasts) are efficiently infected with a particular alphavirus they often fail to release detectable type I IFN into supernatants, tied to global cessation of host gene expression arbitrated by the virus (discussed below and in Chapter 1). While this may reflect some of what occurs locally in these cell types in the host, it should be remembered that the cells are being studied in the absence of a network of other cell types that may be essential to the overall *in vivo* phenotype. Consequently, the findings of these studies should be interpreted with caution.

The complex network of signalling pathways downstream of the PRRs that result in induction of type I IFN and other VSIGs have been elucidated in considerable detail. Canonical cytosolic RNA virus-sensing pathways in non-myeloid cells converge on the activation and dimerization of the constitutively expressed IRF3 transcription factor, which then translocates to the nucleus and activates transcription of a gene subset that includes IFN-β and IRF7 (Grandvaux *et al.*, 2002; Schafer *et al.*, 1998) (Fig. 4.1). Infections of various non-myeloid cell types from mice and primates by wild-type strains of SINV, SFV, CHIKV, VEEV, EEEV or WEEV have been shown to result in the activation and nuclear accumulation of IRF3 with minimal interference by the virus (Breakwell *et al.*, 2007; Burke *et al.*, 2009; Peltier *et al.*, 2013; White *et al.*, 2011). Extrapolation from individual studies with different non-myeloid cell types/alphaviruses suggests that recruitment of IPS-1/CARDIF, activation of TBK1, and TBK1-mediated phosphorylation of IRF3 act to transmit the signal for IRF3 nuclear translocation and promoter binding (Burke *et al.*, 2009; Schilte *et al.*, 2010; White *et al.*, 2011).

Downstream of IRF3 activation and translocation, findings diverge depending upon cell type and virus. In wild-type SINV-infected MEFs (Burke *et al.*, 2009; Garmashova *et al.*, 2006) or neurons (Yin *et al.*, 2009) little/no transcription or translation of the IFN-β gene or other VSIGs are detected, but this inhibition occurs coincident with global host macromolecular synthesis arrest, and thus specific inhibition of the IRF3 pathway, if present, cannot be distinguished. However, CHIKV infection of fibroblasts (White *et al.*, 2011) and VEEV (Yin *et al.*, 2009) or WEEV (Peltier *et al.*, 2013) infection of neurons leads to transcription of IRF3-dependent genes including IFN-β in the early stages of infection, but the mRNA is not translated, again in an environment of generalized protein synthesis arrest. Later in infection, transcription is also curtailed (White *et al.*, 2011).

A non-cytopathic (nc) SINV strain with a mutation in nsP2 that abrogates the virus's ability to arrest host macromolecular synthesis (Frolov *et al.*, 1999) induces a robust type I IFN response in MEFs (Burke *et al.*, 2009). Burke *et al.* (2009) used this virus to determine which of the major cytosolic PRRs would mediate type I IFN production from SINV-infected cells if global macromolecular synthesis did not occur. While MDA5 may interact with ds or highly structured RNA, RIG-I is activated by interaction with uncapped RNA molecules containing 5′-triphosphates (Gitlin *et al.*, 2006; Hornung *et al.*, 2006; Pichlmair *et al.*, 2006). Secretion of type I IFN is impaired, but not abrogated, in ncSINV-infected MDA5$^{-/-}$ and PKR$^{-/-}$ MEFs, confirmed by siRNA-mediated knock-down of MDA5 or PKR expression in control MEFs (Burke *et al.*, 2009). Similarly, type I IFN synthesis was augmented by PKR in SFV-infected MEFs (Barry *et al.*, 2009; Schulz *et al.*, 2010). Surprisingly, siRNA-mediated knockdown of RIG-I expression or overexpression of a dominant-negative protein significantly *increased* type I IFN induction following ncSINV infection (Burke *et al.*, 2009). Consistent with this finding, overexpression of RIG-1 in IFN-α/β signalling deficient human fibroblasts exhibited antiviral activity against CHIKV and VEEV-TC83 (Schoggins and Rice, 2011). Together, these data are suggestive that under certain circumstances RIG-I can mediate antiviral activity (directly or indirectly) against alphaviruses by an IFN-α/β induction-independent mechanism.

Overall, it appears that alphavirus PAMPs can be detected by cytosolic PRRs including MDA5, PKR and possibly RIG-I. The presence of PKR augments type I IFN synthesis, at least in some circumstances, most likely via its role in maintaining IFN-β mRNA integrity (Schulz *et al.*, 2010) and/or its effects on levels of the NF-κB inhibitor, IκB-α (McAllister *et al.*, 2012). However, findings for MEF cultures are not consistent with type I IFN induction profiles for myeloid cell-tropic alphaviruses in cultured DCs or macrophages *in vitro* or *in vivo* (see below). In particular, IFN-β gene expression in fibroblasts and neurons in response to alphavirus infection is extremely delayed compared with myeloid cell responses or detection of type I IFN in sera of infected animals

(Ryman and Klimstra, 2008). To date, there is no clear evidence that TLRs play a role in type I IFN responses to alphavirus infection in non-myeloid cells.

Interferon induction by alphaviruses in myeloid-lineage cells and mice

In a direct comparison of type I IFN induction in mice infected with SINV, CHIKV, VEEV or EEEV propagation-competent viruses, the highest levels of biological type I IFN were produced by VEEV, followed by SINV then CHIKV, while barely detectable levels were produced by NA EEEV (Ryman and Klimstra, 2008), consistent with previous studies examining each virus individually (Aguilar *et al.*, 2005; Charles *et al.*, 2001; Gardner *et al.*, 2008; Griffin, 2006; Klimstra *et al.*, 1999). This response is induced early after infection, generally detectable in sera within 8–12 h post-infection, and appears to play a role in protection of tissues distant from the site of infection prior to viral dissemination, particularly the CNS (Gardner *et al.*, 2009; Konopka *et al.*, 2009). Furthermore, infection of cells within lymphoid tissues and, in particular, myeloid cells, appears to be highly associated with production of type I IFN measured in the sera of infected animals (Gardner *et al.*, 2008, 2009; Tonkin *et al.*, 2010, 2012). Therefore, the interaction of alphaviruses with myeloid cells and magnitude of type I IFN induction measured in serum may be critical to disease outcome. Using a strategy to segregate infected cell transcriptional up-regulation from that of uninfected cells *in vivo*, it was also concluded that both uninfected and infected DLN cells contributed to production of tissue and serum type I IFN induction after VEEV replicon infection (Konopka *et al.*, 2007). For CHIKV- (Rudd *et al.*, 2012), SINV- and VEEV- (unpublished observations) infected mice, the induction of type I IFN was minorly dependent upon IRF3 but majorly dependent upon IRF7.

Interestingly, similar induction profiles are observed when mice are infected with non-propagative replicon particles from which the structural protein genes had been deleted (Tonkin *et al.*, 2012; unpublished observations), or with VEEV or EEEV mutants unable to initiate macromolecular shutoff [viruses described in Garmashova *et al.* (2007a) and Aguilar *et al.* (2008)]: these viruses elicit wild-type levels of serum IFN (unpublished observations). These results illustrate an important point: whether or not a wild-type alphavirus induces a systemic type I IFN response and/or is sensitive to this response, does not correlate with the arrest of cellular gene expression *in vitro*. Therefore, (i) the mechanisms by which wild-type alphaviruses induce type I IFN in MEFs (in which efficient infection generally fails to elicit IFN-β production) differ from those triggered in intact hosts; (ii) EEEV does not produce large amounts of type I IFN *in vivo* regardless of the presence or absence of the transcription-arresting capsid protein (virus vs. replicon); and (iii) VEEV induces the production of robust serum IFN-α/β even in the presence of the capsid protein.

As described above, depending upon the specific alphavirus, important infected cell types in mice include DCs and macrophages during the early stages of infection. Several groups have demonstrated type I IFN induction by alphavirus infection of bone marrow-derived mDCs (Hidmark *et al.*, 2005; Ryman *et al.*, 2002, 2005; Shabman *et al.*, 2007). SINV, VEEV, RRV and SFV each produced biologically active IFN-α/β after infection of these cells, which, with SFV, was shown to be independent of MyD88 but requiring IRF3 (Hidmark *et al.*, 2005). In two of these studies, type I IFN was induced by infection of the cells with partially ultraviolet (UV)-inactivated viruses (Hidmark *et al.*, 2005; Shabman *et al.*, 2007) suggesting that the complete replication cycle was not required. Consistent with results in fibroblasts, IFN-α/β induction was partially impaired in MDA5$^{-/-}$ macrophages infected with SINV (Gitlin *et al.*, 2006), while bone marrow-derived mDCs lacking PKR exhibited delayed IFN-α/β induction in response to SINV infection, although maximum levels were relatively unaffected, suggestive that PKR had a role in the timing of IFN-α/β induction (Ryman *et al.*, 2002). We have more recently completed similar studies with GM-CSF/FLT3 ligand-differentiated cells, which contain a significant proportion of pDCs (Zhan *et al.*, 2012), a major source of type I IFN during some virus infections (Fitzgerald-Bocarsly and Feng, 2007). Type I IFN induction in these cultures following infection with SINV or VEEV was rapid and robust, mirroring *in vivo* induction kinetics; however, the cytokine appeared to be secreted by virus-infected CD11b^{+} macrophage-like cells in these mixed cultures, not by the pDCs themselves (unpublished observations). EEEV failed to induce type I IFN in myeloid cells *in vitro*; however, rates of infection were extremely low due to the block in genome translation initiation (Gardner *et al.*, 2008).

Preliminary comparison of type I IFN responses in fibroblast versus myeloid cells would suggest that macromolecular synthesis shutoff may not be as effective in myeloid-lineage cells, allowing transcription/translation of IFN-α/β and other stress responses. However, our recent experiments with myeloid-lineage cell lines indicate that macromolecular shutoff can be rapid with viruses that can replicate robustly in these cells (e.g. VEEV), and yet substantial type I IFN is still released (unpublished observations). This leads us to speculate that either myeloid-lineage cells can respond more rapidly to virus infection than fibroblasts and essentially 'outrace' the virus or that the production of type I IFN by these cells is not sensitive to global arrest of macromolecular processes. Ultimately, while replication in myeloid cells is implicated in systemic type I IFN induction after alphavirus infection, the particular cell type(s) involved and requirements for productive replication remain to be fully elucidated.

Type I IFN induction of antiviral effectors with activity against alphaviruses

Most alphaviruses rapidly arrest host macromolecular synthesis (transcription and/or translation) and are highly cytopathic in permissive cell lines (see Chapter 1 for details of viral replication). However, if the same cells are primed by exposure to IFN-α/β (or IFN-γ, although this is not as effective, e.g. Gil *et al.*, 2001; Ryman *et al.*, 2002) prior to infection to induce a refractory or antiviral state, global cellular transcription and protein synthesis are no longer overtly altered but instead the production of progeny virions is dramatically reduced and type I IFN may be produced. It is becoming apparent that the arrest of host macromolecular synthesis by the virus and the suppression of virus replication by the host operate in a highly sensitive equilibrium that is dependent upon many factors including the cell type and maturation state, which influences constitutive levels of virus-sensing, IFN signalling and antiviral effector pathway components (Farmer *et al.*, 2013; Frolov *et al.*, 2012; Marfell *et al.*, 2013; Zhang *et al.*, 2007), as well as the particular virus and its ability to antagonize or circumvent these pathways. Thus, the outcome of the infection (i.e. the permissivity of the cell) is determined very soon after (or even before) the virus enters the cell (Frolov *et al.*, 2012).

The cellular responses to secreted type I, II or III IFNs involve separate but overlapping signalling cascades that result in synthesis of hundreds to thousands of proteins encoded by IFN-stimulated genes (ISGs), which exert a multitude of functions including a growing number of antiviral effectors. IFN signalling pathways are triggered when IFN-α/β, IFN-γ or IFN-λ bind distinct, ubiquitously expressed receptors on the surface of cells. The major antiviral induction pathway occurs via dimerization of the IFNAR1/2 receptor subunits upon binding of type I IFN and consequent autophosphorylation of Jak1/Tyk2 kinases that are constitutively associated with each subunit (Fig. 4.2). Activated Jak1/Tyk2 phosphorylates the IFNAR subunits, recruiting and phosphorylating STAT1 and STAT2. STAT1/2 heterodimers associate with IRF9, forming the ISGF3 complex, which translocates to the nucleus and directs the transcription of a large subset of ISGs by binding to the IFN-stimulated response element (ISRE) in the gene promoter. While IFN-λ up-regulates ISRE-containing genes similar to type I IFNs (Witte *et al.*, 2010), IFN-γ signalling involves STAT1 homodimers binding to IFN-γ-activated sequence (GAS) elements (Mirkovitch *et al.*, 1992; Schindler *et al.*, 1992). STAT1-independent type I and II IFN-mediated signalling has been described (Gil *et al.*, 2001) which may occur via phosphatidylinositol (PI) 3′-kinase (PI3K; Uddin *et al.*, 2000a,b) and/or p38/MAP kinase (Li *et al.*, 2004; Platanias, 2003). ISGs may have ISRE and/or GAS sites in their promoters, yielding independent and overlapping effects of the different IFNs. As the type I IFN response has been shown to provide the first and the strongest antiviral protection against alphavirus infection if

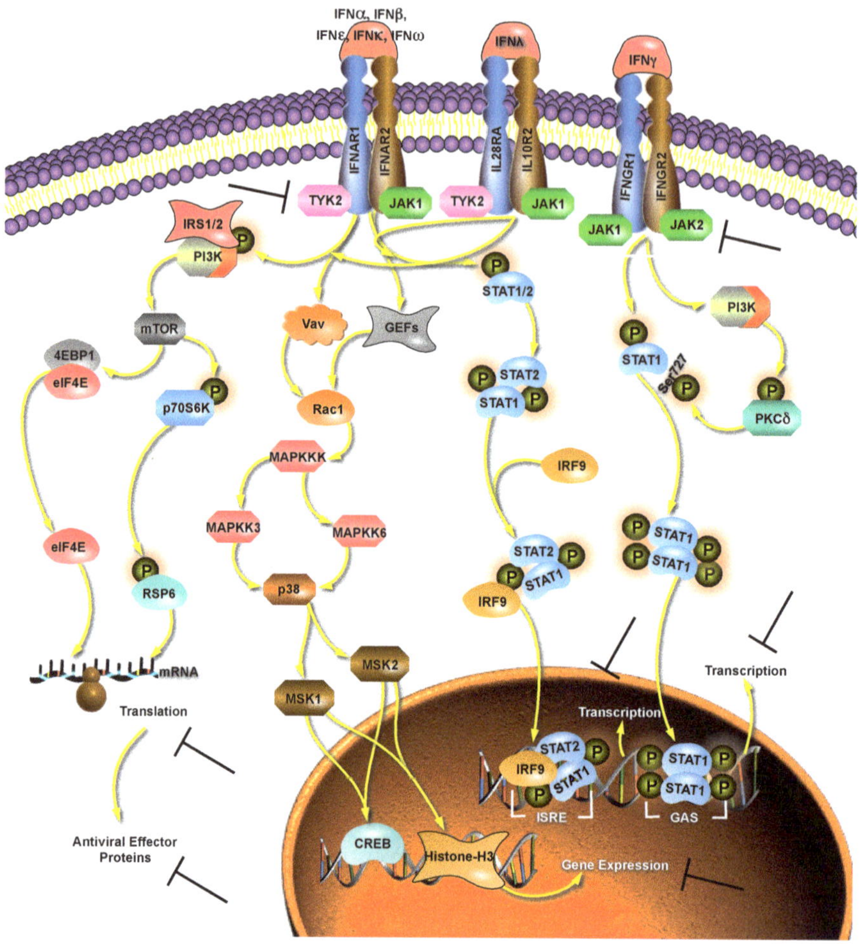

Figure 4.2 Diagram of type I IFN signalling pathways in murine cells and the points of antagonism that have been identified for alphaviruses. Type I IFNs bind a common cell-surface receptor (IFNAR). Interaction of the IFNs with their receptor induces largely overlapping subsets of IFN-stimulated genes (ISGs) that inhibit viral replication. The receptor is composed of two distinct subunits, each associated constitutively with a specific Janus-activated kinase (JAK) family member (Tyk2 and JAK1). The receptor-associated JAKs are activated (autophosphorylated) by ligand-dependent dimerization and phosphorylation of the receptor subunits, then regulate the phosphorylation and activation of STAT proteins. Although a number of STATs can be activated, the major pathways involve STAT1/STAT2 heterodimerization. Type I IFN-activated STAT1/2 dimers join IRF9 to form the ISGF3 complex, which translocates to the nucleus and initiates transcription of ISGs by binding IFN-stimulated response element (ISRE) sites in the promoters of these genes. Alternative signalling cascades regulated by the IFNs have been characterized but are not represented in this diagram. VEEV, CHIKV and SINV have been shown to block type I IFN signalling pathways at or before STAT activation in vitro shown with a ‘T’ blockade symbol (Fros *et al*., 2010; Simmons *et al*., 2009; Yin *et al*., 2009). However, arguably the dominant antagonism of IFN-mediated antiviral effector induction in infected cells results from the arrest of cellular transcription and/or translation by the viruses (Yin *et al*., 2009). Adapted from figures in Randall and Goodbourn (2008) using Protein Lounge software.

not circumvented or antagonized, our discussion of antiviral ISG effectors will concentrate largely on this pathway. It is important also to remember that the ISGs also overlap considerably with genes directly regulated by virus sensing pathways without the requirement for IFN (e.g. by activated IRF3; Grandvaux *et al.*, 2002), typically referred to as VSIGs (Sen and Peters, 2007).

In the past, most studies of the antiviral response have been performed in cell lines that do not produce type I IFN, either due to an inherent defect or because their permissiveness allows rapid and complete arrest of induction by the infecting virus. In this way, the antiviral state can be established and controlled by the exposure of cells to exogenously added IFN. In some more recent studies, attempts have been made to use cells more like those in the animal, both in terms of differentiation and relevance to pathogenesis. However, studies in primary cell types, especially myeloid-lineage cells, are complicated by the inherent resistance of these cells to infection (due to a pre-existing antiviral state), their ability to produce type I IFN and other VSIG products in response to infection, and the possibility of autocrine signalling and antiviral amplification in the infected cell later in infection. Often, pre-infection neutralization of type I IFN or the use of cells derived from IFN response-deficient mice render such cells permissive (e.g. Gardner *et al.*, 2012; Ryman *et al.*, 2000; White *et al.*, 2001), unless they are refractory for an alternative reason [e.g. EEEV in myeloid-lineage cells (Gardner *et al.*, 2008) and CHIKV in murine monocyte/macrophages (Couderc *et al.*, 2008; Gardner *et al.*, 2012)].

The identification of the ISG effector pathways has been tackled from three different angles: what steps in the virus propagation cycle can be blocked/inhibited by host antiviral effectors; what are the effectors and how do they act; and how do relatively IFN-resistant alphaviruses evade or antagonize these effector mechanisms? The first two are addressed below, while the third is addressed in the section entitled 'Antagonism of type I IFN responses by the alphaviruses'.

Identification of steps in alphavirus propagation inhibited by host antiviral effectors

Over the last five decades, researchers have attempted to determine the point(s) at which the alphavirus replication cycle is inhibited in cells that have been pre-exposed to IFN-α/β, or are inherently refractory to infection via IFN-dependent mechanisms (reviewed in Ryman and Klimstra, 2008). To-date, there are no clear reports that attachment, entry or genome uncoating steps are inhibited by type I IFN-dependent responses; indeed there is some evidence to the contrary (Tesfay *et al.*, 2008). However, a strong inhibition of alphavirus replication occurs immediately after uncoating, with lower levels of nsPs translated (Friedman, 1968; Tesfay *et al.*, 2008) and formation of the dsRNA replicative intermediate inhibited (Friedman *et al.*, 1967; Friedman and Sonnabend, 1965) in type I IFN-primed cells. A primary defect appears to be prevention of infectious viral genomic RNAs from forming polyribosomes for efficient translation of the nsP polyproteins (Gardner *et al.*, 2008; Joklik and Merigan, 1966; Tesfay *et al.*, 2008). Constitutively expressed and modestly IFN-inducible protein PKR has been shown to be activated and/or eIF2α has been found to be phosphorylated in SINV (Gorchakov *et al.*, 2004; Ryman *et al.*, 2002, 2005; Sanz *et al.*, 2009; Ventoso *et al.*, 2006), SFV (Barry *et al.*, 2009; Ventoso *et al.*, 2006), CHIKV (Clavarino *et al.*, 2012; White *et al.*, 2011) and EEEV (Aguilar *et al.*, 2008) infected cells (discussed further below). However, the potent IFN-α/β-inducible inhibition of alphavirus genome translation occurs in the absence of PKR or other eIF2α kinases (Tesfay *et al.*, 2008). This anti-alphaviral activity is dependent upon IFN-induced *de novo* RNA polymerase II-mediated transcription of ISG(s) encoding a protein (or proteins), which inhibits translation initiation of capped RNAs that enter the cytoplasmic compartment by infection/transfection but not from the nucleus (Tesfay *et al.*, 2008). Virus delivery of RNA into the cytoplasm or its transcription within that compartment represents a point of distinction between the replication cycle of the alphaviruses and normal host processes. Therefore, this may represent an important point of virus vulnerability to antiviral strategies that may be exploited for therapeutic benefit. Potential type I IFN-inducible effectors of this antiviral activity are discussed below.

Any antiviral effects downstream of genome translation inhibition are difficult to distinguish from the consequences of this early block and therefore have yet to be revealed with the exception of a report that anomalous alphaviral particles are found inside cytoplasmic vacuoles in IFN-α/β primed cells, suggesting that antiviral effectors may also interfere with morphogenesis/egress (Rebello *et al.*, 1993). It is probable that other such mechanisms exist as the type I IFN antiviral system exhibits multiple redundancies of action to protect the host while the virus evolves countering approaches.

IFN-inducible antiviral effectors

Although it has been known for half a century that type I IFN stimulates an antiviral response in many cells *in vitro*, the identification of the effectors of this refractory state has been a slow and painstaking process, often specific to the virus family concerned. In recent years, interest in the alphaviruses has shifted to the human pathogens, VEEV, EEEV, CHIKV and RRV, but it is arguably from the more type I IFN-sensitive SINV that we have learned the most about anti-alphaviral effectors (Chan *et al.*, 2008; Lenschow *et al.*, 2005; MacDonald *et al.*, 2007; Ryman *et al.*, 2002, 2005; Tesfay *et al.*, 2008; Zhang *et al.*, 2007) as this virus clearly exhibits the least ability to withstand IFN responses, particularly *in vivo* (although that is not to say that it lacks any ability to do so).

The first type I IFN-inducible antiviral pathways to be identified were the PKR, RNase L/OAS and Mx pathways (reviewed by Samuel, 1994). However, with the availability of mice and cells triply deficient in the PKR, RNase L and Mx proteins (Zhou *et al.*, 1999) came the discovery that these antiviral effectors represented only the tip of the iceberg in the antiviral response against SINV (Ryman *et al.*, 2002) and that, even in the absence of these proteins, the so-called 'alternative pathway(s)' (Zhou *et al.*, 1999) provided the majority of protection (Ryman *et al.*, 2005). Infection of these triply deficient mice was asymptomatic (Ryman *et al.*, 2002). In an attempt to identify ISG products with antiviral activity against SINV (and potentially other alphaviruses) from amongst the hundreds that are typically induced by type I IFN priming of cells (first shown by Der *et al.*, 1998), microarray analysis of mRNAs from IFN-α/β-treated versus virus-infected primary DC cultures, generated from wild-type, PKR$^{-/-}$ and IFNAR1$^{-/-}$ mice, was used in a subtractive approach to reveal candidate effectors of PKR/RNase L/Mx-independent, IFN-α/β-induced antiviral activity (Ryman *et al.*, 2005): those genes that were up-regulated in the presence of normal IFN-α/β signalling (i.e. in either IFN-α/β-primed or SINV-infected cells with or without PKR), but were not up-regulated in its absence (i.e. in virus-infected IFNAR$^{-/-}$ cells). Forty-five candidate genes were identified using this subtractive approach, five of which have been shown to have antiviral activity *in vitro* and/or *in vivo*: ISG20, ZAP, ISG15, IFIT1 (p56) and viperin (Bick *et al.*, 2003; Lenschow *et al.*, 2005; Zhang *et al.*, 2007) and are discussed further below.

More recently, Schoggins *et al.* (2011) used a lentiviral vector-based overexpression system in STAT1-deficient human fibroblasts (inherently impaired in STAT1-dependent ISGF3 formation; Dupuis *et al.*, 2003) to screen a large set of known ISGs for antiviral activity against VEEV (TC83 vaccine strain) and CHIKV (La Rèunion wild-type strain). Surprisingly, although the sensitivity of the TC83 VEEV vaccine strain to the antiviral effects of type I IFN *in vitro* and *in vivo* are dramatically increased over the wild-type virus (Spotts *et al.*, 1998; White *et al.*, 2001), and CHIKV is quite sensitive to type I IFN inhibition (Couderc *et al.*, 2008; Gardner *et al.*, 2012; Schilte *et al.*, 2010), only a handful of ISG products, from over 300 screened, significantly reduced infectivity of these viruses. Furthermore, this screen did not identify several ISG products with known efficacy against the alphaviruses although they were among the ISGs screened. Interestingly, the proteins identified are mostly known as components of the virus PAMP-sensing pathways [e.g. IRF1, RIG-I, MDA5 and C6ORF150 (cGAMP synthase)], and thus their overexpression may increase the rapidity or magnitude of the global antiviral state, rather than being directly antiviral at the protein level (see below).

PKR and eIF2α pathways

As briefly mentioned above, the PKR kinase, which is constitutively expressed and modestly IFN-inducible, is activated and/or eIF2α is phosphorylated in SINV (Gorchakov *et al.*, 2004; Ryman *et al.*, 2002, 2005; Sanz *et al.*, 2009; Ventoso *et al.*, 2006), SFV (Barry *et al.*, 2009; McInerney *et al.*, 2005; Ventoso *et al.*, 2006), CHIKV (Clavarino *et al.*, 2012; White *et al.*, 2011) and

EEEV (Aguilar *et al.*, 2008) infected cells. The PKR/eIF2α pathway of translation control has primarily been studied as a contributor to cell protein arrest, which is generally considered pro-viral, freeing host translation machinery for synthesis of alphavirus proteins and suppressing translation of IFNs, VSIGs and ISGs. However, although the alphavirus subgenomic RNA possesses 5′ UTR structures that allow it to resist PKR/eIF2α-mediated translation arrest (McInerney *et al.*, 2005; Ventoso, 2012; Ventoso *et al.*, 2006), this pathway appears to play some antiviral role early in infection both *in vitro* and *in vivo* (Barry *et al.*, 2009; Ryman *et al.*, 2002) and, as described above, augments type I IFN production from infected cells even in the absence of global translation shutoff. Thus, the complex and often contradictory interactions between PKR/eIF2α and alphaviruses remain controversial and incompletely understood. Overall, however, potent IFN-α/β-inducible inhibition of alphavirus replication occurs in the absence of PKR or other eIF2α kinases (Gorchakov *et al.*, 2004; Ryman *et al.*, 2002, 2005; Tesfay *et al.*, 2008; White *et al.*, 2011).

IFIT protein family

Although the IFIT family proteins are highly induced in response to type I IFN treatment and by certain viral PAMPs via IRF3 activation (reviewed by Diamond and Farzan, 2013; Zhou *et al.*, 2013), only the IFIT1 protein (p56 or ISG56) confers significant antiviral activity against SINV, and the effects of overexpression in untreated cells, or siRNA-mediated knockdown in IFN-α/β-primed cells *in vitro*, are modest (Zhang *et al.*, 2007), whereas wild-type strains of other alphaviruses may be more resistant to this antiviral effector. Alphavirus RNA genomes lack the 2′ O methylation of the 5′ cap structure used by some viruses to avoid IFIT1-mediated inhibition and masquerade as host mRNAs (Daffis *et al.*, 2010; Diamond and Farzan, 2013; Szretter *et al.*, 2012) and, therefore, may use an alternative mechanism for resistance. IFIT1 is also known to sequester single-stranded viral RNAs with 5′ triphosphates in a complex with other IFIT family members (Pichlmair *et al.*, 2011), a process for which alphaviral negative-strand replicative intermediates could be a target, and is involved in regulation of eIF4F translational activity through interaction with eIF3 (Hui *et al.*, 2005). Whether or not IFIT1 mediates antiviral activity via these mechanisms remains to be tested.

Ubiquitin-like ISG15

ISG15 was one of the first ISGs to be identified and was found to encode a ubiquitin-like protein that functions, at least in part, as a modifier of protein function (Zhao *et al.*, 2013). ISG15 conjugates to target proteins via isopeptide bonding with a series of other type I IFN-inducible enzymes: E1-activating enzyme, UbE1L; E2-conjugating enzyme UbcH8/UbcM8; E3 ligase enzymes Herc5, HHARI, and Efp; and deconjugating enzyme UBP43 (reviewed by Zhao *et al.*, 2013). Evidence implicating ISG15 as a highly IFN-inducible protein with broad-spectrum antiviral activity continues to accumulate, inhibiting by conjugation (or 'ISGylation') and also as a free, unconjugated protein (reviewed by Lenschow, 2010). While ISG15 has substantial antiviral activity against SINV (Giannakopoulos *et al.*, 2009; Lenschow *et al.*, 2005, 2007; Zhang *et al.*, 2007) and CHIKV (Werneke *et al.*, 2011) *in vivo*, it has little to no effect *in vitro* (Zhang *et al.*, 2007). Surprisingly, the mechanism of action by which ISG15 limits SINV infection is dependent upon UbE1L-mediated conjugation (Giannakopoulos *et al.*, 2009) and negatively regulated by UBP43 (Ritchie *et al.*, 2004), but against CHIKV ISG15-mediated antiviral activity is UbE1L-independent (Werneke *et al.*, 2011). Furthermore, ISG15 does not directly inhibit CHIKV replication *in vivo*, but instead seems to perform an immunomodulatory role in keeping with the lack of efficacy *in vitro*.

ISG20 3′–5′ exoribonuclease

ISG20, whose expression is strongly induced by type I IFN, is an antiviral effector with an apparent preference for RNA viruses (Espert *et al.*, 2003; Espert *et al.*, 2005; Jiang *et al.*, 2008; Zhang *et al.*, 2007). ISG20 is a member of the DEDDh subgroup of the DEDD exonuclease superfamily (Espert *et al.*, 2003; Horio *et al.*, 2004), with demonstrated 3′–5′ exonuclease activity, and has a preference for single-stranded RNA substrates. However, 3′-terminal stem–loop structures in the RNA substrate (perhaps such as those in the alphavirus genomic RNAs), strongly reduce its enzymatic activity (Espert *et al.*, 2003). Specifically how ISG20 interferes with virus replication has yet to be revealed. As little/no antiviral effect was observed in cells overexpressing a mutated ISG20 protein defective in exonuclease activity (unpublished data), it is tempting to speculate that ISG20 controls virus replication by degrading viral RNA via its 3′–5′ exoribonuclease activity. However, we have shown indirectly that viral RNA extracted from un-primed or IFN-primed (known to be expressing ISG20) cells and introduced into unprimed cells was equally infectious and therefore intact (Tesfay *et al.*, 2008) and others have suggested that global mRNA is not altered (Espert *et al.*, 2003; Zhou *et al.*, 2011). Identification of ISG20 RNA targets in virus-infected cells thus remains a major challenge for the understanding and exploitation of its mechanism of antiviral action and for the elucidation of ways in which relatively IFN-α/β-resistant viruses might circumvent its activity.

Zinc finger antiviral protein

The zinc finger antiviral protein (ZAP) is a highly IFN-α/β- and IRF3-inducible protein (Ryman *et al.*, 2005; Wang *et al.*, 2010), with strong antiviral activity against multiple alphaviruses including SINV, SFV, RRV and VEEV (Bick *et al.*, 2003; Kerns *et al.*, 2008; Zhang *et al.*, 2007). The antiviral activity of ZAP is achieved by binding to viral RNA (Chen *et al.*, 2012; Guo *et al.*, 2004) and, most likely, exosome-mediated degradation of the viral RNA (Guo *et al.*, 2007; Zhu and Gao, 2008; Zhu *et al.*, 2012). Via this mechanism, ZAP blocks an early step after entry and at or before the synthesis of the viral polyproteins (Bick *et al.*, 2003). It has recently been shown that ZAP is expressed as two distinct isoforms by alternative splicing: the long isoform has the greater anti-alphaviral activity (Kerns *et al.*, 2008) while the short isoform interacts with RIG-I to enhance IFN-β production (Hayakawa *et al.*, 2011). ZAP appears to function synergistically with other ISGs to confer maximal protection against alphavirus infections, 31 possible candidates for which were recently identified in a lentiviral screen (Karki *et al.*, 2012). Thus, ZAP is pivotal in the cellular response to alphaviral infection at several levels, although its activity *in vivo* has yet to be tested.

Viperin

Virus-inhibitory protein, endoplasmic reticulum-associated, interferon-inducible (viperin) also known as cig5 or RSAD2, was first reported as a type I IFN-inducible effector with antiviral activity against alphaviruses by Zhang *et al.* (2007) after it was identified in a global screen (Ryman *et al.*, 2005). Overexpression of viperin modestly inhibits SINV (Zhang *et al.*, 2007) and CHIKV (Schoggins *et al.*, 2011) replication *in vitro*, and SINV- (Zhang *et al.*, 2007) and CHIKV-induced disease *in vivo* (Teng *et al.*, 2012). The mechanisms by which viperin inhibits alphaviruses have yet to be elucidated, but it has been shown to target the assembly and egress of several other viruses (Mattijssen and Pruijn, 2012).

UNC93b1 chaperone

A recent study by Esen *et al.* (2012) demonstrated a significant protective role for the UNC93b1 protein in mice infected intracerebrally with a neuro-adapted strain of SINV, which appeared to be independent of TLR3 or MyD88-mediated responses or altered type I IFN induction. UNC93b1 is a type I IFN-inducible (Schoggins and Rice, 2011) ER protein required to control the trafficking of intracellular TLRs from the ER to endolysosomes (Lee *et al.*, 2013). Macrophages from mice with a loss-of-function mutation in UNC93b1 exhibit profoundly impaired responses to nucleic acid-sensing TLRs as measured by the production of multiple cytokines, including type I IFN production (Brinkmann *et al.*, 2007; Tabeta *et al.*, 2006). However, Esen *et al.* (2012) speculate that the UNC93b1 chaperone protein serves a critical, but TLR3/MyD88-independent, function in the setting of intracellular alphavirus infection, supporting the findings of Schoggins *et al.* (Schoggins and Rice, 2011) that overexpression of this protein in STAT1-deficient human fibroblasts exhibits relatively strong antiviral activity against CHIKV and VEEV-TC83 infections *in vitro*. Further study is required but it is interesting to note that autosomal recessive deficiency in the UNC-93B protein has been linked to the occurrence of herpes simplex virus encephalitis (Casrouge *et al.*, 2006) and might be expected to increase susceptibility to alphavirus infection.

cGAMP synthase

C6ORF150 (murine homologue; E330016A19) has recently been identified as cGAMP synthase (cGAS), which belongs to the nucleotidyltransferase family (Sun *et al.*, 2013). cGAS is a PRR that binds cytosolic DNA (plasmid or viral) and catalyses the synthesis of cyclic guanosine monophosphate-adenosine monophosphate (cyclic GMP-AMP, or cGAMP). cGAMP activates the adaptor protein STING which in turn activates IRF3 and up-regulates IRF3-dependent genes including IFN-β (see 'Type I IFN induction of antiviral effectors with activity against alphaviruses', above). Sun *et al.* (2013) further demonstrated that overexpression of cGAS (C6ORF150) from a DNA plasmid activated IRF3 and induced IFN-β via a STING-dependent mechanism even in the absence of additional cytosolic DNA. It is perhaps, therefore, mostly likely that the anti-alphaviral activity of cGAS overexpression observed by Schoggins *et al.* (2011) results from the activation of this STING-mediated DNA sensing pathway by the lentiviral vector (Yan *et al.*, 2010) and consequent expression of IFN-β and other IRF3-dependent genes, as it is unlikely this DNA sensing protein detects alphavirus RNA replication. However, this remains to be determined.

From data summarized above, it is becoming apparent that the alphaviruses have potential sensitivity to a large number of ISG products and therefore the pathogenic alphaviruses must encode mechanisms to either generally or specifically antagonize and/or avoid these anti-alphaviral effectors. Specific mechanisms of antagonism/avoidance are described above for each effector (where known), while general mechanisms are described below. While the activity of each effector alone does not appear to be critical, in combination their activity could be substantial unless averted, and is likely to be synergistic since minor suppression of virus replication by one effector, reducing arrest of host gene expression, could lead to the expression of many additional effectors.

Antagonism of type I IFN responses by the alphaviruses

The most efficient means by which alphaviruses generally interfere with the type I IFN response are the shutdown of cellular mRNA and protein synthesis via sequential and independent mechanisms, abilities shared by the Old and New World viruses.

Effects of proteins mediating macromolecular synthesis inhibition

As alluded to previously in this chapter and described in Chapter 1, in cells that are highly permissive all of the alphaviruses studied to date are capable of antagonizing induction and/or antiviral activities of stress and innate immune responses, including the production of type I IFN, ISGs and VSIGs in infected cells (Aguilar *et al.*, 2007; Breakwell *et al.*, 2007; Cruz *et al.*, 2010; Frolova *et al.*, 2002; Yin *et al.*, 2009). For Old World viruses such as SINV, SFV and CHIKV, activities of the nsP2 protein have been primarily implicated in this ability (Breakwell *et al.*, 2007; Burke *et al.*, 2009; Frolova *et al.*, 2002; Fros *et al.*, 2010). Furthermore, for SINV-related viruses, a mutation in nsP1 is linked to increased type I IFN induction *in vivo* (Cruz *et al.*, 2010; Heise *et al.*, 2003). In contrast, for the New World viruses, VEEV, WEEV and EEEV, the capsid protein is most prominently involved in these effects (Aguilar *et al.*, 2007; Garmashova *et al.*, 2007a; Peltier *et al.*, 2013).

Host cell transcriptional arrest

The multiple functions of the nsP2 protein in alphavirus replication are reviewed in Chapter 1. For the Old World alphaviruses, a large proportion of the nsP2 protein produced is translocated to the nucleus during infection (Breakwell *et al.*, 2007; Garmashova *et al.*, 2006, 2007a; Peranen *et al.*, 1990; Rikkonen, 1996; Tamm *et al.*, 2008) linked to host cell macromolecular synthesis shutoff and induction of cytopathic effect (cpe) in the cultured cell model systems used for analysis of type I IFN induction and response (Breakwell *et al.*, 2007; Burke *et al.*, 2009; Garmashova *et al.*, 2006, 2007a,b). Old World SINV and CHIKV alphavirus nsP2 mutants that were selected for non-cytopathogenicity (Frolova *et al.*, 2002; Fros *et al.*, 2010) and an SFV mutant with a defective nsP2 nuclear localization signal (Breakwell *et al.*, 2007) do not inhibit host cell transcription or translation activities, replicate poorly compared with their wild-type parents and induce type I IFN release (Breakwell *et al.*, 2007; Burke *et al.*, 2009; Frolov *et al.*, 2009; Frolova *et al.*, 2002; Garmashova *et al.*, 2006). It was recently reported that SINV, SFV and CHIKV nsP2 proteins inhibit cellular transcription by inducing rapid ubiquitinylation and degradation of Rpb1, a catalytic subunit of the RNA polymerase (Pol) II complex (Akhrymuk *et al.*, 2012). In highly permissive BHK-21 and 3T3 fibroblasts this occurs within 6 hours post infection; however, it remains to be determined whether or not this is critical for subversion of antiviral responses in more differentiated cells *in vitro* or *in vivo*.

In the New World alphaviruses the capsid proteins appear to be the major mediators of host cell transcriptional shutoff *in vitro* (Aguilar *et al.*, 2007; Garmashova *et al.*, 2006, 2007a,b). The VEEV capsid protein is also partially localized to the nucleus (Garmashova *et al.*, 2007a), specifically to nuclear pore structures (Garmashova *et al.*, 2007b). A tetrameric complex of capsid protein, importin-α/β, and the CRM1 nuclear export receptor appears to block nucleocytoplasmic trafficking of cellular proteins through the nuclear pore, which is strongly correlated with the development of transcriptional shutoff (Atasheva *et al.*, 2010; Gorchakov *et al.*, 2008). The effects of the capsid protein map to a positively charged amino terminal fragment in VEEV and EEEV and viruses lacking the activity described above (Aguilar *et al.*, 2007; Garmashova *et al.*, 2007a). Although it has been demonstrated that the EEEV capsid mutant virus is highly attenuated *in vivo* (Aguilar *et al.*, 2008), it has not been proven that this results from their inability to suppress the type I IFN response, for example by testing if their virulence is restored in the IFNAR1$^{-/-}$ mice.

Host cell translation arrest

Cessation of translation in alphavirus-infected cells appears to occur both as a consequence of transcription arrest and by independent mechanisms that directly reduce mRNA translation. These are difficult to distinguish when studying Old World nsP2, as both functions are impacted by similar domains within the protein, and the mutation of nsP2 has secondary effects upon the kinetics of virus replication and accumulation of replicative intermediates (Burke *et al.*, 2009; Frolov *et al.*, 2009; Fros *et al.*, 2010). Thus, the shut-off of host mRNA translation is a controversial issue. It has been suggested that Old World viruses (Clavarino *et al.*, 2012; McInerney *et al.*, 2005) almost entirely inhibit translation through a PKR-dependent phosphorylation of eIF2α in MEFs, while others suggest that translation is arrested early and throughout infection via primarily PKR-independent mechanisms, and transcriptional arrest occurs but only in the latter stages of infection (White *et al.*, 2011); unpublished observations). Interestingly, VEEV replicon particles that do not express the capsid protein were found to potently shut off translation in primary cortical neuron cultures in the absence of dramatic host transcription inhibition (Yin *et al.*, 2009) indicating a separate translation-inhibiting activity associated with the replication and/or nsP expression of the New World encephalitic viruses in certain cell types. The capsid proteins of both New World and Old World viruses also appear to contribute to the direct arrest of host translation by activation of PKR and consequent phosphorylation of the eIF2α translation initiation factor (Aguilar *et al.*, 2007; Favre *et al.*, 1996). Since translation inhibition is largely PKR/eIF2α independent, however, clearly the major pathway remains to be elucidated.

As indicated above, most studies have concluded that type I IFN response-antagonism and general disruption of host cell macromolecular synthesis are overlapping activities. However, for the SFV nsP2 mutant with minimal nuclear translocation, effects upon host transcription early after infection were similar to those elicited by the wild-type virus, although the mutant did exhibit a lesser inhibition of cellular DNA synthesis (Rikkonen, 1996) and a reduced rate of viral RNA and protein accumulation (Breakwell *et al.*, 2007). Notably, wild-type SFV also blocked accumulation of the TNF-α protein in cultured cell supernatants (but the mutant did not), implying an effect upon host responses in addition to the IFN pathways, but did not interfere with nuclear translocation of IRF3 or NF-κB transcription factors (Breakwell *et al.*, 2007).

Although it remains to be determined if the IFN induction inhibitory capacity of the alphavirus nsP or capsid proteins in cultured cells is an indirect effect of a global block to transcription and/or translation or if more specific effects are involved, it appears that global arrest of translation is typically evident *before* substantial effects on gene transcription, with the kinetics varying between cell types. Further characterization of the mechanism through which nsP2 and capsid proteins affect gene expression and the careful dissection of the apparently multiple antagonistic activities of nsP2 and possibly capsid (Frolov *et al.*, 2012; Peltier *et al.*, 2013) will be extremely informative regarding this issue. It is important to note that similar domains of nsP2 have been implicated in alphavirus infection-mediated blockade of STAT1-dependent signal transduction of autocrine type I IFN responses (Fros *et al.*, 2010; Simmons *et al.*, 2009; Yin *et al.*, 2009; see below).

Antagonism of STAT1-dependent signalling pathways

Although the resistance of the alphaviruses to the replication-inhibiting activities of the IFN-α/β-induced antiviral state is VEEV > EEEV > CHIKV and SINV, all of these viruses substantially block phosphorylation of STAT1/2 when cells are exposed to IFN-α/β after infection (Fros *et al.*, 2010; Simmons *et al.*, 2009; Yin *et al.*, 2009; unpublished observations). The mechanism by which alphaviruses block STAT1/2 phosphorylation could involve direct interaction of viral nsP with IFN-α/β receptor subunits, upstream activators JAK or Tyk, the STAT1/2 kinases themselves or, conceivably, virus-mediated reduction in the abundance of mediators upstream of the STAT1/2 proteins (e.g. cell surface expression of the IFN-α/β receptor). Surprisingly, virus-mediated blockade of STAT1/2 phosphorylation in neurons (Yin *et al.*, 2009) or Vero cells (Simmons *et al.*, 2009) made only a minor contribution to inhibition of ISG induction in the face of the potent virus-mediated arrest of macromolecular synthesis, even in the absence of VEEV capsid-mediated transcriptional shut-off. While these results are inconclusive with respect to the importance of JAK/STAT pathway blockade in cells capable of producing IFN-α/β in response to infection, it is possible that this mechanism of type I IFN pathway antagonism delays clearance of virus infection in neurons (and presumably other cell types) provided that structural protein-mediated macromolecular shutoff is not highly efficient and the ISG transcription-stimulating effect of IFN-α/β exposure is less prominent. This could reflect a virus-mediated antagonistic effect upon IFN-mediated clearance from neurons such as the non-cytolytic clearance of SINV mediated by IFN-γ (reviewed by Griffin and Metcalf, 2011).

Effects of differential virus-cell surface receptor interactions

Another putative mechanism of type I IFN induction antagonism arises from the alternating mosquito-vertebrate transmission cycle of the alphaviruses, related to the attachment receptor utilized by viruses that have replicated in a mosquito or mosquito cells. Recent studies have established that a number of arthropod-vectored viruses can utilize lectins, including the C-type lectins DC-SIGN or L-SIGN, as attachment receptors during infection of mDCs derived from mammalian hosts (Davis *et al.*, 2006; Klimstra *et al.*, 2003; Miller *et al.*, 2008; Navarro-Sanchez *et al.*, 2003; Tassaneetrithep *et al.*, 2003). Carbohydrate modifications on the envelope proteins of alphavirus particles propagated on insect cells are efficient ligands for these receptors, greatly increasing infectivity for cells expressing these receptors (Klimstra *et al.*, 2003; Shabman *et al.*, 2007). However, the carbohydrate modifications on alphaviruses derived from mammalian cells generally do not promote efficient virus particle attachment to C-type lectins (Klimstra *et al.*, 2003). This is likely due to the more limited carbohydrate processing performed by insect cell enzymes that results in high mannose carbohydrate structures as terminal modifications, as documented for SINV E1 and E2 proteins derived from these cells (Hsieh and Robbins, 1984; Hsieh *et al.*, 1983). Two studies (Shabman *et al.*, 2007, 2008) indicated that infection of murine mDCs by insect cell-derived RRV, VEEV or Barmah Forest viruses elicited less type I IFN than infection with analogous mammalian cell-derived viruses, associated with higher infectivity of the insect cell-derived viruses. Extrapolating from previous studies by Klimstra *et al.* (2003), Shabman and coworkers suggested arthropod-vectored viruses may interact with DCs via carbohydrate modifications on the E1 or E2 proteins and, as a consequence, stimulate less type I IFN induction in the first round of infection, thereby increasing initial replication within the infected host, although increased induction antagonism correlated with increased replication cannot be ruled out.

Another mechanism of antagonism was suggested for RRV particles that entered macrophages by an antibody-dependent enhanced (ADE) infection (Lidbury and Mahalingam, 2000; Mahalingam and Lidbury, 2002). During ADE-mediated RRV infection, low levels of virus-specific neutralizing antibodies or non-neutralizing antibodies attach to virus particles and increase rates of cell infection via efficient antibody attachment to Fc receptors present upon myeloid cells, particularly macrophages (Porterfield, 1986). ADE of RRV is not unique among alphaviruses (Porterfield, 1986) and has been implicated as one mechanism underlying dengue haemorrhagic fever/dengue shock syndrome (Morens and Halstead, 1990). Enhancement of RRV infection of cultured RAW 264.7 macrophages by ADE-mediated entry reduced the accumulation of mRNAs of stress-responsive genes including TNF-α and IFN-β, and also appeared to reduce the abundance of several stress-responsive transcription factor complexes involving STAT1 and NF-κB (Lidbury and Mahalingam, 2000). The effects upon cell stress responses required active virus replication and were ablated by UV-inactivation of the viruses. Again, however, it is difficult to determine whether this is a direct consequence of ADE-mediated entry or a general consequence of increased virus replication and consequent arrest of cellular processes.

Effects of differential cell and tissue tropisms

In addition to the direct effects of alphaviruses upon infected host cells, *in vivo* cell and tissue tropisms have recently been identified as major factors in the avoidance (as opposed to antagonism) of systemic type I IFN responses by NA EEEV (Gardner *et al.*, 2008, 2011, 2013). This appears to be achieved through two features of the virus: (i) natural high-efficiency binding to HS proteoglycans (Gardner *et al.*, 2011, 2013); and (ii) 'failure' to translate the incoming viral genome in myeloid cells (Gardner *et al.*, 2008; Trobaugh *et al.*, 2013). As opposed to Old World alphaviruses where efficient HS binding is a consequence of *in vitro* adaptation (Gardner *et al.*, 2012; Klimstra *et al.*, 1998; Ryman *et al.*, 2007a), evaluation of E2 attachment protein sequences has demonstrated that naturally circulating, non-passaged EEEV isolates bind this glycosaminoglycan efficiently (Gardner *et al.*, 2011, 2013). Binding of HS by EEEV appears to limit virus seeding of lymphoid tissues, suppressing replication and consequent type I IFN induction (Gardner *et al.*, 2011). This, combined with a defect in translation of the EEEV genome in myeloid-lineage cells (Gardner *et al.*, 2008; Trobaugh *et al.*, 2013), results in little/no detectable systemic type I IFN induction (Gardner *et al.*, 2008, 2011; Trobaugh *et al.*, 2013), or prodromal disease, most likely contributing substantially to the extreme neurovirulence of this virus (Gardner *et al.*, 2008). Interestingly, CHIKV appears to infect DCs and macrophages relatively poorly (Gardner *et al.*, 2012; Sourisseau *et al.*, 2007), compared with, for example, SINV or VEEV. Although a type I IFN-dependent response is triggered by exposure of these cells to CHIKV (Her *et al.*, 2010), the limited inherent tropism for these cells is not controlled by antiviral ISGs (Gardner *et al.*, 2012). In the future, this phenomenon may be revealed to be influential to the pathogenesis of CHIKV.

Perspectives

Upon virus infection of a host cell, a cascade of events occurs resulting in the induction of innate immune cytokines including IFN-α/β. This IFN then acts to reduce virus replication in infected cells and increase the resistance of surrounding uninfected cells to initiation of infection. Considerable research is currently devoted to investigation of alphaviral mechanisms of antagonism of IFN-α/β induction, signalling and effector pathways as a strategy through which viruses increase their replicative fitness within vertebrate hosts (reviewed in Ryman and Klimstra, 2008). In the authors' opinion the ability of the alphaviruses to shut-off host macromolecular synthesis as shown *in vitro* does not fully explain their *in vivo* phenotypes and these observations have rarely been confirmed in differentiated cell types relevant to pathogenesis or *in vivo*. Most of the studies implicating particular alphavirus proteins in antagonism of type I IFN transcription and/or translation use highly permissive cultured cell lines such as immortalized BHK-21 or 3T3 fibroblasts (Breakwell *et al.*, 2007; Frolova *et al.*, 2002; Burke *et al.*, 2009) in which alphavirus infection causes substantial shutoff of host transcription and translation by 3–8 hours post infection. In sharp contrast, when primary cultures of heterogeneous, immature bone marrow-derived DCs are infected with wild-type SINV and global transcription analysis performed at 12 hours post infection, over 400 genes were transcriptionally up-regulated ≥ 2-fold over mock (some by greater than 50-fold over mock, including the mRNAs for IFN-β and multiple IFN-α subtypes) and high levels of biologically active IFN-α/β were detected in cell supernatants (Ryman *et al.*, 2005). Similarly, analyses of transcription in brain tissue from mice infected with SINV (Lenschow *et al.*, 2005) or VEEV (Schoneboom *et al.*, 2000b; Sharma and Maheshwari, 2009) indicated robust induction of type I IFN and other stress-responsive genes, and serum IFN-α/β is highly induced within 8–12 hours post infection by both viruses (Charles *et al.*, 2001; Klimstra *et al.*, 1999; Ryman *et al.*, 2002, 2007b). However, the differential virulence of VEEV versus SINV for example is not explained except that, after a few hours delay, VEEV suppresses antiviral responses even in type I IFN-primed cells, whereas SINV does not (Yin *et al.*, 2009), by unknown mechanism(s). For NA EEEV the capacity to antagonize IFN responses may be related to the rapid shutoff of host processes characteristic of infection of established mesenchymal and similar cell lines.

Further research is required to determine to what degree antagonism of IFN-α/β gene induction by alphaviruses is associated with general shutoff of host macromolecular synthesis that is increased as a function of infection and/or replication efficiency or, perhaps, as has been suggested by several recent studies (Frolov *et al.*, 2012; Peltier *et al.*, 2013; Trobaugh *et al.*, 2013), more subtle and specific effects upon particular type I IFN response pathways within infected cells. Clearly, *in vitro* IFN-α/β induction, signalling and effector antagonism/evasion studies need to be placed in the context of pathogenesis in animal models to demonstrate whether or not antagonism occurs in the intact infected host and to what extent this contributes to viral virulence and disease aetiology. A priority for validation of *in vitro* studies is detailed characterization of cell types and pathways involved in virus-induced type I IFN production in mice, tissues and primary cells derived from them relevant to initial virus replication, dissemination within the host and manifestation of virus-specific disease (for example, multiple DC subsets, migratory and tissue-resident macrophages, fibroblasts, osteoblasts, myofibers and neurons). Furthermore, these studies must include scrupulous controls for clonal variation in passaged cells and for virus infection efficiency.

References

Aguilar, P.V., Paessler, S., Carrara, A.S., Baron, S., Poast, J., Wang, E., Moncayo, A.C., Anishchenko, M., Watts, D., Tesh, R.B., *et al.* (2005). Variation in interferon sensitivity and induction among strains of eastern equine encephalitis virus. J. Virol. *79*, 11300–11310.

Aguilar, P.V., Weaver, S.C., and Basler, C.F. (2007). Capsid protein of eastern equine encephalitis virus inhibits host cell gene expression. J. Virol. *81*, 3866–3876.

Aguilar, P.V., Leung, L.W., Wang, E., Weaver, S.C., and Basler, C.F. (2008). A five-amino-acid deletion of the eastern equine encephalitis virus capsid protein attenuates replication in mammalian systems but not in mosquito cells. J. Virol. *82*, 6972–6983.

Akhrymuk, I., Kulemzin, S.V., and Frolova, E.I. (2012). Evasion of the innate immune response: the Old World alphavirus nsP2 protein induces rapid degradation of Rpb1, a catalytic subunit of RNA polymerase II. J. Virol. *86*, 7180–7191.

Atasheva, S., Krendelchtchikova, V., Liopo, A., Frolova, E., and Frolov, I. (2010). Interplay of acute and persistent infections caused by Venezuelan equine encephalitis virus encoding mutated capsid protein. J. Virol. *84*, 10004–10015.

Atkins, G.J., and Lancashire, C.L. (1976). The induction of interferon by temperature-sensitive mutants of Sindbis virus: its relationship to double-stranded RNA synthesis and cytopathic effect. J. Gen. Virol. *30*, 157–165.

Balachandran, S., Roberts, P.C., Kipperman, T., Bhalla, K.N., Compans, R.W., Archer, D.R., and Barber, G.N. (2000). Alpha/beta interferons potentiate virus-induced apoptosis through activation of the FADD/Caspase-8 death signaling pathway. J. Virol. *74*, 1513–1523.

Barnes, B.J., Moore, P.A., and Pitha, P.M. (2001). Virus-specific activation of a novel interferon regulatory factor, IRF-5, results in the induction of distinct interferon alpha genes. J. Biol. Chem. *276*, 23382–23390.

Baron, S., Buckler, C.E., McCloskey, R.V., and Kirschstein, R.L. (1966). Role of interferon during viremia. I. Production of circulating interferon. J. Immunol. *96*, 12–16.

Barry, G., Breakwell, L., Fragkoudis, R., Attarzadeh-Yazdi, G., Rodriguez-Andres, J., Kohl, A., and Fazakerley, J.K. (2009). PKR acts early in infection to suppress Semliki Forest virus production and strongly enhances the type I interferon response. J. Gen. Virol. *90*, 1382–1391.

Bick, M.J., Carroll, J.W., Gao, G., Goff, S.P., Rice, C.M., and MacDonald, M.R. (2003). Expression of the zinc-finger antiviral protein inhibits alphavirus replication. J. Virol. *77*, 11555–11562.

Bradish, C.J., Allner, K., and Maber, H.B. (1972). Infection, interaction and the expression of virulence by defined strains of Semliki forest virus. J. Gen. Virol. *16*, 359–372.

Breakwell, L., Dosenovic, P., Karlsson Hedestam, G.B., D'Amato, M., Liljestrom, P., Fazakerley, J., and McInerney, G.M. (2007). Semliki Forest virus nonstructural protein 2 is involved in suppression of the type I interferon response. J. Virol. *81*, 8677–8684.

Brinkmann, M.M., Spooner, E., Hoebe, K., Beutler, B., Ploegh, H.L., and Kim, Y.M. (2007). The interaction between the ER membrane protein UNC93B and TLR3, 7, and 9 is crucial for TLR signaling. J. Cell Biol. *177*, 265–275.

van den Broek, M.F., Muller, U., Huang, S., Zinkernagel, R.M., and Aguet, M. (1995). Immune defence in mice lacking type I and/or type II interferon receptors. Immunol. Rev. *148*, 5–18.

Burke, C.W., Gardner, C.L., Steffan, J.J., Ryman, K.D., and Klimstra, W.B. (2009). Characteristics of alpha/beta interferon induction after infection of murine fibroblasts with wild-type and mutant alphaviruses. Virology *395*, 121–132.

Casrouge, A., Zhang, S.Y., Eidenschenk, C., Jouanguy, E., Puel, A., Yang, K., Alcais, A., Picard, C., Mahfoufi, N., Nicolas, N., *et al.* (2006). Herpes simplex virus encephalitis in human UNC-93B deficiency. Science *314*, 308–312.

Chan, Y.L., Chang, T.H., Liao, C.L., and Lin, Y.L. (2008). The cellular antiviral protein viperin is attenuated by proteasome-mediated protein degradation in Japanese encephalitis virus-infected cells. J. Virol. *82*, 10455–10464.

Charles, P.C., Walters, E., Margolis, F., and Johnston, R.E. (1995). Mechanism of neuroinvasion of Venezuelan equine encephalitis virus in the mouse. Virology *208*, 662–671.

Charles, P.C., Trgovcich, J., Davis, N.L., and Johnston, R.E. (2001). Immunopathogenesis and immune modulation of Venezuelan equine encephalitis virus-induced disease in the mouse. Virology *284*, 190–202.

Chen, S., Xu, Y., Zhang, K., Wang, X., Sun, J., Gao, G., and Liu, Y. (2012). Structure of N-terminal domain of ZAP indicates how a zinc-finger protein recognizes complex RNA. Nat. Struct. Mol. Biol. *19*, 430–435.

Cheng, T.F., Brzostek, S., Ando, O., Van Scoy, S., Kumar, K.P., and Reich, N.C. (2006). Differential activation of IFN regulatory factor (IRF)-3 and IRF-5 transcription factors during viral infection. J. Immunol. *176*, 7462–7470.

Clavarino, G., Claudio, N., Couderc, T., Dalet, A., Judith, D., Camosseto, V., Schmidt, E.K., Wenger, T., Lecuit, M., Gatti, E., *et al.* (2012). Induction of GADD34 is necessary for dsRNA-dependent interferon-beta production and participates in the control of Chikungunya virus infection. PLoS Pathog. *8*, e1002708.

Couderc, T., Chretien, F., Schilte, C., Disson, O., Brigitte, M., Guivel-Benhassine, F., Touret, Y., Barau, G., Cayet, N., Schuffenecker, I., *et al.* (2008). A mouse model for Chikungunya: young age and inefficient type-I interferon signaling are risk factors for severe disease. PLoS Pathog. *4*, e29.

Cruz, C.C., Suthar, M.S., Montgomery, S.A., Shabman, R., Simmons, J., Johnston, R.E., Morrison, T.E., and Heise, M.T. (2010). Modulation of type I IFN induction by a virulence determinant within the alphavirus nsP1 protein. Virology *399*, 1–10.

Daffis, S., Szretter, K.J., Schriewer, J., Li, J., Youn, S., Errett, J., Lin, T.Y., Schneller, S., Zust, R., Dong, H., *et al.* (2010). 2′-O methylation of the viral mRNA cap evades host restriction by IFIT family members. Nature *468*, 452–456.

Davis, C.W., Nguyen, H.Y., Hanna, S.L., Sanchez, M.D., Doms, R.W., and Pierson, T.C. (2006). West Nile virus discriminates between DC-SIGN and DC-SIGNR for cellular attachment and infection. J. Virol. *80*, 1290–1301.

Der, S.D., Zhou, A., Williams, B.R., and Silverman, R.H. (1998). Identification of genes differentially regulated by interferon alpha, beta, or gamma using oligonucleotide arrays. Proc. Natl. Acad. Sci. U.S.A. *95*, 15623–15628.

Diamond, M.S., and Farzan, M. (2013). The broad-spectrum antiviral functions of IFIT and IFITM proteins. Nat. Rev. Immunol. *13*, 46–57.

Dupuis, S., Jouanguy, E., Al-Hajjar, S., Fieschi, C., Al-Mohsen, I.Z., Al-Jumaah, S., Yang, K., Chapgier, A., Eidenschenk, C., Eid, P., *et al.* (2003). Impaired response to interferon-alpha/beta and lethal viral disease in human STAT1 deficiency. Nat. Genet. *33*, 388–391.

Esen, N., Blakely, P.K., Rainey-Barger, E.K., and Irani, D.N. (2012). Complexity of the microglial activation pathways that drive innate host responses during lethal alphavirus encephalitis in mice. ASN Neuro. *4*, 207–221.

Espert, L., Degols, G., Gongora, C., Blondel, D., Williams, B.R., Silverman, R.H., and Mechti, N. (2003). ISG20, a new interferon-induced RNase specific for single-stranded RNA, defines an alternative antiviral pathway against RNA genomic viruses. J. Biol. Chem. *278*, 16151–16158.

Espert, L., Degols, G., Lin, Y.L., Vincent, T., Benkirane, M., and Mechti, N. (2005). Interferon-induced exonuclease ISG20 exhibits an antiviral activity against human immunodeficiency virus type 1. J. Gen. Virol. *86*, 2221–2229.

Farmer, J.R., Altschaefl, K.M., O'Shea, K.S., and Miller, D.J. (2013). Activation of the type I interferon pathway is enhanced in response to human neuronal differentiation. PLOS ONE *8*, e58813.

Favre, D., Studer, E., and Michel, M.R. (1996). Semliki Forest virus capsid protein inhibits the initiation of translation by upregulating the double-stranded RNA-activated protein kinase (PKR). Biosci. Rep. *16*, 485–511.

Fazakerley, J.K. (2002). Pathogenesis of Semliki Forest virus encephalitis. J. Neurovirol. *8*(Suppl. 2), 66–74.

Finlay, B.B., and McFadden, G. (2006). Anti-immunology: evasion of the host immune system by bacterial and viral pathogens. Cell *124*, 767–782.

Finter, N.B. (1964). Protection of mice by interferon against systemic virus infections. BMJ *2*, 981–985.

Fitzgerald-Bocarsly, P., and Feng, D. (2007). The role of type I interferon production by dendritic cells in host defense. Biochimie *89*, 843–855.

Fragkoudis, R., Breakwell, L., McKimmie, C., Boyd, A., Barry, G., Kohl, A., Merits, A., and Fazakerley, J.K. (2007). The type I interferon system protects mice from Semliki Forest virus by preventing widespread virus dissemination in extraneural tissues, but does not mediate the restricted replication of avirulent virus in central nervous system neurons. J. Gen. Virol. *88*, 3373–3384.

Friedman, R.M. (1968). Inhibition of arbovirus protein synthesis by interferon. J. Virol. *2*, 1081–1085.

Friedman, R.M., and Sonnabend, J.A. (1965). Inhibition by interferon of production of double-stranded Semliki forest virus ribonucleic acid. Nature *206*, 532.

Friedman, R.M., Fantes, K.H., Levy, H.B., and Carter, W.B. (1967). Interferon action on parental Semliki forest virus ribonucleic acid. J. Virol. *1*, 1168–1173.

Frolov, I., Agapov, E., Hoffman, T.A., Jr., Pragai, B.M., Lippa, M., Schlesinger, S., and Rice, C.M. (1999). Selection of RNA replicons capable of persistent noncytopathic replication in mammalian cells. J. Virol. *73*, 3854–3865.

Frolov, I., Garmashova, N., Atasheva, S., and Frolova, E.I. (2009). Random insertion mutagenesis of sindbis virus nonstructural protein 2 and selection of variants incapable of downregulating cellular transcription. J. Virol. *83*, 9031–9044.

Frolov, I., Akhrymuk, M., Akhrymuk, I., Atasheva, S., and Frolova, E.I. (2012). Early events in alphavirus replication determine the outcome of infection. J. Virol. *86*, 5055–5066.

Frolova, E.I., Fayzulin, R.Z., Cook, S.H., Griffin, D.E., Rice, C.M., and Frolov, I. (2002). Roles of nonstructural protein nsP2 and Alpha/Beta interferons in determining the outcome of Sindbis virus infection. J. Virol. *76*, 11254–11264.

Fros, J.J., Liu, W.J., Prow, N.A., Geertsema, C., Ligtenberg, M., Vanlandingham, D.L., Schnettler, E., Vlak, J.M., Suhrbier, A., Khromykh, A.A., *et al.* (2010). Chikungunya virus nonstructural protein 2 inhibits type I/II interferon-stimulated JAK-STAT signaling. J. Virol. *84*, 10877–10887.

Garcia-Sastre, A., and Biron, C.A. (2006). Type 1 interferons and the virus–host relationship: a lesson in detente. Science *312*, 879–882.

Gardner, C.L., Burke, C.W., Tesfay, M.Z., Glass, P.J., Klimstra, W.B., and Ryman, K.D. (2008). Eastern and Venezuelan equine encephalitis viruses differ in their ability to infect dendritic cells and macrophages: impact of altered cell tropism on pathogenesis. J. Virol. *82*, 10634–10646.

Gardner, C.L., Yin, J., Burke, C.W., Klimstra, W.B., and Ryman, K.D. (2009). Type I interferon induction is correlated with attenuation of a South American eastern equine encephalitis virus strain in mice. Virology *390*, 338–347.

Gardner, C.L., Ebel, G.D., Ryman, K.D., and Klimstra, W.B. (2011). Heparan sulfate binding by natural eastern equine encephalitis viruses promotes neurovirulence. Proc. Natl. Acad. Sci. U.S.A. *108*, 16026–16031.

Gardner, C.L., Burke, C.W., Higgs, S.T., Klimstra, W.B., and Ryman, K.D. (2012). Interferon-alpha/beta deficiency greatly exacerbates arthritogenic disease in mice infected with wild-type chikungunya virus but not with the cell culture-adapted live-attenuated 181/25 vaccine candidate. Virology *425*, 103–112.

Gardner, C.L., Choi-Nurvitadhi, J., Sun, C., Bayer, A., Hritz, J., Ryman, K.D., and Klimstra, W.B. (2013). Natural variation in the heparan sulfate binding domain of the eastern equine encephalitis virus E2 glycoprotein alters interactions with cell surfaces and virulence in mice. J. Virol. *87*, 8582–8590.

Gardner, J., Anraku, I., Le, T.T., Larcher, T., Major, L., Roques, P., Schroder, W.A., Higgs, S., and Suhrbier, A. (2010). Chikungunya virus arthritis in adult wild-type mice. J. Virol. *84*, 8021–8032.

Garmashova, N., Gorchakov, R., Frolova, E., and Frolov, I. (2006). Sindbis virus nonstructural protein nsP2 is cytotoxic and inhibits cellular transcription. J. Virol. *80*, 5686–5696.

Garmashova, N., Atasheva, S., Kang, W., Weaver, S.C., Frolova, E., and Frolov, I. (2007a). Analysis of Venezuelan equine encephalitis virus capsid protein function in the inhibition of cellular transcription. J. Virol. *81*, 13552–13565.

Garmashova, N., Gorchakov, R., Volkova, E., Paessler, S., Frolova, E., and Frolov, I. (2007b). The Old World and New World alphaviruses use different virus-specific proteins for induction of transcriptional shutoff. J. Virol. *81*, 2472–2484.

Giannakopoulos, N.V., Arutyunova, E., Lai, C., Lenschow, D.J., Haas, A.L., and Virgin, H.W. (2009). ISG15 Arg151 and the ISG15-conjugating enzyme UbE1L are important for innate immune control of Sindbis virus. J. Virol. *83*, 1602–1610.

Gil, M.P., Bohn, E., O'Guin, A.K., Ramana, C.V., Levine, B., Stark, G.R., Virgin, H.W., and Schreiber, R.D. (2001). Biologic consequences of Stat1-independent IFN signaling. Proc. Natl. Acad. Sci. U.S.A. *98*, 6680–6685.

Gitlin, L., Barchet, W., Gilfillan, S., Cella, M., Beutler, B., Flavell, R.A., Diamond, M.S., and Colonna, M. (2006). Essential role of mda-5 in type I IFN responses to polyriboinosinic:polyribocytidylic acid and encephalomyocarditis picornavirus. Proc. Natl. Acad. Sci. U.S.A. *103*, 8459–8464.

Gorchakov, R., Frolova, E., Williams, B.R., Rice, C.M., and Frolov, I. (2004). PKR-dependent and -independent mechanisms are involved in translational shutoff during Sindbis virus infection. J. Virol. *78*, 8455–8467.

Gorchakov, R., Frolova, E., Sawicki, S., Atasheva, S., Sawicki, D., and Frolov, I. (2008). A new role for ns polyprotein cleavage in Sindbis virus replication. J. Virol. *82*, 6218–6231.

Grandvaux, N., Servant, M.J., tenOever, B., Sen, G.C., Balachandran, S., Barber, G.N., Lin, R., and Hiscott, J. (2002). Transcriptional profiling of interferon regulatory factor 3 target genes: direct involvement in the regulation of interferon-stimulated genes. J. Virol. *76*, 5532–5539.

Grieder, F.B., and Vogel, S.N. (1999). Role of interferon and interferon regulatory factors in early protection against Venezuelan equine encephalitis virus infection. Virology *257*, 106–118.

Grieder, F.B., Davis, N.L., Aronson, J.F., Charles, P.C., Sellon, D.C., Suzuki, K., and Johnston, R.E. (1995). Specific restrictions in the progression of Venezuelan equine encephalitis virus-induced disease resulting from single amino acid changes in the glycoproteins. Virology *206*, 994–1006.

Grieder, F.B., Davis, B.K., Zhou, X.D., Chen, S.J., Finkelman, F.D., and Gause, W.C. (1997). Kinetics of cytokine expression and regulation of host protection following infection with molecularly cloned Venezuelan equine encephalitis virus. Virology *233*, 302–312.

Griffin, D.E. (1976). Role of the immune response in age-dependent resistance of mice to encephalitis due to Sindbis virus. J. Infect. Dis. *133*, 456–464.

Griffin, D.E. (2006). Alphaviruses. In Fields Virology, P.M. Howley and DM Knipe, eds. (Lippincott, Williams and Wilkins, Philadelphia, PA, USA), pp. 1023–1068.

Griffin, D.E., and Metcalf, T. (2011). Clearance of virus infection from the CNS. Curr. Opin. Virol. *1*, 216–221.

Guo, X., Carroll, J.W., Macdonald, M.R., Goff, S.P., and Gao, G. (2004). The zinc finger antiviral protein directly binds to specific viral mRNAs through the CCCH zinc finger motifs. J. Virol. *78*, 12781–12787.

Guo, X., Ma, J., Sun, J., and Gao, G. (2007). The zinc-finger antiviral protein recruits the RNA processing exosome to degrade the target mRNA. Proc. Natl. Acad. Sci. U.S.A. *104*, 151–156.

Hahn, C.S., Lustig, S., Strauss, E.G., and Strauss, J.H. (1988). Western equine encephalitis virus is a recombinant virus. Proc. Natl. Acad. Sci. U.S.A. *85*, 5997–6001.

Hayakawa, S., Shiratori, S., Yamato, H., Kameyama, T., Kitatsuji, C., Kashigi, F., Goto, S., Kameoka, S., Fujikura, D., Yamada, T., *et al.* (2011). ZAPS is a potent stimulator of signaling mediated by the RNA helicase RIG-I during antiviral responses. Nat. Immunol. *12*, 37–44.

Heise, M.T., White, L.J., Simpson, D.A., Leonard, C., Bernard, K.A., Meeker, R.B., and Johnston, R.E. (2003). An attenuating mutation in nsP1 of the Sindbis-group virus S.A.AR86 accelerates nonstructural protein processing and up-regulates viral 26S RNA synthesis. J. Virol. 77, 1149–1156.

Her, Z., Malleret, B., Chan, M., Ong, E.K., Wong, S.C., Kwek, D.J., Tolou, H., Lin, R.T., Tambyah, P.A., Renia, L., *et al.* (2010). Active infection of human blood monocytes by Chikungunya virus triggers an innate immune response. J. Immunol. *184*, 5903–5913.

Hidmark, A.S., McInerney, G.M., Nordstrom, E.K., Douagi, I., Werner, K.M., Liljestrom, P., and Karlsson Hedestam, G.B. (2005). Early alpha/beta interferon production by myeloid dendritic cells in response to UV-inactivated virus requires viral entry and interferon regulatory factor 3 but not MyD88. J. Virol. *79*, 10376–10385.

Hoarau, J.J., Jaffar Bandjee, M.C., Krejbich Trotot, P., Das, T., Li-Pat-Yuen, G., Dassa, B., Denizot, M., Guichard, E., Ribera, A., Henni, T., *et al.* (2010). Persistent chronic inflammation and infection by Chikungunya arthritogenic alphavirus in spite of a robust host immune response. J. Immunol. *184*, 5914–5927.

Honda, K., Takaoka, A., and Taniguchi, T. (2006). Type I interferon [corrected] gene induction by the interferon regulatory factor family of transcription factors. Immunity 25, 349–360.

Horio, T., Murai, M., Inoue, T., Hamasaki, T., Tanaka, T., and Ohgi, T. (2004). Crystal structure of human ISG20, an interferon-induced antiviral ribonuclease. FEBS Lett. *577*, 111–116.

Hornung, V., Ellegast, J., Kim, S., Brzozka, K., Jung, A., Kato, H., Poeck, H., Akira, S., Conzelmann, K.K., Schlee, M., *et al.* (2006). 5′-Triphosphate RNA is the ligand for RIG-I. Science *314*, 994–997.

Hsieh, P., and Robbins, P.W. (1984). Regulation of asparagine-linked oligosaccharide processing. Oligosaccharide processing in *Aedes albopictus* mosquito cells. J. Biol. Chem. *259*, 2375–2382.

Hsieh, P., Rosner, M.R., and Robbins, P.W. (1983). Host-dependent variation of asparagine-linked oligosaccharides at individual glycosylation sites of Sindbis virus glycoproteins. J. Biol. Chem. *258*, 2548–2554.

Hui, D.J., Terenzi, F., Merrick, W.C., and Sen, G.C. (2005). Mouse p56 blocks a distinct function of eukaryotic initiation factor 3 in translation initiation. J. Biol. Chem. *280*, 3433–3440.

Iwasaki, A., and Medzhitov, R. (2004). Toll-like receptor control of the adaptive immune responses. Nat. Immunol. *5*, 987–995.

Jackson, A.C., and Rossiter, J.P. (1997). Apoptotic cell death is an important cause of neuronal injury in experimental Venezuelan equine encephalitis virus infection of mice. Acta Neuropathol. *93*, 349–353.

Jiang, D., Guo, H., Xu, C., Chang, J., Gu, B., Wang, L., Block, T.M., and Guo, J.T. (2008). Identification of three interferon-inducible cellular enzymes that inhibit the replication of hepatitis C virus. J. Virol. *82*, 1665–1678.

Jiang, Z., Zamanian-Daryoush, M., Nie, H., Silva, A.M., Williams, B.R., and Li, X. (2003). Poly(I-C)-induced Toll-like receptor 3 (TLR3)-mediated activation of NFkappa B and MAP kinase is through an interleukin-1 receptor-associated kinase (IRAK)-independent pathway employing the signaling components TLR3-TRAF6-TAK1-TAB2-PKR. J. Biol. Chem. *278*, 16713–16719.

Johnson, R.T., McFarland, H.F., and Levy, S.E. (1972). Age-dependent resistance to viral encephalitis: studies of infections due to Sindbis virus in mice. J. Infect. Dis. *125*, 257–262.

Joklik, W.K., and Merigan, T.C. (1966). Concerning the mechanism of action of interferon. Proc. Natl. Acad. Sci. U.S.A. *56*, 558–565.

Julander, J.G., Siddharthan, V., Blatt, L.M., Schafer, K., Sidwell, R.W., and Morrey, J.D. (2007). Effect of exogenous interferon and an interferon inducer on western equine encephalitis virus disease in a hamster model. Virology *360*, 454–460.

Kang, D.C., Gopalkrishnan, R.V., Wu, Q., Jankowsky, E., Pyle, A.M., and Fisher, P.B. (2002). mda-5: An interferon-inducible putative RNA helicase with double-stranded RNA-dependent ATPase activity and melanoma growth-suppressive properties. Proc. Natl. Acad. Sci. U.S.A. *99*, 637–642.

Karki, S., Li, M.M., Schoggins, J.W., Tian, S., Rice, C.M., and MacDonald, M.R. (2012). Multiple interferon stimulated genes synergize with the zinc finger antiviral protein to mediate anti-alphavirus activity. PLOS ONE 7, e37398.

Kerns, J.A., Emerman, M., and Malik, H.S. (2008). Positive selection and increased antiviral activity associated with the PARP-containing isoform of human zinc-finger antiviral protein. PLoS Genet. *4*, e21.

Klimstra, W.B., Ryman, K.D., and Johnston, R.E. (1998). Adaptation of Sindbis virus to BHK cells selects for use of heparan sulfate as an attachment receptor. J. Virol. *72*, 7357–7366.

Klimstra, W.B., Ryman, K.D., Bernard, K.A., Nguyen, K.B., Biron, C.A., and Johnston, R.E. (1999). Infection of neonatal mice with sindbis virus results in a systemic inflammatory response syndrome. J. Virol. *73*, 10387–10398.

Klimstra, W.B., Nangle, E.M., Smith, M.S., Yurochko, A.D., and Ryman, K.D. (2003). DC-SIGN and L-SIGN can act as attachment receptors for alphaviruses and distinguish between mosquito cell- and mammalian cell-derived viruses. J. Virol. *77*, 12022–12032.

Konopka, J.L., Penalva, L.O., Thompson, J.M., White, L.J., Beard, C.W., Keene, J.D., and Johnston, R.E. (2007). A two-phase innate host response to alphavirus infection identified by mRNP-tagging in vivo. PLoS Pathog. 3, e199.

Konopka, J.L., Thompson, J.M., Whitmore, A.C., Webb, D.L., and Johnston, R.E. (2009). Acute infection with venezuelan equine encephalitis virus replicon particles catalyzes a systemic antiviral state and protects from lethal virus challenge. J. Virol. *83*, 12432–12442.

Kumar, S., Jaffar-Bandjee, M.C., Giry, C., Connen de Kerillis, L., Merits, A., Gasque, P., and Hoarau, J.J. (2012). Mouse macrophage innate immune response to Chikungunya virus infection. Virol. J. *9*, 313.

Lazear, H.M., Lancaster, A., Wilkins, C., Suthar, M.S., Huang, A., Vick, S.C., Clepper, L., Thackray, L., Brassil, M.M., Virgin, H.W., *et al.* (2013). IRF-3, IRF-5, and IRF-7 coordinately regulate the type I IFN response in myeloid dendritic cells downstream of MAVS signaling. PLoS Pathog. *9*, e1003118.

Lee, B.L., Moon, J.E., Shu, J.H., Yuan, L., Newman, Z.R., Schekman, R., and Barton, G.M. (2013). UNC93B1 mediates differential trafficking of endosomal TLRs. eLife *2*, e00291.

Lenschow, D.J. (2010). Antiviral Properties of ISG15. Viruses *2*, 2154–2168.

Lenschow, D.J., Giannakopoulos, N.V., Gunn, L.J., Johnston, C., O'Guin, A.K., Schmidt, R.E., Levine, B., and Virgin, H.W.t. (2005). Identification of interferon-stimulated gene 15 as an antiviral molecule during Sindbis virus infection in vivo. J. Virol. *79*, 13974–13983.

Lenschow, D.J., Lai, C., Frias-Staheli, N., Giannakopoulos, N.V., Lutz, A., Wolff, T., Osiak, A., Levine, B., Schmidt, R.E., Garcia-Sastre, A., *et al.* (2007). IFN-stimulated gene 15 functions as a critical antiviral molecule against influenza, herpes, and Sindbis viruses. Proc. Natl. Acad. Sci. U.S.A. *104*, 1371–1376.

Levitt, N.H., Miller, H.V., and Edelman, R. (1979). Interaction of alphaviruses with human peripheral leukocytes: in vitro replication of Venezuelan equine encephalomyelitis virus in monocyte cultures. Infect. Immun. *24*, 642–646.

Li, Y., Sassano, A., Majchrzak, B., Deb, D.K., Levy, D.E., Gaestel, M., Nebreda, A.R., Fish, E.N., and Platanias, L.C. (2004). Role of p38alpha Map kinase in Type I interferon signaling. J. Biol. Chem. *279*, 970–979.

Lidbury, B.A., and Mahalingam, S. (2000). Specific ablation of antiviral gene expression in macrophages by antibody-dependent enhancement of Ross River virus infection. J. Virol. *74*, 8376–8381.

McAllister, C.S., Taghavi, N., and Samuel, C.E. (2012). Protein kinase PKR amplification of interferon beta induction occurs through initiation factor eIF-2alpha-mediated translational control. J. Biol. Chem. *287*, 36384–36392.

MacDonald, G.H., and Johnston, R.E. (2000). Role of dendritic cell targeting in Venezuelan equine encephalitis virus pathogenesis. J. Virol. *74*, 914–922.

MacDonald, M.R., Machlin, E.S., Albin, O.R., and Levy, D.E. (2007). The zinc finger antiviral protein acts synergistically with an interferon-induced factor for maximal activity against alphaviruses. J. Virol. *81*, 13509–13518.

McFadden, G., Mohamed, M.R., Rahman, M.M., and Bartee, E. (2009). Cytokine determinants of viral tropism. Nat. Rev. Immunol. *9*, 645–655.

McGreal, E.P., Martinez-Pomares, L., and Gordon, S. (2004). Divergent roles for C-type lectins expressed by cells of the innate immune system. Mol. Immunol. *41*, 1109–1121.

McInerney, G.M., Kedersha, N.L., Kaufman, R.J., Anderson, P., and Liljestrom, P. (2005). Importance of eIF2alpha phosphorylation and stress granule assembly in alphavirus translation regulation. Mol. Biol. Cell *16*, 3753–3763.

McKnight, K.L., Simpson, D.A., Lin, S.C., Knott, T.A., Polo, J.M., Pence, D.F., Johannsen, D.B., Heidner, H.W., Davis, N.L., and Johnston, R.E. (1996). Deduced consensus sequence of Sindbis virus strain AR339: mutations contained in laboratory strains which affect cell culture and *in vivo* phenotypes. J. Virol. *70*, 1981–1989.

Mahalingam, S., and Lidbury, B.A. (2002). Suppression of lipopolysaccharide-induced antiviral transcription factor (STAT-1 and NF-kappa B) complexes by antibody-dependent enhancement of macrophage infection by Ross River virus. Proc. Natl. Acad. Sci. U.S.A. *99*, 13819–13824.

Marcus, P.I., and Carver, D.H. (1967). Intrinsic interference: a new type of viral interference. J. Virol. *1*, 334–343.

Marcus, P.I., and Fuller, F.J. (1979). Interferon induction by viruses. II. Sindbis virus: interferon induction requires one-quarter of the genome--genes G and A. J. Gen. Virol. *44*, 169–177.

Marfell, B.J., O'Brien, R., and Griffin, J.F. (2013). Global gene expression profiling of monocyte-derived macrophages from red deer (*Cervus elaphus*) genotypically resistant or susceptible to *Mycobacterium avium* subspecies *paratuberculosis* infection. Dev. Comp. Immunol. *40*, 210–217.

Mattijssen, S., and Pruijn, G.J. (2012). Viperin, a key player in the antiviral response. Microbes Infect. *14*, 419–426.

Meylan, E., Tschopp, J., and Karin, M. (2006). Intracellular pattern recognition receptors in the host response. Nature *442*, 39–44.

Miller, J.L., de Wet, B.J., Martinez-Pomares, L., Radcliffe, C.M., Dwek, R.A., Rudd, P.M., and Gordon, S. (2008). The mannose receptor mediates dengue virus infection of macrophages. PLoS Pathog. *4*, e17.

Mirkovitch, J., Decker, T., and Darnell, J.E., Jr. (1992). Interferon induction of gene transcription analyzed by in vivo footprinting. Mol. Cell. Biol. *12*, 1–9.

Morens, D.M., and Halstead, S.B. (1990). Measurement of antibody-dependent infection enhancement of four dengue virus serotypes by monoclonal and polyclonal antibodies. J. Gen. Virol. *71*(Pt 12), 2909–2914.

Morrison, T.E., Oko, L., Montgomery, S.A., Whitmore, A.C., Lotstein, A.R., Gunn, B.M., Elmore, S.A., and Heise, M.T. (2011). A mouse model of chikungunya virus-induced musculoskeletal inflammatory disease: evidence of arthritis, tenosynovitis, myositis, and persistence. Am. J. Pathol. *178*, 32–40.

Navarro-Sanchez, E., Altmeyer, R., Amara, A., Schwartz, O., Fieschi, F., Virelizier, J.L., Arenzana-Seisdedos, F., and Despres, P. (2003). Dendritic-cell-specific ICAM3-grabbing non-integrin is essential for the productive infection of human dendritic cells by mosquito-cell-derived dengue viruses. EMBO Rep. *4*, 723–728.

Paz, S., Sun, Q., Nakhaei, P., Romieu-Mourez, R., Goubau, D., Julkunen, I., Lin, R., and Hiscott, J. (2006). Induction of IRF-3 and IRF-7 phosphorylation following activation of the RIG-I pathway. Cell. Mol. Biol. (Noisy-le-grand) *52*, 17–28.

Peltier, D.C., Lazear, H.M., Farmer, J.R., Diamond, M.S., and Miller, D.J. (2013). Neurotropic arboviruses induce interferon regulatory factor 3-mediated neuronal responses that are cytoprotective, interferon independent, and inhibited by Western equine encephalitis virus capsid. J. Virol. *87*, 1821–1833.

Peranen, J., Rikkonen, M., Liljestrom, P., and Kaariainen, L. (1990). Nuclear localization of Semliki Forest virus-specific nonstructural protein nsP2. J. Virol. *64*, 1888–1896.

Pichlmair, A., Schulz, O., Tan, C.P., Naslund, T.I., Liljestrom, P., Weber, F., and Reis e Sousa, C. (2006). RIG-I-mediated antiviral responses to single-stranded RNA bearing 5′-phosphates. Science *314*, 997–1001.

Pichlmair, A., Lassnig, C., Eberle, C.A., Gorna, M.W., Baumann, C.L., Burkard, T.R., Burckstummer, T., Stefanovic, A., Krieger, S., Bennett, K.L., *et al.* (2011). IFIT1 is an antiviral protein that recognizes 5′-triphosphate RNA. Nat. Immunol. *12*, 624–630.

Platanias, L.C. (2003). The p38 mitogen-activated protein kinase pathway and its role in interferon signaling. Pharmacol. Ther. *98*, 129–142.

Pollara, G., Kwan, A., Newton, P.J., Handley, M.E., Chain, B.M., and Katz, D.R. (2005). Dendritic cells in viral pathogenesis: protective or defective? Int. J. Exp. Pathol. *86*, 187–204.

Porterfield, J.S. (1986). Antibody-dependent enhancement of viral infectivity. Adv. Virus Res. *31*, 335–355.

Postic, B., Schleupner, C.J., Armstrong, J.A., and Ho, M. (1969). Two variants of Sindbis virus which differ in interferon induction and serum clearance. I. The phenomenon. J. Infect. Dis. *120*, 339–347.

Rabinovich, S., and Liu, C. (1968). Interferon content of organs and serum after intravenous inoculation of mice with Sindbis virus. Exp. Med. Surg. *26*, 117–123.

Randall, R.E., and Goodbourn, S. (2008). Interferons and viruses: an interplay between induction, signalling, antiviral responses and virus countermeasures. J. Gen. Virol. *89*, 1–47.

Rebello, M.C., Fonseca, M.E., Marinho, J.O., and Rebello, M.A. (1993). Interferon action on Mayaro virus replication. Acta Virol. *37*, 223–231.

Rikkonen, M. (1996). Functional significance of the nuclear-targeting and NTP-binding motifs of Semliki Forest virus nonstructural protein nsP2. Virology *218*, 352–361.

Ritchie, K.J., Hahn, C.S., Kim, K.I., Yan, M., Rosario, D., Li, L., de la Torre, J.C., and Zhang, D.E. (2004). Role of ISG15 protease UBP43 (USP18) in innate immunity to viral infection. Nature Med. *10*, 1374–1378.

Rudd, P.A., Wilson, J., Gardner, J., Larcher, T., Babarit, C., Le, T.T., Anraku, I., Kumagai, Y., Loo, Y.M., Gale, M., Jr., *et al.* (2012). Interferon response factors 3 and 7 protect against Chikungunya virus hemorrhagic fever and shock. J. Virol. *86*, 9888–9898.

Ryman, K.D., and Klimstra, W.B. (2008). Host responses to alphavirus infection. Immunol. Rev. *225*, 27–45.

Ryman, K.D., Klimstra, W.B., Nguyen, K.B., Biron, C.A., and Johnston, R.E. (2000). Alpha/beta interferon protects adult mice from fatal Sindbis virus infection and is an important determinant of cell and tissue tropism. J. Virol. *74*, 3366–3378.

Ryman, K.D., White, L.J., Johnston, R.E., and Klimstra, W.B. (2002). Effects of PKR/RNase L-dependent and alternative antiviral pathways on alphavirus replication and pathogenesis. Viral Immunol. *15*, 53–76.

Ryman, K.D., Meier, K.C., Nangle, E.M., Ragsdale, S.L., Korneeva, N.L., Rhoads, R.E., MacDonald, M.R., and Klimstra, W.B. (2005). Sindbis virus translation is inhibited by a PKR/RNase L-independent effector induced by alpha/beta interferon priming of dendritic cells. J. Virol. *79*, 1487–1499.

Ryman, K.D., Gardner, C.L., Burke, C.W., Meier, K.C., Thompson, J.M., and Klimstra, W.B. (2007a). Heparan sulfate binding can contribute to the neurovirulence of neuroadapted and nonneuroadapted Sindbis viruses. J. Virol. *81*, 3563–3573.

Ryman, K.D., Gardner, C.L., Meier, K.C., Biron, C.A., Johnston, R.E., and Klimstra, W.B. (2007b). Early restriction of alphavirus replication and dissemination contributes to age-dependent attenuation of systemic hyperinflammatory disease. J. Gen. Virol. *88*, 518–529.

Ryman, K.D., Meier, K.C., Gardner, C.L., Adegboyega, P.A., and Klimstra, W.B. (2007c). Non-pathogenic Sindbis virus causes hemorrhagic fever in the absence of alpha/beta and gamma interferons. Virology *368*, 273–285.

Samuel, C.E. (1994). Interferon-induced proteins and their mechanisms of action. Hokkaido Igaku Zasshi *69*, 1339–1347.

Sanz, M.A., Castello, A., Ventoso, I., Berlanga, J.J., and Carrasco, L. (2009). Dual mechanism for the translation of subgenomic mRNA from Sindbis virus in infected and uninfected cells. PLOS ONE *4*, e4772.

Schafer, S.L., Lin, R., Moore, P.A., Hiscott, J., and Pitha, P.M. (1998). Regulation of type I interferon gene expression by interferon regulatory factor-3. J. Biol. Chem. *273*, 2714–2720.

Schilte, C., Couderc, T., Chretien, F., Sourisseau, M., Gangneux, N., Guivel-Benhassine, F., Kraxner, A., Tschopp, J., Higgs, S., Michault, A., *et al.* (2010). Type I IFN controls chikungunya virus via its action on nonhematopoietic cells. J. Exp. Med. *207*, 429–442.

Schilte, C., Buckwalter, M.R., Laird, M.E., Diamond, M.S., Schwartz, O., and Albert, M.L. (2012). Cutting edge: independent roles for IRF-3 and IRF-7 in hematopoietic and nonhematopoietic cells during host response to Chikungunya infection. J. Immunol. *188*, 2967–2971.

Schindler, C., Shuai, K., Prezioso, V.R., and Darnell, J.E., Jr. (1992). Interferon-dependent tyrosine phosphorylation of a latent cytoplasmic transcription factor. Science *257*, 809–813.

Schoggins, J.W., and Rice, C.M. (2011). Interferon-stimulated genes and their antiviral effector functions. Curr. Opin. Virol. *1*, 519–525.

Schoggins, J.W., Wilson, S.J., Panis, M., Murphy, M.Y., Jones, C.T., Bieniasz, P., and Rice, C.M. (2011). A diverse range of gene products are effectors of the type I interferon antiviral response. Nature *472*, 481–485.

Schoneboom, B.A., Catlin, K.M., Marty, A.M., and Grieder, F.B. (2000a). Inflammation is a component of neurodegeneration in response to Venezuelan equine encephalitis virus infection in mice. J. Neuroimmunol. *109*, 132–146.

Schoneboom, B.A., Lee, J.S., and Grieder, F.B. (2000b). Early expression of IFN-alpha/beta and iNOS in the brains of Venezuelan equine encephalitis virus-infected mice. J. Interferon Cytokine Res. *20*, 205–215.

Schulz, O., Pichlmair, A., Rehwinkel, J., Rogers, N.C., Scheuner, D., Kato, H., Takeuchi, O., Akira, S., Kaufman, R.J., and Reis e Sousa, C. (2010). Protein kinase R contributes to immunity against specific viruses by regulating interferon mRNA integrity. Cell Host Microbe *7*, 354–361.

Sen, G.C., and Peters, G.A. (2007). Viral stress-inducible genes. Adv. Virus Res. *70*, 233–263.

Seymour, R.L., Rossi, S.L., Bergren, N.A., Plante, K.S., and Weaver, S.C. (2013). the role of innate versus adaptive immune responses in a mouse model of o'nyong-nyong virus infection. Am. J. Trop. Med. Hyg. *88*, 1170–1179.

Shabman, R.S., Morrison, T.E., Moore, C., White, L., Suthar, M.S., Hueston, L., Rulli, N., Lidbury, B., Ting, J.P., Mahalingam, S., *et al.* (2007). Differential induction of type I interferon responses in myeloid dendritic cells by mosquito and mammalian-cell-derived alphaviruses. J. Virol. *81*, 237–247.

Shabman, R.S., Rogers, K.M., and Heise, M.T. (2008). Ross River virus envelope glycans contribute to type I interferon production in myeloid dendritic cells. J. Virol. *82*, 12374–12383.

Sharma, A., and Maheshwari, R.K. (2009). Oligonucleotide array analysis of Toll-like receptors and associated signalling genes in Venezuelan equine encephalitis virus-infected mouse brain. J. Gen. Virol. *90*, 1836–1847.

Sherman, L.A., and Griffin, D.E. (1990). Pathogenesis of encephalitis induced in newborn mice by virulent and avirulent strains of Sindbis virus. J. Virol. *64*, 2041–2046.

Silva, A.M., Whitmore, M., Xu, Z., Jiang, Z., Li, X., and Williams, B.R. (2004). Protein kinase R (PKR) interacts with and activates mitogen-activated protein kinase kinase 6 (MKK6) in response to double-stranded RNA stimulation. J. Biol. Chem. *279*, 37670–37676.

Silverman, M.A., Misasi, J., Smole, S., Feldman, H.A., Cohen, A.B., Santagata, S., McManus, M., and Ahmed, A.A. (2013). Eastern equine encephalitis in children, Massachusetts and New Hampshire, USA, 1970–2010. Emerg. Infect. Dis. *19*, 194–201; quiz 352.

Simmons, J.D., White, L.J., Morrison, T.E., Montgomery, S.A., Whitmore, A.C., Johnston, R.E., and Heise, M.T. (2009). Venezuelan equine encephalitis virus disrupts STAT1 signaling by distinct mechanisms independent of host shutoff. J. Virol. *83*, 10571–10581.

Sourisseau, M., Schilte, C., Casartelli, N., Trouillet, C., Guivel-Benhassine, F., Rudnicka, D., Sol-Foulon, N.,

Le Roux, K., Prevost, M.C., Fsihi, H., *et al.* (2007). Characterization of reemerging chikungunya virus. PLoS Pathog. *3*, e89.

Spotts, D.R., Reich, R.M., Kalkhan, M.A., Kinney, R.M., and Roehrig, J.T. (1998). Resistance to alpha/beta interferons correlates with the epizootic and virulence potential of Venezuelan equine encephalitis viruses and is determined by the 5′ noncoding region and glycoproteins. J. Virol. *72*, 10286–10291.

Sun, L., Wu, J., Du, F., Chen, X., and Chen, Z.J. (2013). Cyclic GMP-AMP synthase is a cytosolic DNA sensor that activates the type I interferon pathway. Science *339*, 786–791.

Szretter, K.J., Daniels, B.P., Cho, H., Gainey, M.D., Yokoyama, W.M., Gale, M., Jr., Virgin, H.W., Klein, R.S., Sen, G.C., and Diamond, M.S. (2012). 2′-O methylation of the viral mRNA cap by West Nile virus evades ifit1-dependent and -independent mechanisms of host restriction in vivo. PLoS Pathog. *8*, e1002698.

Tabeta, K., Hoebe, K., Janssen, E.M., Du, X., Georgel, P., Crozat, K., Mudd, S., Mann, N., Sovath, S., Goode, J., *et al.* (2006). The Unc93b1 mutation 3d disrupts exogenous antigen presentation and signaling via Toll-like receptors 3, 7 and 9. Nat. Immunol. *7*, 156–164.

Tamm, K., Merits, A., and Sarand, I. (2008). Mutations in the nuclear localization signal of nsP2 influencing RNA synthesis, protein expression and cytotoxicity of Semliki Forest virus. J. Gen. Virol. *89*, 676–686.

Tassaneetrithep, B., Burgess, T.H., Granelli-Piperno, A., Trumpfheller, C., Finke, J., Sun, W., Eller, M.A., Pattanapanyasat, K., Sarasombath, S., Birx, D.L., *et al.* (2003). DC-SIGN (CD209) mediates dengue virus infection of human dendritic cells. J. Exp. Med. *197*, 823–829.

Teng, T.S., Foo, S.S., Simamarta, D., Lum, F.M., Teo, T.H., Lulla, A., Yeo, N.K., Koh, E.G., Chow, A., Leo, Y.S., *et al.* (2012). Viperin restricts chikungunya virus replication and pathology. J. Clin. Invest. *122*, 4447–4460.

Tesfay, M.Z., Yin, J., Gardner, C.L., Khoretonenko, M.V., Korneeva, N.L., Rhoads, R.E., Ryman, K.D., and Klimstra, W.B. (2008). Alpha/beta interferon inhibits cap-dependent translation of viral but not cellular mRNA by a PKR-independent mechanism. J. Virol. *82*, 2620–2630.

Tonkin, D.R., Jorquera, P., Todd, T., Beard, C.W., Johnston, R.E., and Barro, M. (2010). Alphavirus replicon-based enhancement of mucosal and systemic immunity is linked to the innate response generated by primary immunization. Vaccine *28*, 3238–3246.

Tonkin, D.R., Whitmore, A., Johnston, R.E., and Barro, M. (2012). Infected dendritic cells are sufficient to mediate the adjuvant activity generated by Venezuelan equine encephalitis virus replicon particles. Vaccine *30*, 4532–4542.

Trgovcich, J., Aronson, J.F., and Johnston, R.E. (1996). Fatal Sindbis virus infection of neonatal mice in the absence of encephalitis. Virology *224*, 73–83.

Trgovcich, J., Ryman, K., Extrom, P., Eldridge, J.C., Aronson, J.F., and Johnston, R.E. (1997). Sindbis virus infection of neonatal mice results in a severe stress response. Virology *227*, 234–238.

Trobaugh, D.W., Gardner, C.L., Haddow, A., Wang, E., Weaver, S.C., Ryman, K.D., and Klimstra, W.B. (2014). RNA viruses can hijack vertebrae microRNAs to suppress innate immunity. Nature *747*, 245–248.

Uddin, S., Lekmine, F., Sharma, N., Majchrzak, B., Mayer, I., Young, P.R., Bokoch, G.M., Fish, E.N., and Platanias, L.C. (2000a). The Rac1/p38 mitogen-activated protein kinase pathway is required for interferon alpha-dependent transcriptional activation but not serine phosphorylation of Stat proteins. J. Biol. Chem. *275*, 27634–27640.

Uddin, S., Majchrzak, B., Wang, P.C., Modi, S., Khan, M.K., Fish, E.N., and Platanias, L.C. (2000b). Interferon-dependent activation of the serine kinase PI 3′-kinase requires engagement of the IRS pathway but not the Stat pathway. Biochem. Biophys. Res. Commun. *270*, 158–162.

Ventoso, I. (2012). Adaptive changes in alphavirus mRNA translation allowed colonization of vertebrate hosts. J. Virol. *86*, 9484–9494.

Ventoso, I., Sanz, M.A., Molina, S., Berlanga, J.J., Carrasco, L., and Esteban, M. (2006). Translational resistance of late alphavirus mRNA to eIF2alpha phosphorylation: a strategy to overcome the antiviral effect of protein kinase PKR. Genes Dev. *20*, 87–100.

Vilcek, J. (1964). Production of interferon by newborn and adult mice infected with Sindbis virus. Virology *22*, 651–652.

Vogel, P., Kell, W.M., Fritz, D.L., Parker, M.D., and Schoepp, R.J. (2005). Early events in the pathogenesis of eastern equine encephalitis virus in mice. Am. J. Pathol. *166*, 159–171.

Wang, F., Barrett, J.W., Ma, Y., Dekaban, G.A., and McFadden, G. (2009). Induction of alpha/beta interferon by myxoma virus is selectively abrogated when primary mouse embryo fibroblasts become immortalized. J. Virol. *83*, 5928–5932.

Wang, N., Dong, Q., Li, J., Jangra, R.K., Fan, M., Brasier, A.R., Lemon, S.M., Pfeffer, L.M., and Li, K. (2010). Viral induction of the zinc finger antiviral protein is IRF3-dependent but NF-kappaB-independent. J. Biol. Chem. *285*, 6080–6090.

Weaver, S.C., Kang, W., Shirako, Y., Rumenapf, T., Strauss, E.G., and Strauss, J.H. (1997). Recombinational history and molecular evolution of western equine encephalomyelitis complex alphaviruses. J. Virol. *71*, 613–623.

Werneke, S.W., Schilte, C., Rohatgi, A., Monte, K.J., Michault, A., Arenzana-Seisdedos, F., Vanlandingham, D.L., Higgs, S., Fontanet, A., Albert, M.L., *et al.* (2011). ISG15 is critical in the control of Chikungunya virus infection independent of UbE1L mediated conjugation. PLoS Pathog. *7*, e1002322.

Wesselingh, S.L., Levine, B., Fox, R.J., Choi, S., and Griffin, D.E. (1994). Intracerebral cytokine mRNA expression during fatal and nonfatal alphavirus encephalitis suggests a predominant type 2 T cell response. J. Immunol. *152*, 1289–1297.

White, L.J., Wang, J.G., Davis, N.L., and Johnston, R.E. (2001). Role of alpha/beta interferon in Venezuelan equine encephalitis virus pathogenesis: effect of an attenuating mutation in the 5′ untranslated region. J. Virol. *75*, 3706–3718.

White, L.K., Sali, T., Alvarado, D., Gatti, E., Pierre, P., Streblow, D., and Defilippis, V.R. (2011). Chikungunya virus induces IPS-1-dependent innate immune activation and protein kinase R-independent translational shutoff. J. Virol. *85*, 606–620.

Witte, K., Witte, E., Sabat, R., and Wolk, K. (2010). IL-28A, IL-28B, and IL-29: promising cytokines with type I interferon-like properties. Cytokine Growth Factor Rev. *21*, 237–251.

Yan, N., Regalado-Magdos, A.D., Stiggelbout, B., Lee-Kirsch, M.A., and Lieberman, J. (2010). The cytosolic exonuclease TREX1 inhibits the innate immune response to human immunodeficiency virus type 1. Nat. Immunol. *11*, 1005–1013.

Yin, J., Gardner, C.L., Burke, C.W., Ryman, K.D., and Klimstra, W.B. (2009). Similarities and differences in antagonism of neuron alpha/beta interferon responses by Venezuelan equine encephalitis and Sindbis alphaviruses. J. Virol. *83*, 10036–10047.

Yoneyama, M., Kikuchi, M., Natsukawa, T., Shinobu, N., Imaizumi, T., Miyagishi, M., Taira, K., Akira, S., and Fujita, T. (2004). The RNA helicase RIG-I has an essential function in double-stranded RNA-induced innate antiviral responses. Nat. Immunol. *5*, 730–737.

Zhan, Y., Vega-Ramos, J., Carrington, E.M., Villadangos, J.A., Lew, A.M., and Xu, Y. (2012). The inflammatory cytokine, GM-CSF, alters the developmental outcome of murine dendritic cells. Eur. J. Immunol. *42*, 2889–2900.

Zhang, Y., Burke, C.W., Ryman, K.D., and Klimstra, W.B. (2007). Identification and characterization of interferon-induced proteins that inhibit alphavirus replication. J. Virol. *81*, 11246–11255.

Zhao, C., Collins, M.N., Hsiang, T.Y., and Krug, R.M. (2013). Interferon-induced ISG15 pathway: an ongoing virus–host battle. Trends in microbiology *21*, 181–186.

Zhou, A., Paranjape, J.M., Der, S.D., Williams, B.R., and Silverman, R.H. (1999). Interferon action in triply deficient mice reveals the existence of alternative antiviral pathways. Virology *258*, 435–440.

Zhou, X., Michal, J.J., Zhang, L., Ding, B., Lunney, J.K., Liu, B., and Jiang, Z. (2013). Interferon induced IFIT family genes in host antiviral defense. Int. J. Biol. Sci. *9*, 200–208.

Zhou, Z., Wang, N., Woodson, S.E., Dong, Q., Wang, J., Liang, Y., Rijnbrand, R., Wei, L., Nichols, J.E., Guo, J.T., *et al.* (2011). Antiviral activities of ISG20 in positive-strand RNA virus infections. Virology *409*, 175–188.

Zhu, Y., and Gao, G. (2008). ZAP-mediated mRNA degradation. RNA Biol. *5*, 65–67.

Zhu, Y., Wang, X., Goff, S.P., and Gao, G. (2012). Translational repression precedes and is required for ZAP-mediated mRNA decay. EMBO J. *31*, 4236–4246.

Alphaviruses and their Role in Elucidating Antiviral Responses in Mosquitoes

5

Melanie McFarlane and Alain Kohl

Abstract

The interaction between mosquito vectors and arthropod-borne viruses (arboviruses) such as alphaviruses of the *Togaviridae* family has long been of central interest to virologists. Indeed, the infection-induced processes and outcomes appear to differ considerably between vertebrate and arthropod hosts. Control of arbovirus infection in mosquitoes by host immune responses is beginning to be elucidated. Much of this progress is due to research using alphaviruses. Additionally, the detailed understanding of their replicative processes, and also the sophisticated molecular tools available to manipulate these viruses, have been used to great advantage. This chapter will give insights into the interactions of alphaviruses with vector immunity, and the nature of these responses.

Introduction

Arboviruses are a large group of viruses relevant to both human and veterinary medicine. They cause a wide range of symptoms ranging from mild to severe disease. This group of viruses is composed of a number of different virus families and genera, including the medically important *Alphavirus* genus (*Togaviridae*). The nature of the disease caused by *Alphaviruses* is discussed elsewhere (see Chapters 7 and 8) and will not be mentioned in this chapter. The absolute requirement of these viruses to be transmitted to a mammalian host via the bite of an infected vector means that the viruses have two very different selection pressures, that of the mammalian immune system and the immune system of the arthropod vector. Historically, it was assumed that the interaction between the virus and mosquito vector was largely non-pathogenic; however, it is now clear that pathogenic infections of the mosquito vector can occur. An increasing body of evidence has shown that there is a complex interaction between the virus and the arthropod immune system, which acts to control virus infection and dissemination in the insect. Infection of the vector results in a targeted innate immune response that can vary depending on the infecting virus. There are three main signalling pathways that may be activated in response to virus infection in insects; IMD, Toll and JAK/STAT pathways. Other cellular pathways play a role in modulation of virus infection including apoptosis, melanisation and autophagy. Moreover, insects have an extremely effective RNA interference response, which is thought to be the major immune pathway in the mosquito vector for control of virus infection. Here we will discuss the roles of these pathways in alphavirus–mosquito interactions specifically.

A historical perspective on antiviral response studies in mosquito cells using alphaviruses

Immune and host responses to viruses including arboviruses in vertebrates have been investigated in great detail. The field has benefited from contributions from a number of research areas, which have allowed questions relating to virus detection, immune pathways, antibody production and specificity to be addressed. In comparison, less is known about the vector innate immune mechanisms that control arbovirus replication.

Nonetheless, in the context of this chapter we wish to point out some pioneering work by the group of Dennis Brown, which suggested that antiviral responses against alphaviruses do take place following infection of mosquito cells. This came at a time when insect immunity and insect cell responses were very poorly understood, yet the results and tools, such as cell lines, from these studies turned out to be important to the field many years later. Infection of mosquito cells in culture by alphaviruses (and other arboviruses) is generally characterized by an early acute phase of infection with high levels of virus production to give a culture of persistently infected cells with low levels of virus production. In particular, the *Ae. albopictus*-derived cell lines U4.4, C6/36 and C7/10 that were used in these studies turned out to have interesting properties that were in part explained decades later, as explained further below.

Early studies with Sindbis virus, a well characterized alphavirus, showed that culture medium of persistently infected mosquito cells contained antiviral activity which was virus specific. This could also act during the acute phase of infection and was a protein or had a protein component (Riedel and Brown, 1979). Antibodies against Sindbis virus did not eliminate this activity; however, it needs to be stressed that at that point it was not clear whether this activity was derived from the host cell or the virus. Later work showed that this activity was effective in U4.4 and C6/36 cells when pretreated with supernatant containing the antiviral activity, and that it affected viral RNA synthesis (Condreay and Brown, 1988). However, this was not the case with C7/10 cells, which do not produce this activity. The antiviral protein was later purified as a 3.2-kDa hydrophobic protein and was found to inhibit Sindbis virus in a specific manner and also induced cell cycle arrest (Luo and Brown, 1993). At that moment in time, it was difficult to assess whether this protein was a defensive response or whether this was an effect due to normal cellular processes. Nonetheless, these experiments produced important early observations and it was later observed that this antiviral protein also induced the synthesis of a 55-kDa protein in parallel with resistance to the virus appearing in the infected cells (Luo and Brown, 1994). Thus, the first hints of an antiviral response in mosquito cells were appearing. The cell lines used in these studies turned out be, alongside others such as the *Ae. aegypti*-derived Aag2 cells, very useful for many of the studies we will describe in this chapter.

Control of alphavirus replication by small RNA pathways

The major immune pathways in vector species, such as mosquitoes, implicated in control of virus replication are thought to be the small RNA mediated RNA interference pathways. There are four small interfering RNA (siRNA) pathways described in invertebrates: the exogenous small RNA (exo-siRNA), the endogenous small RNA (endo-siRNA), micro RNA (miRNA) and PIWI interacting RNAi (piRNA) pathways. Two of these pathways, namely the exo-siRNA and piRNA, have been shown to have antiviral activities in invertebrates and so will be discussed in detail in this chapter.

Exogenous RNAi pathway

Much of what we know about the RNAi pathway components and their function comes from studies in the model organism *Drosophila melanogaster* (Fig. 5.1). The exogenous siRNA pathway is initiated by the detection of long double-stranded RNA molecules, which in the case of alphaviruses, most likely arise from replication intermediates (Siu *et al.*, 2011). In the *Drosophila* model, these double-stranded RNA molecules are recognized by the initiator protein Dicer 2 (Dcr-2) (Aliyari and Ding, 2009; Ding and Voinnet, 2007; Kemp and Imler, 2009; Takeuchi and Akira, 2008). Dcr-2 has endonuclease activity and so, along with the dsRNA binding protein R2D2, cleaves the RNA molecules into small double-stranded RNA fragments (Aliyari *et al.*, 2008). The small RNA molecules are 21 nucleotides in length and are termed siRNAs or, when they are of virus origin, viRNAs (virus-induced small interfering RNAs) (Aliyari *et al.*, 2008). The small RNA fragments are then guided into the RNA induced

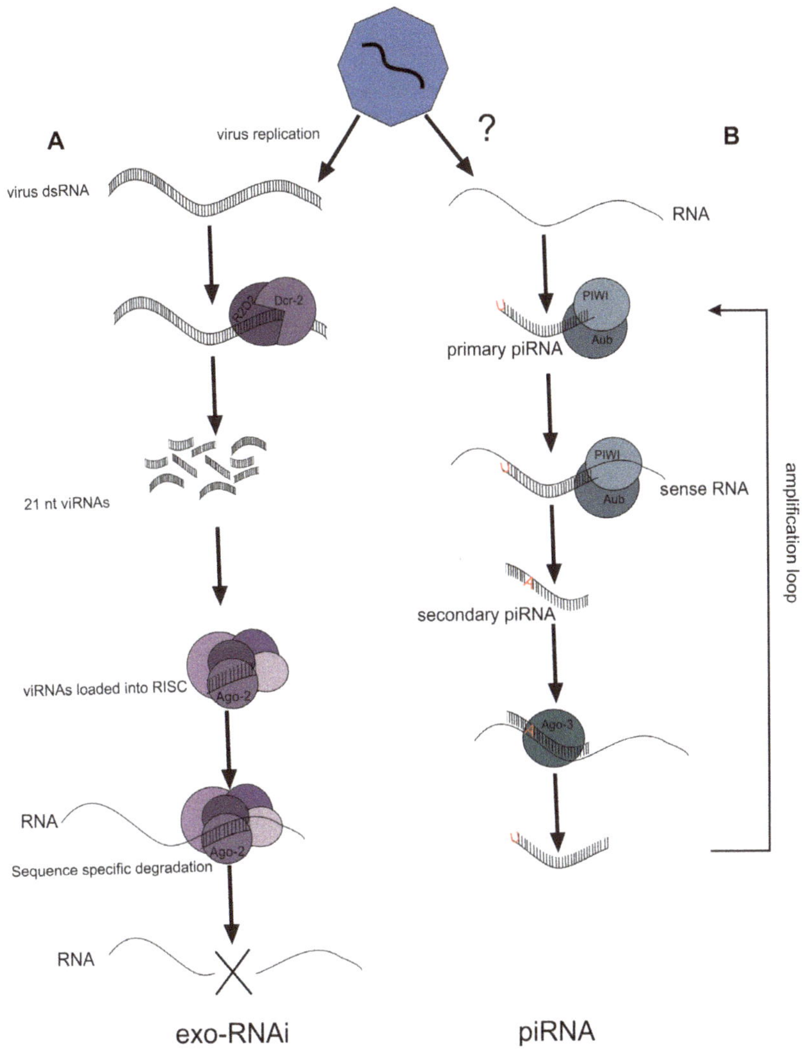

Figure 5.1 Schematic diagram of exo-RNAi and piRNA pathways in insects. The exogenous RNAi pathway (A) is initiated by the presence of virus-specific double-stranded RNA. Short 21-nucleotide RNA fragments are produced, which are then used for sequence specific degradation of virus RNAs and genomes. In *Drosophila* the piRNA pathway (B) is initiated by antisense transposon RNAs but it is unknown how the pathway is initiated during virus infection. Short fragments of 25–29 nucleotides are produced in a 'ping-pong' mechanism where there is a bias for uridine at position one of the primary antisense piRNA and adenine at position 10 of the secondary sense piRNA. These also result in sequence-specific silencing of virus RNAs and genomes. dsRNA: double-stranded RNA; Dcr-2: Dicer-2; nt: nucleotide; viRNA: virus induced small interfering RNA; RISC: RNA-induced silencing complex; Ago-2: Argonaute-2; Aub: Auberine; Ago-3: Argonaute-3.

silencing complex (RISC) where they are bound and unwound by Argonaute 2 (Ago-2) and one of the double strands is degraded by a second protein called C3PO (Ender and Meister, 2010; Liu *et al.*, 2009). The remaining strand, bound by Ago-2, is used as a template to identify complementary mRNA molecules or virus genomes which can then be degraded (Okamura *et al.*, 2004). The slicing by Ago-2 results in the sequence specific degradation of RNA molecules and thus 'silencing' of target proteins (see Fig. 5.1A for an overview). Bioinformatic studies revealed that the major components of these pathways are also present in *Aedes aegypti* (Campbell *et al.*, 2008a).

The first report of RNAi control of alphavirus replication and dissemination was published in 2004 where Keene and colleagues reported that injection of dsRNA targeting virus sequences into *Anopheles gambiae* mosquitoes inhibited O'nyong-nyong virus spread. Moreover, they showed that knocking down Ago-2 significantly increased virus titres when compared with control mosquitoes (Keene *et al.*, 2004). The key role of Ago-2 (Campbell *et al.*, 2008b; Schnettler *et al.*, 2013a) and Dcr-2 (Campbell *et al.*, 2008b) in defence against alphaviruses was also confirmed in subsequent studies. This provided direct evidence that RNAi is used to efficiently control arbovirus replication and dissemination in mosquito vectors. Additionally, Khoo *et al.* (2010) showed that the infection potential of the *Ae. aegypti* midgut by Sindbis virus was enhanced when Dcr-2 was knocked down. They also argued that impairment of the RNAi pathway aided the virus to escape the midgut barrier, thus allowing for enhanced virus dissemination (Khoo *et al.*, 2010). Furthermore, it was illustrated that engineered inhibition of the RNAi pathway resulted in increased mortality after virus infection (Cirimotich *et al.*, 2009; Myles *et al.*, 2008). Injection of *Ae. aegypti* mosquitoes with Sindbis virus or O'nyong-nyong virus expressing an RNA silencing suppressor protein, B2, from the pathogenic insect viruses Flock house virus and Nodamura virus, which binds both long dsRNAs and siRNAs, thus protecting the virus from RNAi-mediated silencing, decreased mosquito survival (Myles *et al.*, 2008). A similar study was performed by infecting *Aedes aegypti*, in addition to *Aedes albopictus* and *Culex tritaeniorhynchus*, with Sindbis virus expressing the same suppressor protein from Flock house virus and resulted in similar findings (Cirimotich *et al.*, 2009).

Data from these studies suggest that RNAi is required to control virus replication in order to protect the vector from pathology associated with an overwhelming virus infection and thus indirectly aids the maintenance of the arbovirus transmission cycle. This was also suggested to correlate with the absence of silencing suppressor proteins in alphaviruses. In contrast, it is thought that alphaviruses, such as Semliki Forest virus, employ an evasion strategy to evade rather than actively inhibit the RNAi mechanism. The first report of an RNAi evasion mechanism was outlined in 2011 by Siu and colleagues, where they identified a possible virus decoy mechanism. Deep sequencing data revealed the presence of viRNAs originating from throughout the Semliki Forest virus genome. There was, however, a bias towards production of viRNAs from certain areas of the virus genome (Siu *et al.*, 2011). Upon further investigation it was shown that the viRNAs present with the highest number of virus reads did not act to silence the virus efficiently, in contrast to the areas with fewer reads which were more effective at inhibiting Semliki Forest virus replication (Siu *et al.*, 2011). It was hypothesized from these data that the high expression of viRNAs, which cannot silence the virus, is an evasion mechanism whereby these viRNAs saturate the RISC complex, leading to virus escape from repression (Siu *et al.*, 2011). These observations will need to be further confirmed in the future by analysing the small RNAs present in the RISC.

Taken together these findings strongly suggest that the exo-RNAi pathway is a major component of the mosquito antiviral response system. The initial findings provoked a surge in the number of studies demonstrating that exo-RNAi was involved in the control of arboviruses in vector species. Studies in a number of vectors, including various mosquito species and midges, and their derived cell lines (Table 5.1), have revealed that the RNAi pathway controls the replication of a number of viruses, including the prototype alphaviruses Sindbis and Semliki Forest virus as well as the medically important chikungunya (*Togaviridae*), dengue and West Nile viruses (*Flaviviridae*), and the veterinary important bluetongue virus (*Reoviridae*), Rift Valley fever virus and Schmallenberg virus (*Bunyaviridae*) (Attarzadeh-Yazdi *et al.*, 2009; Brackney *et al.*, 2009; Campbell *et al.*, 2008b; Leger *et al.*, 2013; Morazzani *et al.*, 2012; Sanchez-Vargas *et al.*, 2009; Schnettler *et al.*, 2013a,b). In *Drosophila melanogaster* the RNAi pathway has also been demonstrated to limit replication of a DNA virus, invertebrate iridescent virus 6 (Bronkhorst *et al.*, 2012). However, whether this antiviral activity is mediated by the exogenous or endogenous RNAi pathway is not clear as both pathways share the

Table 5.1 Mosquito cell lines used for recent innate immunity studies with or on alphaviruses

Cell line	Origin	Pathway	Reference(s)
U4.4	*Aedes. albopictus*	RNAi, Toll, IMD, JAK/STAT	Attarzadeh-Yazdi *et al.* (2009), Fragkoudis *et al.* (2008), Morazzani *et al.* (2012), Siu *et al.* (2011)
C6/36	*Aedes albopictus*	RNAi (defective in exoRNAi)	Morazzani *et al.* (2012)
C7/10	*Aedes albopictus*	RNAi (defective in exoRNAi)	Morazzani *et al.* (2012)
Aag2	*Aedes aegypti*	RNAi	Siu *et al.* (2011)

key proteins Dcr-2 and Ago-2. This is the first report of the RNAi pathway controlling replication of a DNA virus, indicating that RNAi is used to control not only alphaviruses but all major classes of viruses replicating in insects.

piRNA pathway

The second small RNA pathway that has been identified as having antiviral function in vector species is the PIWI interacting RNA pathway or the piRNA pathway. The PIWI interacting RNAs (piRNAs) are larger in size than siRNAs and can vary in size between 25–29 nt in length. piRNAs were originally described and implicated in control of transposons in the germline. Recently, however, the presence of piRNAs has been described in *Drosophila melanogaster* somatic cells where they can target viruses (Brennecke *et al.*, 2007; Wu *et al.*, 2010). The origin of piRNAs is unclear, although they are thought to be derived from precursor RNA molecules which are transcribed from regions in the genome that contain transposable elements. These RNAs are in an antisense polarity and are called primary piRNAs which can then bind to sense mRNAs and result in mRNA cleavage. The short cleaved RNAs can then act as secondary piRNAs and cleave the expressed antisense RNAs, resulting in an amplification cycle of the piRNA molecules that is often referred to as a 'ping-pong' mechanism. The production of primary and secondary piRNA molecules results in a specific signature appearing on the RNA molecules. Primary piRNAs have a bias for uridine at position 1 on the RNA and, because the molecules are partially overlapping secondary piRNAs, have a bias for adenine at position 10 (Saito and Siomi, 2010; Senti and Brennecke, 2010; Siomi *et al.*, 2010, 2011; van Rij and Berezikov, 2009). In *Drosophila melanogaster*, the proteins involved in the piRNA pathway are two PIWI proteins, an Ago-3 protein and an Aubergine (Aub) protein. In mosquitoes there have been several more PIWI family members identified but no Aub protein has yet been identified (Campbell *et al.*, 2008a). This suggests that the pathways are slightly different between the different species and perhaps the expansion of the PIWI family has occurred to fulfil a specific role in mosquitoes. The importance of the piRNA pathway as an antiviral defence mechanism in mosquitoes is only now coming to light and again these discoveries owe much to studies with alphaviruses. Our current understanding of this pathway is summarized in Fig. 5.1B.

There have been several reports of the presence of piRNA molecules produced in mosquitoes and their derived cell lines. The first report of PIWI interacting RNA pathway control of alphavirus replication was in 2004 where Keene *et al* reported that knockdown of Ago-3 in *Anopheles gambiae* resulted in higher O'nyong-nyong virus titres compared with control mosquitoes (Keene *et al.*, 2004). At this time the pathway was not known to be antiviral but the proteins were assumed to be involved in development due to their expression in the *Drosophila* germline. It was not until several years later that virus derived piRNAs were reported to occur in *Drosophila* somatic sheet cells, suggesting an antiviral role for this pathway in adult insects could be possible (Wu *et al.*, 2010).

Following these observations Vodovar *et al.* (2012) showed that infection of *Ae. albopictus* derived U4.4 and C6/36 cells with Sindbis virus produced small RNAs of between 25–29 nt in length and displayed the piRNA signature of uridine on position 1 of the antisense strand and adenine on position 10 of the sense strand. A similar result was obtained when Aag2 cells were infected with Sindbis virus (Vodovar *et al.*, 2012).

Moreover, Morazzani *et al.* (2012) demonstrated that piRNAs derived from chikungunya virus, are present not only in mosquito derived cell lines but in whole mosquitoes. Additionally, it was shown that when the siRNA pathway activity is dampened by the presence of a silencing suppressor protein the levels of piRNAs that are produced in the mosquito increase (Morazzani *et al.*, 2012). The authors suggested that it is active virus replication and thus double-stranded RNA that may be inducing the PIWI-interacting response rather than single-stranded RNA. However, experiments performed by Schnettler *et al* (2013a) demonstrated that transfection of long double-stranded RNA did not elicit a piRNA response. The antiviral activity of the piRNA pathway was, however, demonstrated when the production of SFV-derived piRNAs was lost following knockdown of the seven PIWI proteins in Aag2 cells. Additionally, knockdown of PIWI 4 alone resulted in increased SFV replication and thus higher virus titres (Schnettler *et al.*, 2013a). This demonstrated for the first time that the piRNA pathway does in fact have an antiviral role in mosquito cell lines.

Similar studies showing the production of virus specific piRNAs have been done using LaCrosse virus infection of C6/36 cells (Vodovar *et al.*, 2012), dengue and West Nile virus infection of *Ae. aegypti* (Hess *et al.*, 2011) and Rift Valley Fever virus infection of mosquito derived cell lines (Leger *et al.*, 2013). Interestingly, Leger *et al.* (2013) suggested that the expression of piRNAs became more prominent during the persistent phase of the Rift Valley Fever virus infection, whereas the expression of 21 nt siRNAs was predominant in the acute phase of the virus replication cycle (Leger *et al.*, 2013). This hints that there may be some switch between the two immune pathways during virus infection and that this may be why the mosquito can become persistently infected without obvious adverse pathology associated with persistent infection. If this were to be shown for other arboviruses, including alphaviruses, this may explain the presence of two small RNA pathways in insects that seem to act antivirally. Perhaps the exo-RNAi pathway controls the initial high level of virus replication in the acute stage of infection and once the replication has been controlled to sufficient, non-pathogenic levels the piRNA pathway steps in to control viral replication to prevent it increasing again and thereby frees the siRNA pathway to monitor and control any other incoming virus infections. However, it may also be that pathways target virus families depending on genetic make-up and replication strategies.

Immune signalling pathways and extracellular mechanisms involved in antiviral defences in mosquitoes

Despite the prominent role of small RNA pathways in mosquito antiviral defences, it is becoming clear that immune pathways, previously linked to responses against bacterial and fungal pathogens, play a role in antiviral activities. Again, this owes much to studies in *Drosophila* where the roles of the key pathways have been elucidated (Imler, 2013). These are the Immune Deficiency (IMD), Toll and JAK/STAT pathways, and they display a degree of specificity according to the pathogen they ultimately mediate activities against. The Toll and IMD pathways stimulate activation of antimicrobial peptides and target fungi and Gram-positive bacteria or Gram-negative bacteria, respectively. The JAK/STAT pathway has also been involved in antiviral signalling in *Drosophila* (Ferrandon *et al.*, 2007; Kemp and Imler, 2009; Kounatidis and Ligoxygakis, 2012). The signalling cascades within these pathways have been elucidated in some detail, and have influenced research in mosquitoes (Fig. 5.2). Initial work with dengue virus in *Aedes aegypti* demonstrated a role for the JAK/STAT and Toll in responses against this virus (Souza-Neto *et al.*, 2009; Xi *et al.*, 2008). This was a step forward from earlier studies with microarrays using Sindbis virus and *Aedes aegypti*, which suggested an involvement of the Toll pathway in antiviral responses but did not take the observations further to demonstrate this (Sanders *et al.*, 2005). Similarly, studies with O'nyong-nyong virus in *Anopheles gambiae* showed differential regulation of a number of host genes, though none had any obvious link to these signalling pathways (Sim *et al.*, 2005). Nonetheless, these studies were the first to demonstrate the profound effects alphavirus infection has on

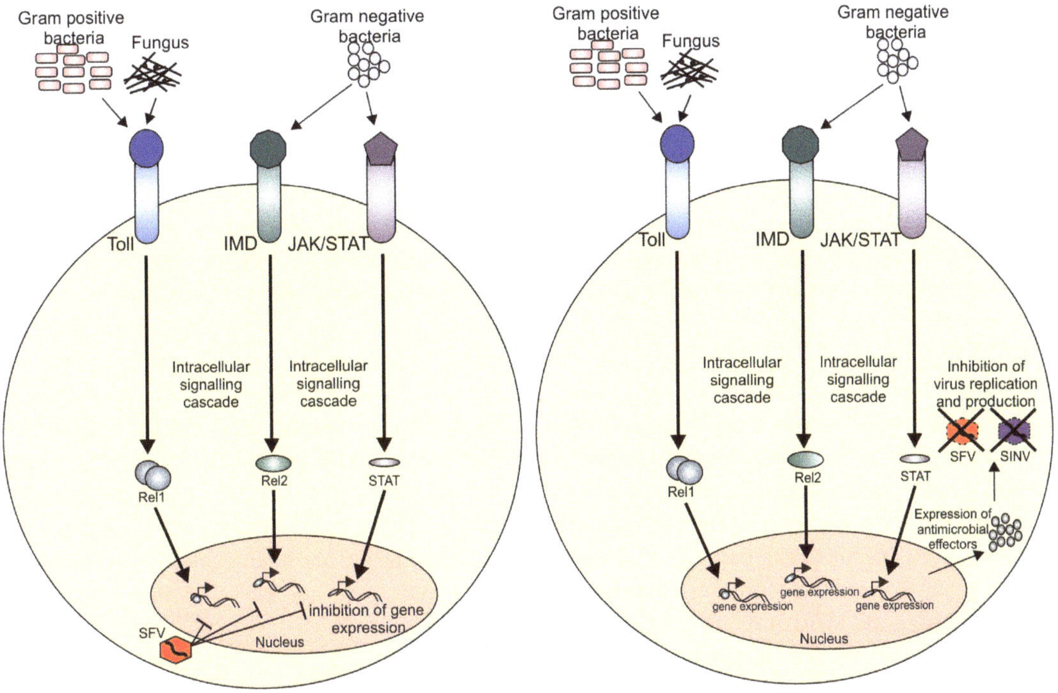

Figure 5.2 Schematic diagram of innate signalling pathways in insects. Three key signalling pathways in arthropods, including mosquito vectors, are the JAK/STAT, IMD and Toll pathways. The Toll signalling cascade is activated by Gram-positive bacteria and fungi. JAK/STAT and IMD are activated by Gram-negative bacteria; how viruses may activate these pathways is not clear. The pathways themselves have not been fully characterized in mosquitoes but are expected to be similar to those discovered in *Drosophila melanogaster*. Activation results in expression of antimicrobial peptides which can act antivirally. Semliki Forest virus (SFV) has been shown to reduce immune signalling by repression of host gene expression. Sindbis (SINV) virus replication has been shown to be inhibited by activation of the IMD pathway and SFV replication has been shown to be reduced by activation of one or both of the JAK/STAT and IMD pathways.

the vector. Recent detailed work on the impact of these signalling pathways on alphavirus infection has shown interesting differences to what is known from flaviviruses.

There have, until now, only been a few studies investigating the effect of these immune pathways on alphavirus replication (Table 5.1). Fragkoudis *et al.* (2008) used reporter constructs, which become activated upon stimulation of the IMD, Toll and JAK/STAT pathways in order to investigate the interactions of the signalling pathways and Semliki Forest virus. Infection of *Aedes albopictus* derived U4.4 cells with Semliki Forest virus did not activate any of the signalling pathways; however, virus infection did result in a significantly reduced level of stimulation of the pathways in response to bacterial challenge, in the case of JAK/STAT and IMD pathways, or transfection of active Toll receptor, for the Toll pathway, by targeting overall host gene expression (Fragkoudis *et al.*, 2008). Moreover, it was shown that prior stimulation of the JAK/STAT and/or IMD pathways (the assay did not allow a clear distinction), but not the Toll pathway, significantly reduced Semliki Forest virus replication (Fragkoudis *et al.*, 2008). Similarly, investigations in the *Drosophila* model implicated the IMD pathway in the control of Sindbis virus replication, with antimicrobial peptides implicated as effectors (Avadhanula *et al.*, 2009; Huang *et al.*, 2013). Virus replication and production was increased in flies mutant for IMD pathway components (Avadhanula *et al.*, 2009). Additionally, the presence, but not necessarily replication, of virus RNA resulted in an increase in Diptericin and Metchnikowin antimicrobial peptide (AMP) expression. Furthermore, analysis of Rel2, the relish homologue, in *Aedes albopictus*-derived C6/36 cells, following Sindbis virus infection, showed Rel2 translocation into the nuclear fraction, indicating that the IMD pathway is also activated in mosquitoes in response to Sindbis virus infection (Avadhanula *et al.*, 2009). However, transcriptome profiling of O'nyong-nyong infected *Anopheles gambiae* mosquitoes followed by gene silencing studies of pathway components indicated no role of the Toll and JAK/STAT pathways and at best only a minor role for the IMD pathway; on the other hand, other antiviral effectors (lysozymes, a galectin, and proteins involved in LPS-sensing) were described as having an antiviral effect in this study (Waldock *et al.*, 2012). The disparity between the different studies is possibly due to the method of study or host systems. Indeed, the interactions between these pathways and virus infection are likely to be complex and there may in fact be differences between different viruses and species of mosquito.

Significantly, O'nyong-nyong virus is the only alphavirus known to infect *Anopheles gambiae* mosquitoes and these mosquitoes have been shown to be highly refractory to the closely related chikungunya virus (Brault *et al.*, 2004; Vanlandingham *et al.*, 2005) and are generally not known to be efficient arbovirus vectors. This difference may be linked to differences in the virus–host interactions, including with the immune signalling pathways. It has been reported that generation of a chikungunya virus clone expressing O'nyong-nyong virus non-structural protein 3 (nsP3) facilitates infection of *Anopheles gambiae* mosquitoes with the chimeric chikungunya virus (Saxton-Shaw *et al.*, 2013). This suggests that nsP3 is the determinant of vector specificity of alphaviruses. How nsP3 defines the specificity is unclear; however, it may be interesting to speculate that nsP3 is required to evade detection by immune pathways in *Anopheles gambiae*, but the existence of fundamental differences between aedine and anopheline immune responses which would impact on arboviruses should not be ruled out as a factor. Attention will also have to be paid to tissue specificity and mode of infection by either injection or viraemic bloodmeal, as these factors may influence gene expression and pathway activation. Future studies in the field will need to take these experimental factors into consideration.

Intriguingly, the role of extracellular phenoloxidase activity in antiviral responses to arboviruses had not been described previously. This response system consists of a cascade which ultimately results in the production of melanin that accumulates at wound sites and around invading pathogens. Recent work from our group has shown by using Semliki Forest virus, that this pathway also displays antiviral activity in cell culture and *in vivo* (Rodriguez-Andres *et al.*, 2012). Thus, defences against alphaviruses and potentially other viruses grip at different levels inside and outside cells. Further investigations into these emerging antiviral responses are required to elucidate the precise interactions and associations between the virus and the host immune system.

The role of apoptosis as mosquito antiviral response

Apoptosis, or programmed cell death, is the mechanism by which cells *in vitro* or *in vivo* activate intracellular pathways resulting in death in response to a number of stimuli, including virus infection (Barber, 2001; Clarke and Clem, 2003; Clem, 2005; McLean *et al.*, 2008). The role of apoptosis as a mosquito antiviral response is unclear but recent work has begun to shed light on this mechanism. Proteins involved in *Aedes aegypti* apoptotic processes are becoming increasingly well characterized. These include effectors as well as inhibitors of the pathway (Bryant *et al.*, 2010; Liu and Clem, 2011; Wang and Clem, 2011) and this will allow dissection of the role of individual components in detail. First indicators of a role of apoptosis in mosquito antiviral responses did not come from work on alphaviruses but stemmed from work on the flaviviruses dengue and West Nile. Early work with West Nile virus showed the induction of apoptosis in the mosquito midgut epithelium following infection in a refractory strain of *Culex pipiens* (Vaidyanathan and Scott, 2006). Recent studies with mosquito strains competent or refractory for dengue virus have also linked the refractory state to induction of apoptosis (Liu *et al.*, 2013). Work on alphaviruses and cell death in mosquito cells is beginning to give interesting hints, as in the case of Sindbis virus infection of *Aedes aegypti*, that apoptosis is required for enhanced infection and dissemination (Wang *et al.*, 2012). However, those studies were carried out by silencing components of the apoptosis pathway (either pro- or anti-apoptotic genes) before bloodmeal infection: it is not clear whether this is relevant during natural infection when apoptosis has not been described. Whether this absence of apoptosis or cell death in mosquito cells is due to other cellular control mechanisms that limit virus replication remains unclear, but again, recent work allows us to speculate in that direction. Studies on the role of antiviral RNAi in Sindbis virus infection have demonstrated cytopathic effects in C6/36 cells (Morazzani *et al.*, 2012), which are deficient for the exogenous RNAi pathway due to a homozygous frameshift mutation targeting the Dcr-2 protein (Brackney *et al.*, 2010; Morazzani *et al.*, 2012; Scott *et al.*, 2010). Those findings suggest that host responses may result in avoidance of apoptosis and thus further virus spread and replication, which may be deleterious and result in the death of the mosquito. However, a viral inhibitory activity preventing cell death cannot be excluded.

Thus, there is a link between excessive viral replication linked to loss of RNAi and cell death, but also a link between apoptosis and competence that may depend on mosquito strains (refractory vs. competent), which may be more difficult to reproduce in cell culture. Nonetheless, given our lack of understanding of what defines competence, recently published data makes this also worthwhile to investigate in the case of alphaviruses. Understanding the factors underlying competence is key to better understanding these virus–vector interactions, but can also inform public health measures, for example by informing on risks associated with local mosquito populations. Efforts into understanding these processes are therefore highly important.

Conclusions

The complexities of mosquito immune responses to arbovirus infections are beginning to be unravelled. Our knowledge, in particular, of antiviral mechanisms against alphaviruses has grown considerably and will continue to grow. This is particularly true for RNAi pathways, but the effector mechanisms induced by innate immune signalling pathways, such as IMD and JAK/STAT, also need to be determined. Work on alphaviruses is also influencing research on other arboviruses and thus our general understanding of arbovirus–vector interactions. A number of key questions about arbovirus–vector interactions remain unsolved – for example the mechanism(s) of persistence and potential role of immunity in this process, how immunity shapes transmission, and the contribution of individual tissues in the various immune responses. Moreover, the hierarchy and timing of antiviral activities needs to be determined. These challenges now need to be met, and alphaviruses will undoubtedly be at the forefront of this research. The studies carried out on vectors may allow development of new control or preventative measures to limit virus infection and/or replication before transmission occurs.

References

Aliyari, R., and Ding, S.W. (2009). RNA-based viral immunity initiated by the Dicer family of host immune receptors. Immunol. Rev. *227*, 176–188.

Aliyari, R., Wu, Q., Li, H.W., Wang, X.H., Li, F., Green, L.D., Han, C.S., Li, W.X., and Ding, S.W. (2008). Mechanism of induction and suppression of antiviral immunity directed by virus-derived small RNAs in Drosophila. Cell Host Microbe *4*, 387–397.

Attarzadeh-Yazdi, G., Fragkoudis, R., Chi, Y., Siu, R.W., Ulper, L., Barry, G., Rodriguez-Andres, J., Nash, A.A., Bouloy, M., Merits, A., *et al.* (2009). Cell-to-cell spread of the RNA interference response suppresses Semliki Forest virus (SFV) infection of mosquito cell cultures and cannot be antagonized by SFV. J. Virol. *83*, 5735–5748.

Avadhanula, V., Weasner, B.P., Hardy, G.G., Kumar, J.P., and Hardy, R.W. (2009). A novel system for the launch of alphavirus RNA synthesis reveals a role for the Imd pathway in arthropod antiviral response. PLoS Pathog. *5*, e1000582.

Barber, G.N. (2001). Host defense, viruses and apoptosis. Cell Death Differ. *8*, 113–126.

Brackney, D.E., Beane, J.E., and Ebel, G.D. (2009). RNAi targeting of West Nile virus in mosquito midguts promotes virus diversification. PLoS Pathog. *5*, e1000502.

Brackney, D.E., Scott, J.C., Sagawa, F., Woodward, J.E., Miller, N.A., Schilkey, F.D., Mudge, J., Wilusz, J., Olson, K.E., Blair, C.D., *et al.* (2010). C6/36 *Aedes albopictus* cells have a dysfunctional antiviral RNA interference response. PLoS Negl. Trop. Dis. *4*, e856.

Brault, A.C., Foy, B.D., Myles, K.M., Kelly, C.L., Higgs, S., Weaver, S.C., Olson, K.E., Miller, B.R., and Powers,

A.M. (2004). Infection patterns of o'nyong nyong virus in the malaria-transmitting mosquito, *Anopheles gambiae*. Insect Mol. Biol. *13*, 625–635.

Brennecke, J., Aravin, A.A., Stark, A., Dus, M., Kellis, M., Sachidanandam, R., and Hannon, G.J. (2007). Discrete small RNA-generating loci as master regulators of transposon activity in Drosophila. Cell *128*, 1089–1103.

Bronkhorst, A.W., van Cleef, K.W., Vodovar, N., Ince, I.A., Blanc, H., Vlak, J.M., Saleh, M.C., and van Rij, R.P. (2012). The DNA virus Invertebrate iridescent virus 6 is a target of the Drosophila RNAi machinery. Proc. Natl. Acad. Sci. U.S.A. *109*, E3604–3613.

Bryant, B., Ungerer, M.C., Liu, Q., Waterhouse, R.M., and Clem, R.J. (2010). A caspase-like decoy molecule enhances the activity of a paralogous caspase in the yellow fever mosquito, *Aedes aegypti*. Insect Biochem. Mol. Biol. *40*, 516–523.

Campbell, C.L., Black, W.C.t., Hess, A.M., and Foy, B.D. (2008a). Comparative genomics of small RNA regulatory pathway components in vector mosquitoes. BMC Genomics *9*, 425.

Campbell, C.L., Keene, K.M., Brackney, D.E., Olson, K.E., Blair, C.D., Wilusz, J., and Foy, B.D. (2008b). Aedes aegypti uses RNA interference in defense against Sindbis virus infection. BMC Microbiol. *8*, 47.

Cirimotich, C.M., Scott, J.C., Phillips, A.T., Geiss, B.J., and Olson, K.E. (2009). Suppression of RNA interference increases alphavirus replication and virus-associated mortality in *Aedes aegypti* mosquitoes. BMC Microbiol. *9*, 49.

Clarke, T.E., and Clem, R.J. (2003). Insect defenses against virus infection: the role of apoptosis. Int. Rev. Immunol. *22*, 401–424.

Clem, R.J. (2005). The role of apoptosis in defense against baculovirus infection in insects. Curr. Top. Microbiol. Immunol. *289*, 113–129.

Condreay, L.D., and Brown, D.T. (1988). Suppression of RNA synthesis by a specific antiviral activity in Sindbis virus-infected *Aedes albopictus* cells. J. Virol. *62*, 346–348.

Ding, S.W., and Voinnet, O. (2007). Antiviral immunity directed by small RNAs. Cell *130*, 413–426.

Ender, C., and Meister, G. (2010). Argonaute proteins at a glance. J. Cell Sci. *123*, 1819–1823.

Ferrandon, D., Imler, J.L., Hetru, C., and Hoffmann, J.A. (2007). The Drosophila systemic immune response: sensing and signalling during bacterial and fungal infections. Nat. Rev. Immunol. *7*, 862–874.

Fragkoudis, R., Chi, Y., Siu, R.W., Barry, G., Attarzadeh-Yazdi, G., Merits, A., Nash, A.A., Fazakerley, J.K., and Kohl, A. (2008). Semliki Forest virus strongly reduces mosquito host defence signaling. Insect Mol. Biol. *17*, 647–656.

Hess, A.M., Prasad, A.N., Ptitsyn, A., Ebel, G.D., Olson, K.E., Barbacioru, C., Monighetti, C., and Campbell, C.L. (2011). Small RNA profiling of Dengue virus–mosquito interactions implicates the PIWI RNA pathway in antiviral defense. BMC Microbiol. *11*, 45.

Huang, Z., Kingsolver, M.B., Avadhanula, V., and Hardy, R.W. (2013). An antiviral role for antimicrobial peptides during the arthropod response to alphavirus replication. J. Virol. *87*, 4272–4280.

Imler, J.L. (2013). Overview of Drosophila immunity: a historical perspective. Dev. Comp. Immunol. *42*, 3–15.

Keene, K.M., Foy, B.D., Sanchez-Vargas, I., Beaty, B.J., Blair, C.D., and Olson, K.E. (2004). RNA interference acts as a natural antiviral response to O'nyong-nyong virus (Alphavirus; Togaviridae) infection of *Anopheles gambiae*. Proc. Natl. Acad. Sci. U.S.A. *101*, 17240–17245.

Kemp, C., and Imler, J.L. (2009). Antiviral immunity in drosophila. Curr. Opin. Immunol. *21*, 3–9.

Khoo, C.C., Piper, J., Sanchez-Vargas, I., Olson, K.E., and Franz, A.W. (2010). The RNA interference pathway affects midgut infection- and escape barriers for Sindbis virus in *Aedes aegypti*. BMC Microbiol. *10*, 130.

Kounatidis, I., and Ligoxygakis, P. (2012). Drosophila as a model system to unravel the layers of innate immunity to infection. Open Biol. *2*, 120075.

Leger, P., Lara, E., Jagla, B., Sismeiro, O., Mansuroglu, Z., Coppee, J.Y., Bonnefoy, E., and Bouloy, M. (2013). Dicer-2- and Piwi-mediated RNA interference in Rift Valley fever virus-infected mosquito cells. J. Virol. *87*, 1631–1648.

Liu, B., Behura, S.K., Clem, R.J., Schneemann, A., Becnel, J., Severson, D.W., and Zhou, L. (2013). P53-mediated rapid induction of apoptosis conveys resistance to viral infection in Drosophila melanogaster. PLoS Pathog. *9*, e1003137.

Liu, Q., and Clem, R.J. (2011). Defining the core apoptosis pathway in the mosquito disease vector Aedes aegypti: the roles of iap1, ark, dronc, and effector caspases. Apoptosis *16*, 105–113.

Liu, Y., Ye, X., Jiang, F., Liang, C., Chen, D., Peng, J., Kinch, L.N., Grishin, N.V., and Liu, Q. (2009). C3PO, an endoribonuclease that promotes RNAi by facilitating RISC activation. Science *325*, 750–753.

Luo, T., and Brown, D.T. (1993). Purification and characterization of a Sindbis virus-induced peptide which stimulates its own production and blocks virus RNA synthesis. Virology *194*, 44–49.

Luo, T., and Brown, D.T. (1994). A 55-kDa protein induced in Aedes albopictus (mosquito) cells by antiviral protein. Virology *200*, 200–206.

McLean, J.E., Ruck, A., Shirazian, A., Pooyaei-Mehr, F., and Zakeri, Z.F. (2008). Viral manipulation of cell death. Curr. Pharm. Des. *14*, 198–220.

Morazzani, E.M., Wiley, M.R., Murreddu, M.G., Adelman, Z.N., and Myles, K.M. (2012). Production of virus-derived ping-pong-dependent piRNA-like small RNAs in the mosquito soma. PLoS Pathog. *8*, e1002470.

Myles, K.M., Wiley, M.R., Morazzani, E.M., and Adelman, Z.N. (2008). Alphavirus-derived small RNAs modulate pathogenesis in disease vector mosquitoes. Proc. Natl. Acad. Sci. U.S.A. *105*, 19938–19943.

Okamura, K., Ishizuka, A., Siomi, H., and Siomi, M.C. (2004). Distinct roles for Argonaute proteins in small RNA-directed RNA cleavage pathways. Genes Dev. *18*, 1655–1666.

Riedel, B., and Brown, D.T. (1979). Novel antiviral activity found in the media of Sindbis virus-persistently infected mosquito (*Aedes albopictus*) cell cultures. J. Virol. *29*, 51–60.

van Rij, R.P., and Berezikov, E. (2009). Small RNAs and the control of transposons and viruses in Drosophila. Trends Microbiol. *17*, 163–171.

Rodriguez-Andres, J., Rani, S., Varjak, M., Chase-Topping, M.E., Beck, M.H., Ferguson, M.C., Schnettler, E., Fragkoudis, R., Barry, G., Merits, A., *et al.* (2012). Phenoloxidase activity acts as a mosquito innate immune response against infection with Semliki Forest virus. PLoS Pathog. *8*, e1002977.

Saito, K., and Siomi, M.C. (2010). Small RNA-mediated quiescence of transposable elements in animals. Dev. Cell *19*, 687–697.

Sanchez-Vargas, I., Scott, J.C., Poole-Smith, B.K., Franz, A.W., Barbosa-Solomieu, V., Wilusz, J., Olson, K.E., and Blair, C.D. (2009). Dengue virus type 2 infections of Aedes aegypti are modulated by the mosquito's RNA interference pathway. PLoS Pathog. *5*, e1000299.

Sanders, H.R., Foy, B.D., Evans, A.M., Ross, L.S., Beaty, B.J., Olson, K.E., and Gill, S.S. (2005). Sindbis virus induces transport processes and alters expression of innate immunity pathway genes in the midgut of the disease vector, Aedes aegypti. Insect Biochem. Mol. Biol. *35*, 1293–1307.

Saxton-Shaw, K.D., Ledermann, J.P., Borland, E.M., Stovall, J.L., Mossel, E.C., Singh, A.J., Wilusz, J., and Powers, A.M. (2013). O'nyong nyong virus molecular determinants of unique vector specificity reside in non-structural protein 3. PLoS Negl. Trop. Dis. *7*, e1931.

Schnettler, E., Donald, C.L., Human, S., Watson, M., Siu, R.W., McFarlane, M., Fazakerley, J.K., Kohl, A., and Fragkoudis, R. (2013a). Knockdown of piRNA pathway proteins results in enhanced Semliki Forest virus production in mosquito cells. J. Gen. Virol. *94*, 1680–1689.

Schnettler, E., Ratinier, M., Watson, M., Shaw, A.E., McFarlane, M., Varela, M., Elliott, R.M., Palmarini, M., and Kohl, A. (2013b). RNA interference targets arbovirus replication in Culicoides cells. J. Virol. *87*, 2441–2454.

Scott, J.C., Brackney, D.E., Campbell, C.L., Bondu-Hawkins, V., Hjelle, B., Ebel, G.D., Olson, K.E., and Blair, C.D. (2010). Comparison of Dengue virus type 2-specific small RNAs from RNA interference-competent and -incompetent mosquito cells. PLoS Negl. Trop. Dis. *4*, e848.

Senti, K.A., and Brennecke, J. (2010). The piRNA pathway: a fly's perspective on the guardian of the genome. Trends Genet. *26*, 499–509.

Sim, C., Hong, Y.S., Vanlandingham, D.L., Harker, B.W., Christophides, G.K., Kafatos, F.C., Higgs, S., and Collins, F.H. (2005). Modulation of *Anopheles gambiae* gene expression in response to o'nyong-nyong virus infection. Insect Mol. Biol. *14*, 475–481.

Siomi, M.C., Miyoshi, T., and Siomi, H. (2010). piRNA-mediated silencing in Drosophila germlines. Semin. Cell Dev. Biol. *21*, 754–759.

Siomi, M.C., Sato, K., Pezic, D., and Aravin, A.A. (2011). PIWI-interacting small RNAs: the vanguard of genome defence. Nat. Rev. Mol. Cell Biol. *12*, 246–258.

Siu, R.W., Fragkoudis, R., Simmonds, P., Donald, C.L., Chase-Topping, M.E., Barry, G., Attarzadeh-Yazdi, G., Rodriguez-Andres, J., Nash, A.A., Merits, A., *et al.* (2011). Antiviral RNA interference responses induced by Semliki Forest Virus infection of mosquito cells: characterization, origin, and frequency-dependent functions of virus-derived small interfering RNAs. J. Virol. *85*, 2907–2917.

Souza-Neto, J.A., Sim, S., and Dimopoulos, G. (2009). An evolutionary conserved function of the JAK-STAT pathway in anti-dengue defense. Proc. Natl. Acad. Sci. U.S.A. *106*, 17841–17846.

Takeuchi, O., and Akira, S. (2008). RIG-I-like antiviral protein in flies. Nat. Immunol. *9*, 1327–1328.

Vaidyanathan, R., and Scott, T.W. (2006). Apoptosis in mosquito midgut epithelia associated with West Nile virus infection. Apoptosis *11*, 1643–1651.

Vanlandingham, D.L., Hong, C., Klingler, K., Tsetsarkin, K., McElroy, K.L., Powers, A.M., Lehane, M.J., and Higgs, S. (2005). Differential infectivities of o'nyong-nyong and chikungunya virus isolates in Anopheles gambiae and Aedes aegypti mosquitoes. Am. J. Trop. Med. Hyg. 72, 616–621.

Vodovar, N., Bronkhorst, A.W., van Cleef, K.W., Miesen, P., Blanc, H., van Rij, R.P., and Saleh, M.C. (2012). Arbovirus-derived piRNAs exhibit a ping-pong signature in mosquito cells. PLOS ONE 7, e30861.

Waldock, J., Olson, K.E., and Christophides, G.K. (2012). Anopheles gambiae antiviral immune response to systemic O'nyong-nyong infection. PLoS Negl. Trop. Dis. *6*, e1565.

Wang, H., and Clem, R.J. (2011). The role of IAP antagonist proteins in the core apoptosis pathway of the mosquito disease vector *Aedes aegypti*. Apoptosis *16*, 235–248.

Wang, H., Gort, T., Boyle, D.L., and Clem, R.J. (2012). Effects of manipulating apoptosis on Sindbis virus infection of Aedes aegypti mosquitoes. J. Virol. *86*, 6546–6554.

Wu, Q., Luo, Y., Lu, R., Lau, N., Lai, E.C., Li, W.X., and Ding, S.W. (2010). Virus discovery by deep sequencing and assembly of virus-derived small silencing RNAs. Proc. Natl. Acad. Sci. U.S.A. *107*, 1606–1611.

Xi, Z., Ramirez, J.L., and Dimopoulos, G. (2008). The *Aedes aegypti* toll pathway controls Dengue virus infection. PLoS Pathog. *4*, e1000098.

Animal Models of Alphavirus-induced Inflammatory Disease

6

Lara J. Herrero, Adam Taylor, Pierre Roques, Brett A. Lidbury and Suresh Mahalingam

Abstract

Animal models that reflect key aspects of human disease are invaluable tools for understanding the mechanisms of disease pathogenesis and development of treatment strategies. In particular, animal models play important roles in expanding the current understanding of the pathobiology of alphavirus infection. Alphaviruses are arthropod-borne viruses that are able to infect a variety of vertebrate hosts. In humans, infection can result in extensive morbidity and mortality. As the global distribution of alphaviruses expands, with recent outbreaks in new areas of the world, there is an overwhelming need to improve our understanding of alphavirus pathogenesis with the key objective of anti-alphaviral drug design. This chapter focuses on the application of alphaviral animal models with particular emphasis on how these models have expanded the understanding of disease mechanisms and have provided a foundation for the development of new treatments.

Mouse models of alphavirus disease

Mouse models of Ross River virus-induced inflammatory disease

Ross River virus (RRV) is an 'Old World' alphavirus of the Semliki Forest group transmitted by both fresh and saltwater mosquitoes. Human disease comprises a range of symptoms including myalgia, lethargy and rash, with prominent polyarthritis of most concern to the care of patients.

Early development of mouse models for the elucidation of Ross River virus disease

RRV was first isolated in 1963 by Doherty and colleagues near the Ross River in North Queensland (Doherty *et al.*, 1963). Similar to other alphaviruses, RRV showed *in vitro* infectivity for BHK-21 and Vero cell lines (Taylor and Marshall, 1975). Studies 20 years later demonstrated RRV infectivity for a range of human and murine monocytes and macrophages, including via the antibody-dependent enhancement (ADE) pathway, with infection not achieved in T-cell and B-cell lines (Linn *et al.*, 1996). The relationship with monocytes and macrophages was subsequently shown to be a feature of cellular responses to RRV infection for humans, and central to mouse models as the primary cellular mediator of disease (Lidbury *et al.*, 2000, 2008).

Initial studies to model RRV disease in mice were reported via consecutive articles from 1973 (Mims *et al.*, 1973; Murphy *et al.*, 1973), with a subsequent 1975 publication reporting earlier studies on the adaptation of RRV field isolates to laboratory mice and cell lines (Seay *et al.*, 1981). These studies were performed with what was later identified as the low virulence Nelson Bay strain (NB) of RRV; NB field isolates showed 86%–100% morbidity in day-old outbred mice, but low (5%–24%) mortality (Seay *et al.*, 1981). Recovery in surviving mice was complete by day 14 post infection, with residual symptoms persisting for an additional 10 days, occasionally associated with muscle wasting.

The initial mouse pathogenesis studies for RRV were described in two distinct parts, with a focus on RRV associated pathology in the brain of mice comprising the first section (Mims *et al.*, 1973), and the pathogenic effects of infection on muscle, periosteum/perichondrium and brown fat the focus of the second report (Murphy *et al.*, 1973). Post RRV infection, in spite of persistent cortical and general neurological lesions, mice survived and recovered from the symptoms of RRV disease. At day 13 post infection, by which time mice had recovered from symptoms, a generalized monocyte infiltration was noted in association with 'cortical thinning' in the mouse brain. This and other observations suggested that such lesions were not responsible for the paralysis observed between days 7–10 post RRV infection. In light of this, it was subsequently proposed that paralysis was a result of severe necrosis in the muscle. Further studies on muscle collected from RRV-infected mice showed that virus could be detected at day 3 post infection by immunofluorescent antibody (IFA) techniques and by electron microscopy budding from myocytes at day 4 post infection. By day 7 post infection, severe 'necrotic' changes in the muscle were noted, with only a small number of inflammatory cells reported in association with this tissue damage. 'Confluent necrosis' was observed at day 10 post infection, corresponding with the most severe clinical signs. Once the peak clinical signs were reached, '... regeneration and repair ...' were noted in the muscle tissue, returning the muscle to a completely normal appearance by day 34 post infection, coinciding with the recovery of mice from all symptoms (Murphy *et al.*, 1973). Examination of the perichondrium and periosteum at day 3 post infection also revealed 'prominent' foci of infection, with only occasional necrotic cells noted. Only very high doses of the mouse-virulent RRV strain T48 induced signs of severe necrosis in periosteal and perichondrial cells at day 3 post infection. These observations led Murphy *et al.* to conclude, '... there was no involvement of joints in the mice that paralleled the striking involvement of joints in the human disease'. This suggested that no correlation existed between the mouse model and human disease. The authors also noted the late appearance of monocytes in the brown fat (at day 13 post infection), suggesting that monocyte involvement may vary depending on the tissue, and day 3 may have been too early to observe significant inflammatory changes in perichondrium/periosteum. In spite of the doubts raised by these studies over the relevance of a mouse model to study human RRV disease (particularly joint pathology), these reports (Mims *et al.*, 1973; Murphy *et al.*, 1973) were the first to note the association of RRV-induced pathology with monocyte infiltrates, providing early evidence for the later elucidation of the central role of monocytes/macrophages in RRV disease of the mouse.

The early observations reported for muscle pathology, and the associations with mononuclear infiltrates post RRV infection, were confirmed by studies in mice by Seay and colleagues (Seay *et al.*, 1981). RRV replication was detected in the serum and muscle of 7-day-old mice by day 1, with virus detectable in muscle up until day 9 post infection, but only until day 4 post infection in serum (as detected via plaque assay). In week-old mice, foci of muscle necrosis were observed in association with local mononuclear infiltrates at day 5 post infection, which by day 7 saw a prominent mononuclear infiltrate and significant muscle necrosis. In further agreement with previous observations, 'confluent necrosis' and inflammation almost totally replaced muscle tissue at day 11 post infection. Seay *et al.* (1981) also noted the reduction in inflammatory infiltrate by day 15 post infection, which by day 65 manifested as the complete recovery of muscle structure and function. These studies were repeated in 4-week-old mice, with no clinical signs of infection detected, which confirmed the age-related resistance to RRV disease that had been previously observed (Mims *et al.*, 1973; Murphy *et al.*, 1973). No RRV antigen was detectable in muscle by IFA techniques for 4-week-old mice, whereas antigen was detected in muscle from week old mice between days 3–7 post infection, but not after day 7. It appears, therefore, that while detectable infection was required for muscle pathology and disease to occur, peak disease symptoms reported for young mice were dependent upon the development and persistence of an RRV-induced mononuclear infiltrate, and that recovery from muscle necrosis could only occur with the gradual disappearance of monocytes from the muscle. As humoral and cell-mediated responses to RRV were similar for the two age groups, and immunosuppressive therapies had no impact on age-related resistance to disease, Seay *et al.* (1981) concluded that RRV-induced myositis was a result of 'viral lysis' of muscle fibres, rather than immune-mediated pathology.

Subsequent studies by Seay and Wolinsky (1982, 1983) on murine brain pathology post RRV infection further emphasized the earlier interpretation of viral replication as crucial to pathology, rather than immunopathogenesis. Again, while RRV could be detected in brain tissue during early infection, the early infiltrate was noted to consist primarily of neutrophils and monocytes, with monocytes eventually being the only infiltrating cell population present by day 8 (whereby virus was not detectable) post infection. Interestingly, Seay and Wolinsky presented evidence at the electron microscope level showing 'debris-filled' macrophages and '... cytoplasmic processes ... were inserted into myelin sheaths ...'. Yet again, Seay and Wolinsky concluded that immune responses were not responsible for pathology, based on negative results from studies with cyclophosphamide. In this context, cells of the monocyte/macrophage lineage, which are insensitive to the action of cyclophosphamide, were not considered to be a component of the immune response, although they did report that the monocyte infiltrate was reduced in cyclophosphamide-treated animals (which was most likely due to impaired T-cell function).

A murine model that represents human RRV disease, and the cellular basis of RRV pathology

With the collective early RRV studies largely focused on very young mice, and the observation from these studies that depletion or neutralization of the immune response did not reverse disease, the prevailing opinion until relatively recently was that immunopathology was not involved in RRV disease, and furthermore, the mouse model did not inform human RRV disease. Murine disease was thought to be caused by brain pathology due to viral lysis, while human disease was believe to be primarily focused on joint tissue and skeletal muscle.

Investigations on disease post RRV infection in older outbred mice (14- to 17-day-old Swiss or CD-1 strains) showed a nexus between the mouse and human. Generalized symptoms that included ruffled fur, hunching and failure to thrive were observed for mice inoculated with RRV between the ages of 14 and 17 days, but more importantly for parallels with human disease, hind-limb paresis and paralysis was observed between days 8 and 11 post infection, with declining or non-detectable RRV titres. A key clue to mechanisms behind this response was that the prior depletion of macrophages by silica or carrageenan (via intraperitoneal injection) ameliorated RRV disease (Lidbury *et al.*, 2000), in contrast to other studies that have shown silica treatment to enhance virus-induced morbidity/mortality in mice (Zisman *et al.*, 1970). Further investigation of hind-limb muscle from symptomatic mice showed a strong F4/80$^+$ infiltrating cell population, denoting monocyte/macrophage phenotype (Lidbury *et al.*, 2000), and confirming previous observations (Mims *et al.*, 1973; Murphy *et al.*, 1973; Seay *et al.*, 1981; Seay and Wolinsky, 1982, 1983). Macrophage-depleted mice showed no RRV symptoms, and this physical state corresponded to an absence of F4/80$^+$ cells in the hind-limb skeletal muscle. Histologically, the muscle tissue fibres from mice with RRV disease showed thinning, with loss of striation and centralized nuclei, in addition to the heavy F4/80$^+$ cell presence between the muscle fibres. Validating the earlier findings of Marshall *et al.* (unpublished) and Seay *et al.* (1981) mice recovered from these RRV symptoms, regained full motor activity, with cellular infiltrates disappearing, and muscle tissue returning to normal morphology within 30–40 days (unpublished data). Similar to other studies, immunosuppressive agents were also used to unravel the nature of the immunology associated with RRV disease; both cyclophosphamide and sub-lethal g-irradiation resulted in the abolition of antibody production and prolonged *in vivo* virus growth, but only marginal changes in morbidity and mortality for infected mice were observed (Lidbury *et al.*, 2000).

Together, these data firmly establish the macrophage as the cellular mediator of RRV disease in the mouse model, and show clearly that while cyclophosphamide- and g-irradiation-sensitive progenitor T and B cells had only very minor roles (if any), contrary to previous opinion the disease in the animal model can be considered to be immune-mediated. Furthermore, the observations that virus concentration was either falling or not detected at the peak of clinical disease, suggests that direct viral cell or tissue damage was not the explanation, but that RRV replication had an earlier role in eliciting macrophage activity.

As reported previously by Seay *et al.* (1981), mice were resistant to RRV disease if inoculated at 4 weeks of age, uncovering a paradox for the mouse model, namely an inverse age relationship to RRV disease when compared to humans (Seay *et al.*, 1981). While there are exceptions, clinical RRV disease is only normally seen in adults from the ages of 20–60 years (Fraser, 1986). In agreement with earlier reports, Lidbury *et al.* (2000) showed that inoculating outbred mice at ages greater than 19 days old led to no disease phenotype, with or without immunosuppressive therapy, confirming the age-associated resistance in outbred mice to RRV disease. Later studies on the inbred strain C57BL/6 showed disease phenotype could be induced by RRV infection up until the age of 28 days (Morrison *et al.*, 2006).

What was quite clear from the above-cited studies was that monocyte/macrophage infiltrates were a key correlate of muscle pathology, particularly as the kinetics of monocyte/macrophage infiltration mirrored the appearance of severe clinical disease and muscle necrosis in mice. Furthermore, a feature of recovery from RRV symptoms in mice was the clearance of monocytes/macrophages from the damaged muscle, which subsequently recovered its normal structure and function post disease. With a mouse model reflective of human disease available, and the primary cellular mechanism identified, opportunities were presented to further dissect disease mechanisms associated with RRV infection in mice. Importantly, results from studies of human RRV disease were used to examine the validity of the mouse model and consistency in terms of the features of disease identified.

Parallels between the mouse model and human Ross River virus disease

Studies on humans with RRV disease (RRVD) have shown in both the virus-associated exanthem and synovium of epidemic polyarthritis (EPA) patients a predominant mononuclear cell infiltrate (Clarris *et al.*, 1975; Fraser *et al.*, 1983; Hazelton *et al.*, 1985), with functional natural killer (NK) cells also detected in one study. In spite of the persistent nature of arthritis/arthralgia in a significant number of individuals with RRV infection, the virus has traditionally been very difficult to detect in synovial samples from EPA sufferers, with the initial successes of RRV isolation restricted to serum samples collected during the first 7 days of disease (Aaskov *et al.*, 1985; Tesh *et al.*, 1981). RRV antigen was also successfully detected by immunofluorescence in synovial monocytes and macrophages during the early phase of illness (Fraser *et al.*, 1981) and in basal epidermal and eccrine duct epithelia 3 days after the onset of RRV exanthem (Fraser *et al.*, 1983). A study by Soden *et al.* (2000) used molecular techniques, with partial success, to detect RRV RNA in synovial tissue from a cohort of RRVD patients more than one month after the onset of symptoms, providing the first evidence of persistent infection in the inflamed synovium of an RRVD sufferer. It is feasible, therefore, that the long-term presence of RRV, or RRV antigens, is required in the synovium for the chronic arthritis/arthralgia symptoms seen in some EPA patients. However, as pointed out by Soden *et al.* (2000) despite the many technical challenges inherent in detecting RNA in primary cell samples, the association of persistent virus with arthritis remains inconclusive as RRV was not detected in the majority of RRVD patients who exhibited synovial inflammation.

Consistently observed in studies of human RRVD was the predominance of a synovial monocyte/macrophage infiltrate, unlike other arthritic conditions which feature neutrophils (Fraser *et al.*, 1981). Early studies on synovial samples from RRVD patients noted that the '... synovial fluid was devoid of neutrophil leucocytes ...'(Clarris *et al.*, 1975), and there was a '... paucity of neutrophils ...' (Fraser *et al.*, 1981), suggesting a different mechanism of pathogenesis for infectious (viral) arthritis when compared to rheumatoid arthritis (Soden *et al.*, 2000). The chemokine basis of this cell infiltrate profile has been explored, based on *in vitro* studies of RRV-infected primary human synovial fibroblasts and RAW 264.7 murine macrophages (Mateo *et al.*, 2000). Infection of synovial fibroblasts was characterized by the pronounced up-regulation of mRNA for IL-8 and MCP-1, as well as TNF-a and granulocyte–macrophage colony-stimulating factor (GM-CSF). For RAW 264.7 mouse macrophage cell line, acute infection of naïve cells resulted in elevated mRNA for IL-8 and MIP-2.

Extension of the RRVD mouse model into synovial pathology and the involvement of immune factors and pathways in disease

With the establishment of a mouse model that resembles human RRVD, further studies were conducted to further elucidate cellular tropism for RRV infection, and examine in finer detail the cellular aspects of immunopathology. Repeating earlier observations (Lidbury *et al.*, 2000; Mims *et al.*, 1973; Murphy *et al.*, 1973), hind-limb skeletal muscle was the primary target of RRV infection, as was bone and joints of the hind-limbs from both outbred CD-1 mice and adult C57BL/6J mice (Morrison *et al.*, 2006). Also in line with earlier observations, severe inflammation of these tissues was detected post infection (Fig. 6.1). In addition to inflammatory macrophages, NK cells, and $CD4^+$ and $CD8^+$ T cells were identified within the inflammatory infiltrates. To determine the functional role of these adaptive immune cells, C57BL/6J RAG-$1^{-/-}$ mice (that lack functional T and B lymphocytes) were infected with RRV and clinical disease and inflammation monitored. RAG-$1^{-/-}$ mice and wild-type controls developed similar disease, accompanied by macrophage and NK cell infiltration into hind-limb tissues, and muscle pathology. Using gene knockout technology, previous studies that applied whole body g-irradiation or cyclophosphamide treatment to ablate adaptive immunity were confirmed, namely, that the adaptive immune response has a minor role, if any, in the development of RRVD in mice (Morrison *et al.*, 2006).

The identification of hind-limb bone and joint tissue infection by RRV, and the detection of inflammation involving macrophages and NK cells, firmly established the mouse as a suitable model of human myalgia and arthritis following infection (Clarris *et al.*, 1975; Fraser *et al.*, 1983; Hazelton *et al.*, 1985). Additional mouse studies that investigated macrophage products involved in RRVD, further confirmed the relevance to human disease. This investigation repeated previous

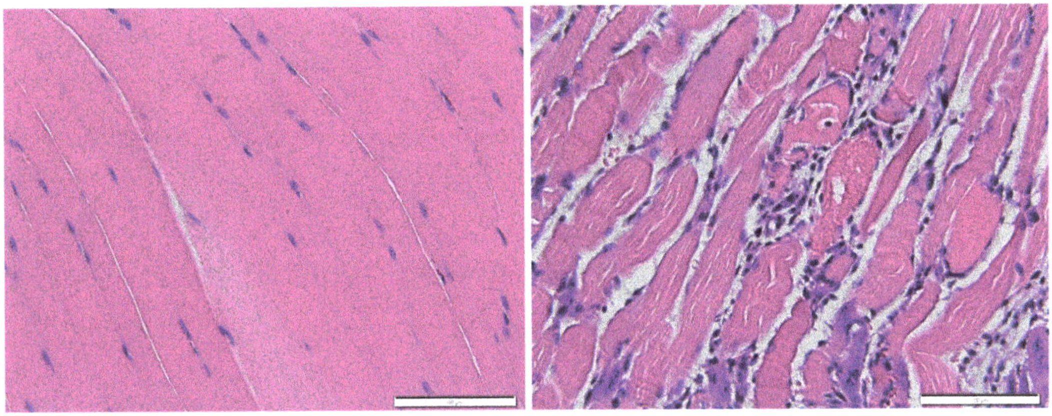

Figure 6.1 Muscle pathology in RRV infected mice. Representative pictures showing the pathology of skeletal muscle tissues from mock infected and RRV infected. C57BL/6 mice were injected subcutaneously with 10^4 PFU of RRV or mock infected with PBS alone. Quadriceps were removed at peak disease, fixed in 4% paraformaldehyde, sectioned at 5 μm and stained with haematoxylin and eosin. Magnification is ×200.

macrophage depletion studies (Lidbury *et al.*, 2000), this time using highly macrophage specific liposome-encapsulated clodronate for pre-infection macrophage depletion (Lidbury *et al.*, 2008). Macrophage-depleted mice showed a significant reduction in inflammatory infiltrate for muscle and ankle joint tissues on histological analysis, as well as reduced clinical disease. Reductions in joint/muscle tissue levels of tumour necrosis factor-α (TNF-α), interferon-g (IFN-γ), reactive nitrogen intermediates (RNI) and macrophage chemoattractant protein-1 (MCP-1) mRNA and protein were associated with macrophage depletion in mice. These four pro-inflammatory proteins were elevated in synovial fluid collected from patients with active RRV polyarthritis, compared to control synovial samples from osteoarthritis patients, hence further demonstrating relevance of the mouse RRVD model for human disease (Lidbury *et al.*, 2008).

Innate immune response to RRV in the mouse: the complement system and viral avoidance of type I interferon response

The role of monocyte/macrophage infiltrates in RRV-induced joint and muscle tissue damage indicates a primacy of innate immune system activity for disease post infection. Using this logic, studies of the complement system, which has been associated with other inflammatory arthritides, attracted deeper investigation. RRV disease severity in mice was enhanced by complement, with complement activation products detected in inflamed tissues and serum post RRV infection. As demonstrated in C3 deficient ($C3^{-/-}$) mice, complement had whole organism relevance, with less severe disease observed compared to wild-type control mice. Interestingly, the extent and composition of the inflammatory infiltrates were similar for $C3^{-/-}$ mice and controls, although significantly less hind-limb skeletal muscle destruction was observed for C3-deficient mice, suggesting no interruption to inflammatory cell chemotaxis but an impact on the effector capacity of infiltrating cells. Synovial fluid from RRV-infected patients had detectable complement activation (Morrison *et al.*, 2007). As an extension, studies using mice deficient in complement receptor 3 (CR3) ($CD11b^{-/-}$), showed less severe disease symptoms and decreased tissue destruction compared to RRV-infected control mice, without impacting viral replication (Morrison *et al.*, 2008). Whether the classical or alternative complement system was of greater importance to RRVD was answered soon after, with the outcome of mouse studies showing it was neither of these pathways, but the mannose-binding lectin (MBL) pathway that was central to the development of RRV-induced disease (Gunn *et al.*, 2012). MBL-deficient mice displayed an amelioration of RRVD, reduced tissue damage and complement deposition compared with controls. A reduction in clinical disease was not observed in infected mice deficient for key components of the classical or alternative complement activation pathways. Elevated serum MBL levels were found in RRV disease patients, with human disease severity correlating with serum and synovial fluid MBL concentration.

Concluding remarks and towards a cure for RRVD

While available since the early 1970s, for much of this time the mouse model of disease following RRV infection was seen to be of little relevance to understanding human RRV disease, or arthrogenic alphavirus disease in general which, as emphasized by the recent re-emergence and spread of chikungunya virus, has global health importance. Since 2000, dramatic progress has been made to establish a mouse model that, as described above, has consistently shown relevance to human disease associated with RRV infection, for example the role of complement activation. Such progress has been further vindicated by successful studies with the monocyte chemotactic protein (MCP) inhibitor, *Bindarit* (Rulli *et al.*, 2009). Bindarit-treated mice had less severe RRV disease, reduced tissue damage and lower intensity of inflammatory cell recruitment. Bindarit-mediated down-regulation of MCP reduced cell infiltration, leading to lower levels of local hind-limb joint/muscle NF-κB and TNF activity, thus ameliorating tissue damage. Therefore, a successful mouse model that has parallels with human disease allows for the thorough pre-clinical investigation of potential therapeutics such as Bindarit. Conversely, fundamental mouse studies identified MCP as a key mediator of mouse and human RRVD, allowing the connection to this MCP inhibitor as of possible benefit. With a number of other innate mechanisms identified in mice, it can be expected that other drug candidates will emerge of benefit to sufferers of alphavirus-induced polyarthritis and associated disease symptoms.

Mouse models of chikungunya virus-induced inflammatory disease

Chikungunya virus (CHIKV) is an 'Old World' alphavirus of the Semliki Forest group, transmitted by *Aedes* mosquitoes. Patients infected with CHIKV often present with a high fever, headache, myalgia and highly debilitating migratory polyarthralgia. Arthropathies can persist in some patients and more atypical clinical manifestations, such as neurological complications, are frequently associated with underlying comorbidities.

Early development of a small animal model permissive to CHIKV infection

CHIKV was first isolated from febrile patients in Tanganyika in 1952 following a large outbreak of the disease. In 1956 Ross performed the first *in vivo* studies of CHIKV infection and associated disease, analysing the susceptibility to CHIKV infection of a number of small animals (Ross, 1956). Although no virus was recovered from the blood or brains of infected guinea pigs or rabbits, inoculated newborn Albino Swiss mice became viraemic and CHIKV was successfully passaged from the organs of these mice. Suckling mice inoculated intracerebrally with sera from acute phase CHIKV patients showed unexpectedly high mortality, often succumbing to infection less than 48 hours after inoculation. Although used largely as a means of viral propagation, mice intracerebrally inoculated with CHIKV were observed closely for the development of disease signs. Serial passage of CHIKV resulted in a gradual extension of the viral incubation period until infected mice sickened on day 3 with manifestations such as limb extension and spasmodic tremor and died 4–7 days post infection. All mice younger than 12 days died, 60% of 12-day-old mice died, and 80% of mice older than 12 days survived. However, continued passage (over 100 serial passages) of CHIKV failed to increase its virulence towards adult mice. This work, therefore, first identified the difference in susceptibility to CHIKV infection between juvenile and adult mice. The responsiveness to CHIKV infection in juvenile mice, although poorly characterized, is an advantage still used in a number of modern mouse models of CHIKV and alphaviral disease in general (Couderc *et al.*, 2008; Morrison *et al.*, 2006).

Mouse models of acute CHIKV disease

Since its isolation in 1952, mice have been used in a variety of setting to better understand the pathogenesis of CHIKV infection. However, only recently have the greatest advances in our knowledge of CHIKV-associated inflammatory disease and pathogenesis *in vivo* been made. The large outbreaks throughout 2005–2007 and the re-emergence of CHIKV as a pathogen of considerable human importance drove the development of a number of animal models to better understand the pathophysiology of CHIKV disease. Several studies have since been able to accurately simulate one or more specific aspects of human CHIKV disease in mice: the classical symptoms associated with acute CHIKV infection in humans being fever, maculopapular erythematous rash, debilitating polyarthralgia and myalgia. Initially, *in vivo* studies were hampered by the asymptomatic nature of infection of various adult mouse strains inoculated with CHIKV by intraperitoneal, intradermal or subcutaneous injection (Couderc *et al.*, 2008; Gardner *et al.*, 2010; Suckling *et al.*, 1978a). The mature innate immune system in adult mice is thought to offer protection from fatal CHIKV infection. It has also been suggested that a strong interferon response to alphaviral infection reduces disease susceptibility in adult mice (Grieder and Vogel, 1999; Levy, 2007; Ryman *et al.*, 2000; Sourisseau *et al.*, 2007). Thus, early efforts to model acute CHIKV disease in mice looked to immunologically immature neonate mice (Couderc *et al.*, 2008; Schilte *et al.*, 2010; Werneke *et al.*, 2011; Ziegler *et al.*, 2008). Using CHIKV isolates from patients infected during the Réunion Island outbreak (CHIKV-21), Couderc *et al.* (2008) observed that neonate mice were permissive to CHIKV infection. Specifically, neonatal C57BL/6 mice inoculated with

CHIKV by intradermal injection of 10^6 PFU exhibited age-dependent lethality: 6-day-old mice died before day 12, whereas more than half of 9-day-old infected mice recovered. Interestingly, both 6- and 9-day-old infected mice all developed flaccid paralysis on day 6–7 post infection and CHIKV was found to disseminate to the periphery in neonates, including muscle and joint tissues. Furthermore, as alphaviruses stimulate the host interferon response and are interferon sensitive, similar CHIKV inoculations were made in adult mice with a partially or totally abrogated type-I IFN pathway. To question the nature of adult mouse non-permissiveness, Couderc *et al.* (2008) infected IFN-α/βR knockout (IFN-α/βR$^{-/-}$) adult mice with CHIKV. Infected mice all developed disease manifestations characterized by loss of muscle tone, lethargy and death at day 3 post infection. Strikingly, adult mice with only a partially abrogated type-I IFN pathway (IFN-α/βR$^{+/-}$) developed no signs of disease upon infection with CHIKV. However, infectious virus was detected in the muscle and joints of IFN-α/βR$^{+/-}$ infected adult mice at day 3 post infection just as in IFN-α/βR$^{-/-}$ infected mice. Specifically, CHIKV was predominantly observed in the epimysium and to a lower extent in the perimysium and endomysium. The IFN-α/βR$^{+/-}$ non-lethal adult mouse model of CHIKV infection reflects the mild disease commonly observed in human adults. These data also suggest that a dose effect of type-I IFN receptor gene controls the level of permissiveness of adult mice towards CHIKV. These studies confirm the importance of the innate immune system in protection of adult mice from fatal CHIKV infection. The dissemination of CHIKV to muscle and the joints, the symptomatic tissues in human disease, in each of these mouse models was an indication of the link between CHIKV tropism and disease. It also indicated the validity of the models to replicate human disease and provided further understanding of the pathogenic properties of this arthritogenic alphavirus. These models and derivatives were used to further dissect the type I IFN antiviral response during CHIKV infection (Schilte *et al.*, 2010; Werneke *et al.*, 2011).

Focusing on modelling the acute arthritic symptoms of CHIKV infection in humans, Ziegler *et al.* (2008) used neonatal mice, strains ICR and CD-1. Inoculating mice subcutaneously with CHIKV strain LR 2006-OPY1, Ziegler *et al.* (2008) observed arthritogenic disease manifestations in infected mice typified by lethargy, difficulty walking and dragging of the hind-limbs. Importantly however, compared with other neonate mouse models, there was a reduction of the high mortality associated with infection, and most animals eventually recovered. Other acute signs of CHIKV disease in this model included loss of balance and a reduced weight gain. All mice were viraemic during the initial 6–7 days post infection. High levels of CHIKV (~10^7 PFU) were found in skeletal leg muscle where replication competent virus persisted up to day 9–10 post infection. Histological analysis revealed major changes to the pathology of the skeletal muscle and joint tissue in infected mice. In muscle, there was severe myositis with signs of focal necrosis and inflammatory cell infiltration followed by fibrosis and dystrophic calcification in the joint cartilage. Signs of necrosis and inflammation in the muscle of infected mice were similar to the tissue damage observed in the biopsies of quadriceps from an acute CHIKV patient, although there was little sign of collagenosis in the mouse muscle (Ozden *et al.*, 2007).

Modern techniques such as *in vivo* imaging were used to visualize viral tropism and pinpoint areas of viral dissemination and replication during acute infection *in vivo*. A recombinant CHIKV construct able to express luciferase during viral replication was injected in the right rear footpad of weaning (3–4 weeks old) CD-1 mice ($10^{5.5}$ $TCID_{50}$) (Ziegler *et al.*, 2011). Replication of CHIKV-luciferase at the point of inoculation was observed in mice as early as 12 hours post infection and continued up to day 7 post infection. Viral dissemination to the hind-limb muscle was also observed at day 3 post infection and remained visible up to day 5. This model is, however, restricted by the limit of detection offered by the current imaging system and so requires further optimization. For example, replication in deep tissue or diffuse replication will not produce a strong enough signal to be detected.

Looking specifically at the site of infection, Thangamani *et al.* (2010) used mice to model the early events during transmission of CHIKV. The effects of infection by mosquito bite or intradermal needle inoculation on immune activation and cytokine expression were examined in 2-week-old outbred CD-1 mice. A skewing of host immunity towards a local Th2 cytokine profile was observed in mosquito bitten mice compared to a significant up-regulation of Th1 cytokines after needle inoculation. Down-regulation of a cutaneous antiviral Th1 response following acute infection by mosquito may promote viral replication and dissemination of CHIKV in the host. Modelling the exact route of viral inoculation has also highlighted the significant role of mosquito saliva during CHIKV transmission.

Proteomic analysis has been used by Dhanwani *et al.* (2011, 2014) in a number of studies of acute CHIKV infection in mice to analyse the host response in CHIKV-infected tissues. Not only has this work further characterized the pathophysiology of CHIKV disease but it has also established a novel permissive neonate (2- to 3-day-old) mouse model of acute CHIKV disease (Dhanwani *et al.*, 2011, 2014). Upon subcutaneous inoculation with 10^6 PFU Indian strain CHIKV DRDE-06 in the loose skin of their back, mice developed acute hind-limb disease. Infected mice showed signs of illness such as reduced body weight, lethargy, hind-limb dragging and hair loss on days 8–10 post infection. The locomotor behaviour of CHIKV infected mice was severely affected, implicating CHIKV-induced severe myopathic symptoms in infected newborn mice, mimicking the clinical presentation of CHIKV disease in human patients. Although infected mice exhibited 10–20% mortality, in the majority of diseased mice manifestations subsided by day 12 post infection and animals eventually recovered from the disease. CHIKV antigen was detected in the brain and muscle tissues of infected mice at peak disease. In muscle, significantly elevated levels of the pro-inflammatory cytokines IL-6, TNF-α, MCP-1, MCP-3 and RANTES were observed. Moreover, histological analysis of infected skeletal muscle revealed atrophied and necrotic muscle fibres in addition to a major infiltration of neutrophils and monocytoid cells in mice at peak disease. Together these data confirm the validity of the mouse model, which is able to replicate the CHIKV-induced inflammation and myopathy of human disease. Tissues permissive to CHIKV replication were subjected to proteomic profiling to identify host factors that potentially contribute to CHIKV disease in mice. Interestingly, a variety of candidate proteins with roles in structural, signalling, metabolism, and inflammation pathways were differentially expressed during CHIKV infection. In-depth studies are required to further elucidate the function of candidate proteins in CHIKV pathogenesis and to validate the host–CHIKV interactions that may provide novel targets for therapeutic intervention. However, in the short term, differential expression of candidate genes in CHIKV-infected tissues could reveal clinical biomarkers that allow for early detection of CHIKV disease in patients.

These neonatal mouse models have provided important information on CHIKV tropism during acute infection and were able to replicate muscle pathologies observed in human CHIKV disease. However, these studies often overlooked the full extent of musculoskeletal disease in mice and failed to report signs of joint pathologies often associated with CHIKV infection in patients. The use of immunologically immature neonate or immune-deficient adult mice in these studies restricts their ability to accurately replicate the immunopathogenesis of CHIKV infection in adult patients. Furthermore, neonate models regularly use lethality to measure CHIKV disease rather than scoring rheumatic manifestations. Thus, development of a permissive mouse model of acute CHIKV infection to better represent the human clinical manifestations of disease led researchers to use juvenile (over 14 days old) mice (Morrison *et al.*, 2011; Rulli *et al.*, 2011). Using a clinical isolate of CHIKV, Morrison *et al.* (2011) inoculated 14-day-old C57BL6/J mice with 100 PFU CHIKV SL15649 in the left rear footpad. All mice developed a prominent gross swelling and oedema of the left foot measurable by calipers. Swelling began 2 days after inoculation and persisted for at least 1 week. Although it was able to disseminate from the site of inoculation, almost 100-fold more CHIKV was present in the left foot versus the non-inoculated right foot. Similarly, no swelling was observed in the right foot, and inflammatory infiltrates, although present, were less extensive in the right foot verses the left. Rheumatic disease signs are a prominent feature of CHIKV infection of humans. Consistent with this, mice developed a mixed cell inflammatory arthritis characterized by the presence of natural killer cells, monocytes and neutrophils in a number of joints in the foot. Furthermore, infected mice displayed chronic active tenosynovitis and, in skeletal muscle tissues, necrotizing myositis. To further characterize the inflammatory response in musculoskeletal tissues during CHIKV infection, Rulli *et al.* examined the profile of proinflammatory factors in the serum of acutely infected CHIKV patients versus infected mice. Twenty-four-day-old C57BL/6 mice were infected subcutaneously with 10^4 PFU of CHIKV Réunion isolate. High levels of MCP-1, TNF–α, and IFN–γ were found in both the serum of CHIKV infected patients and the ankle and skeletal muscle tissue of CHIKV infected mice. Consistent with previous mouse models of CHIKV disease, severe inflammation was also observed in the synovial tissue and skeletal muscle of the infected mice as well as extensive disruption of striated muscle fibres. Strikingly, treatment of these mice with the indazole derivative and MCP-1 synthesis inhibitor Bindarit ameliorated disease and reduced the inflammatory cell infiltrate observed in CHIKV infected muscle and joint tissue. Bindarit-treated mice had reduced tissue damage and lower intensity of inflammatory cell recruitment through down-regulation of MCP-1. The up-regulation of proinflammatory factors in response to CHIKV infection implicates these soluble factors in the immunopathogenesis of CHIKV disease in both the mouse model and human patients. The contributory role of these proinflammatory factors in CHIKV disease presents them as a potential target for therapeutic intervention, as seen with Bindarit. Together this work also highlights the importance of the immune infiltrate and, particularly, infiltrating inflammatory monocytes to CHIKV-associated musculoskeletal disease. Mouse studies with the arthritogenic alphavirus Ross River virus (RRV) have similarly shown that this virus replicates in the muscle, joint, and bone, leading to macrophage-driven inflammation of these tissues (Chen *et al.*, 2014; Morrison *et al.*, 2006).

To develop a wild-type adult mouse model of CHIKV-induced acute rheumatic disease, C57BL/6 mice at least 6 weeks old were inoculated with 10^4 $CCID_{50}$ Réunion Island isolate or mouse adapted Asian strain CHIKV (Gardner *et al.*, 2010). Again, by moving the site of infection to the ventral side of each hind foot, clearly visible foot swelling developed that peaked 6–8 days post infection (Fig. 6.2). Peripheral viraemia lasted 4–5 days, while local titres in the feet peaked at day 1 post infection, with viable virus detectable until day 9. Virus was also found in muscle tissue, the spleen and lymph nodes; however, no virus was detected in the brain. Histological analysis of CHIKV infected feet at peak swelling revealed clear signs of arthritis and tenosynovitis with immense infiltration of mononuclear cells in and around the synovial membranes and connective tissues surrounding the tendons (Fig. 6.2). Skeletal muscle showed prodigious cellular infiltrates and necrotic muscle fibres in CHIKV inoculated mice. Elevated levels of TNF-α, MCP-1, IFN-γ, IL-6 and IFN-α/β were observed in sera of infected mice and likely have a role in macrophage recruitment/activation in addition to promoting inflammation. As expected, macrophage depletion in mice treated with macrophage specific liposome-encapsulated clodronate produced a significant reduction in foot swelling in CHIKV-infected mice. However, clodronate treatment also significantly prolonged viraemia, suggesting that macrophages are also required for the clearance of CHIKV. This adult model of acute CHIKV disease is able to reproduce the self-limiting arthritis, tenosynovitis, and myositis seen in patients infected with CHIKV, and findings correlate with data from other clinical and *in vivo* studies (Labadie *et al.*, 2010). Using the adult mouse model of acute CHIKV disease, studies have closely examined the interferon response and the role of the innate immune system during CHIKV infection (Gardner *et al.*, 2012; Rudd *et al.*, 2012). STAT1-dependent and type I IFN receptor signalling-dependent responses were shown to control CHIKV replication in muscle and joint tissue. Similarly CHIKV infection of adult mice deficient in interferon response factors 3 and 7 was lethal. Also using the established adult mouse model of acute CHIKV disease,

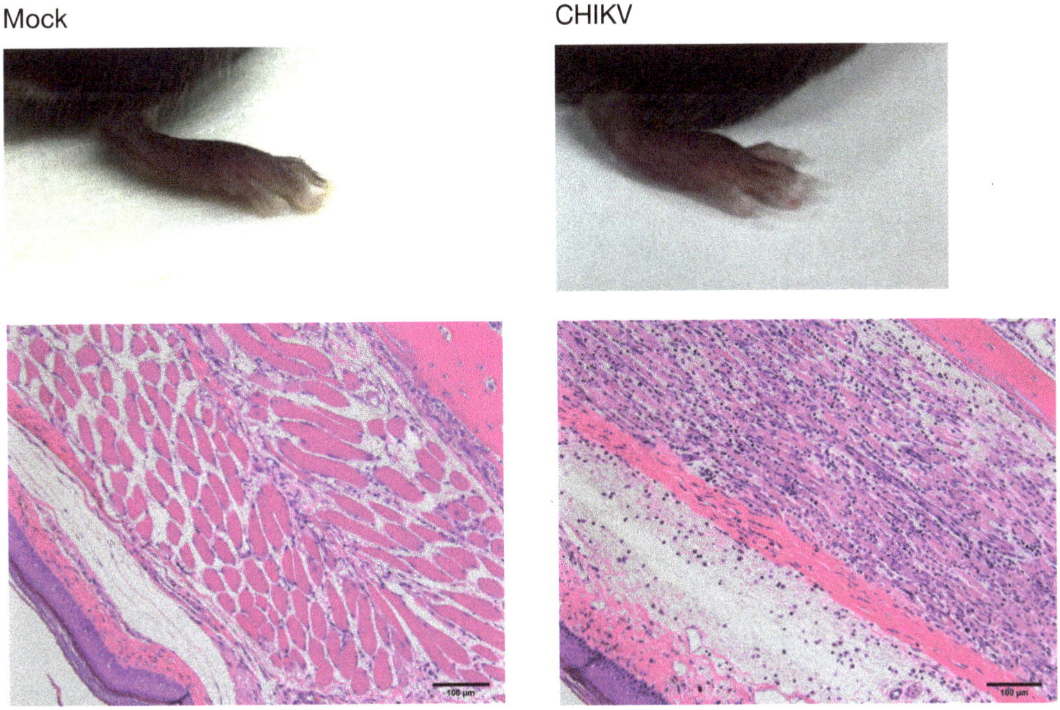

Figure 6.2 Chikungunya virus–induced inflammation. Twenty-eight-day-old C57BL/6 mice were infected subcutaneously in the footpad with 10^4 plaque-forming units CHIKV or mock infected with diluent alone. Foot swelling was monitored daily by measuring foot width and breadth. Peak inflammation was observed 3–5 days post infection at which point feet and ankles were removed, fixed in 4% paraformaldehyde, sectioned at 5 μm and stained with haematoxylin and eosin. Magnification is ×100.

studies comparing the gene expression profiles of CHIKV-induced arthritis and rheumatoid arthritis observed multiple shared inflammatory processes (Nakaya *et al.*, 2012). It was thus proposed that therapies developed for treatment of rheumatoid arthritis may find application in the treatment of CHIKV-induced arthritis and potentially other alphaviral arthritides.

Mouse models of chronic CHIKV disease

Studies suggest that over 60% of patients infected with CHIKV experience chronic symptoms of CHIKV disease (Borgherini *et al.*, 2008; Sissoko *et al.*, 2009). The prolonged symptoms include myalgia, arthralgia or arthritis, which often restrict the patient's movements and can be incapacitating. The symptoms can occur years after the initial infection. Currently no animal model is able to reproduce the chronic rheumatic manifestations associated with chikungunya and so the cause of chronic CHIKV disease remains unclear. There is however a growing body of evidence from studies performed in acute CHIKV mouse models that attempt to elucidate why a considerable number of CHIKV patients experience prolonged rheumatic pain. Using the adult mouse model of acute CHIKV disease an extensive histological examination was performed after the resolution of foot swelling (Gardner *et al.*, 2010). At day 14 post infection, as foot swelling subsided, tibial skeletal muscle, liver, spleen and lymph node histology also returned to normal. Mononuclear cell infiltrates were, however, observed in subcutaneous and peritendinous connective tissues from days 14–21. In addition, small focal infiltrates were also observed in the muscle tissue of the feet and perisynovial tissues. At days 30–40 post infection the infiltrates had generally resolved. In a separate acute model, signs of inflammation were observed in the joint tissues of CHIKV SL15649-inoculated 14-day-old

C57BL6/J mice up to 21 days after inoculation, whereas a regenerative response was observed in skeletal muscle tissue (Morrison *et al.*, 2011). In addition, CHIKV RNA was detected by qPCR in the left and right ankles at 21 days post infection. These findings indicate that while viable CHIKV is cleared from the periphery following acute infection, viral RNA, or fragments of the CHIKV genome, reside in joint tissues for at least 3 weeks post infection and this persistence of viral RNA is associated with an inflammatory infiltrate within joint tissues. It is possible that the persistence of viral RNA drives chronic inflammatory responses. Similar studies in macaques also observed long-term CHIKV infection in the joints, muscles, lymphoid organs and liver, which again may explain the long-lasting CHIKV disease symptoms observed in humans (Labadie *et al.*, 2010). In addition, this study identified the role of macrophages as a cellular sink for CHIKV during the late stages of infection *in vivo*. Further research is required to determine whether viral RNA persistence in these tissues is due to continued viral replication or evasion of host innate or adaptive immune responses. However, both acute animal models point towards viral persistence, macrophage tropism and inflammation having roles in perpetuating CHIKV disease.

Neurological mouse models of CHIKV disease

Since the 1960s, reports have described the neurological manifestations associated with CHIKV infection (Nimmannitya *et al.*, 1969). Early models of CHIKV disease examined the neurological tropism of CHIKV in mice and the pathology of the CNS upon infection. Investigating experimental encephalitides, Suckling *et al.* (1978a) used 28-day-old Swiss A_2G mice inoculated either intracerebrally or intraperitoneally with CHIKV. While mice inoculated intraperitoneally developed no clear signs of infection, mice inoculated intracerebrally presented with disease signs of ruffled fur, lethargy, paralysis and ultimately death at 6 days post infection. Interestingly, histological analysis of the brains of mice infected intracerebrally revealed a gross necrosis of the cerebrum and histopathic lesions. Lesions frequently associated with CHIKV infection included microgliosis and leptomeningeal and perivascular monocytic infiltration.

In humans, neurological complications arising during infection, such as encephalitis or encephalopathy, are considered atypical clinical manifestations of CHIKV disease and were first reported on a large scale during the Réunion Island outbreak (Borgherini *et al.*, 2007; Chandak *et al.*, 2009; Tournebize *et al.*, 2009). Although rare, neurological involvement was frequently observed in adults with underlying illnesses such as diabetes, epilepsy and hypertension. Children too presented with neurological manifestations of chikungunya such as encephalitis, febrile seizures, meningeal syndrome and acute encephalopathy (Robin *et al.*, 2008). Furthermore, the Réunion Island outbreak recorded the first CHIKV-associated fatalities. The scale of this outbreak and severity of the disease brought CHIKV back into the spotlight and again highlighted the lack of a permissive small animal model able to replicate the CHIKV-associated neuropathy found in human disease.

The neurological sequelae associated with severe cases of atypical CHIKV infection during the Réunion outbreak included encephalitis, malaise, meningoencephalitis, syndrome of meningeal irritation and cerebellar syndrome (Economopoulou *et al.*, 2009). Severe neurological signs of CHIKV disease in humans were first reported during the 2005/2006 La Réunion Island outbreak, suggesting this virus strain may have enhanced neurovirulence characteristics. A number of studies have investigated the potential for neurological involvement during CHIKV infection in mice. Neuropathy was examined in BALB/c mice intranasally infected with the Ross strain of CHIKV (Powers and Logue, 2007). CHIKV antigen was detected in neurons of the anterior olfactory nucleus in brain sections taken at day 5 post infection. At day 5, haematoxylin and eosin staining revealed signs of neuronal necrosis with contracted nuclei, cytoplasmic eosinophilia and nuclear debris. Necrotic neurons were also prevalent in the pyriform cortex. Investigations continue to examine the variation in strain-specific neurovirulence determinants of CHIKV. A murine challenge model developed to test vaccine candidates has also been used to examine the neuropathogenesis of CHIKV variant strains. Five-week-old inbred C57BL/6 mice or outbred NIH Swiss mice were infected intranasally with either the Réunion Island or neurovirulent Ross strain of CHIKV. Mice were inoculated intranasally to produce more severe disease for comparative attenuation evaluations of CHIKV vaccine candidates (Wang *et al.*, 2008). Intraperitoneal inoculation of CHIKV in mice yielded poor viraemia and mice remained free of disease signs. Intranasal inoculation of both strains produced viraemia lasting 2–3 days. Infection of C57BL/6 mice with the Ross strain of CHIKV produced detectable signs of disease. Disease manifestations included ruffled fur, hunched posture, inactivity and death at day 7–8 post infection. High viral titres were recovered from the brain on days 3 and 7 post infection. Brain pathology as early as day 3 post infection revealed mononuclear cell infiltration and diffuse areas of neuronal degeneration in CHIKV-infected mice. On day 7 post infection multifocal inflammation and liquefactive necrosis was observed in the cerebral cortex and hippocampus. Perivascular cuffs were located diffusely throughout the cerebral cortex largely composed of lymphocytic cells and severe spongiform changes were found in the superficial cerebral cortex.

Mouse models of Semliki Forest virus-induced inflammatory disease

Semliki Forest virus (SFV) is an 'Old World' alphavirus of the Semliki Forest complex transmitted by several mosquito species of the *Aedes* genus. To date there have been very few clinical symptoms reported, with most human SFV infections being asymptomatic. See Chapter 8 for further description of human SFV disease.

Early development of a mouse model to study Semliki Forest virus

Semliki Forest virus (SFV) was first isolated from mosquitoes in Uganda in 1942 (Smithburn and Haddow, 1944). Initial studies into the pathogenic properties of SFV in mice demonstrated that intracerebral inoculation of adult mice with SFV caused a rapidly fatal encephalitis. Using the original virus isolate, Seamer *et al.* (1967) examined the progression of disease in infected mice of various ages. A high mortality rate was observed in all mice inoculated either intracerebrally or peripherally with the L10 strain of SFV. However, prolonged survival was observed in older mice. Age-related resistance to infection is a known feature of a number of mouse models of alphaviral disease. Disease signs in infected mice prior to death included paralysis. The course of infection was followed in 25-day-old mice inoculated with $10^{1.6}$ PFU L10 in the footpad. Inoculated mice showed signs of weight loss by day 2 post infection and paralysis/death was observed between days 4–7 post infection. Peak viraemia was observed on day 2 post infection and productive viral replication was found in the spleen, liver and brain. Histopathology revealed normal spleen and liver tissues; however, multiple lesions were found throughout the CNS. Cellular infiltration was seen in the brain beginning on day 3 post infection. Inflammatory lesions and neuronal degeneration were observed in infected mice, which developed into a generalized encephalitis prior to death. Signs of myelitis were also detected in infected mice. This early work demonstrated the neurovirulence of SFV and particularly the L10 strain.

A new strain of SFV, A7, was isolated from mosquitoes in Mozambique in 1961 (McIntosh *et al.*, 1961). Although A7 showed no difference in replication kinetics in BHK cells compared to the L10 strain, A7 was less virulent than L10 *in vivo*. The replication kinetics during the early stages of infection in weanling mice were also identical for both strains (Bradish *et al.*, 1971; Pusztai *et al.*, 1971). There were however differences in the extent of viral replication in the brains of infected mice between virulent and avirulent strains of SFV. Whereas virulent strain titres would continue to rise in the brain until death, avirulent strains replicated in the brain at later times post infection and titres were found to decline. The difference in virulence between strains of SFV is now thought to be due in part to the competency of the strains to replicate in neuronal cells (Gates *et al.*, 1985). *In vivo* virulent strains, able to replicate quickly and to high titres in neuronal cells, cause rapid and extensive damage to neurons in the CNS of infected mice (Balluz *et al.*, 1993). Generally, infection of a developing CNS with

SFV is highly destructive and ultimately fatal in mice. For example, neonate mice inoculated with the avirulent A7 strain cannot survive infection. However, infection of mice over 17 days old with avirulent A7 strain is largely non-lethal. In spite of this, older mice develop neurological disease pathology over the course of infection including virus-induced demyelination in the CNS. This has allowed the development of mouse models to accurately simulate the clinical symptoms associated with SFV infection in humans by using morbidity as a measure of disease rather than mortality. Studies using animal models of SFV disease have thus sought to understand the immunopathogenesis and virus–host interactions responsible for viral-induced encephalitis. *In vivo* studies of SFV disease will also advance the understanding of the pathogenesis of closely related encephalitic New World alphaviruses, as well as other viral encephalitides.

Clinical studies of virus-induced encephalitis and demyelinating diseases are restricted by the inaccessibility of the CNS and ethics of human examination. Therefore mouse models of SFV disease provide a platform, not only to investigate the pathogenesis of encephalitic viruses, but also the mechanisms that drive associated neuropathy in other demyelinating diseases. An inflammatory disease of the CNS, the demyelination associated with multiple sclerosis has been likened to the neuronal sequela of SFV infected mice. While studying the extent of demyelination in SFV infected mice, Fleming *et al.* (1982) and Illavia *et al.* (1982) noted a high degree of demyelination of the optic nerve. Tremain and Ikeda (1983) attempted to model these observations using electrophysiological recordings on infected mice to measure the visual latencies of both single cell and field potential responses from the superior colliculus. Infecting Swiss A^2G mice at 4 and 6 weeks of age, visual latencies of both single cell and field potential responses of infected mice were found to be longer than those of the control mice. In addition, the collicular neurons of infected mice revealed a loss of sensitivity to light and a significant reduction in flicker fusion frequency. With the same flash stimuli as in the mouse experiments, visual evoked potentials were recorded from patients with the demyelinating disease multiple sclerosis. As with SFV infected mice, patients showed a significant increase in peak latency of the visual evoked potential and a significantly reduced flicker fusion frequency (Tremain and Ikeda, 1983). Therefore, this mouse model of SFV-induced demyelination also has application in the study of non-viral demyelinating diseases such as multiple sclerosis.

Mouse models of Semliki Forest virus encephalomyelitis

The relatively early isolation of SFV has made it one of the most comprehensively studied alphaviruses. With demyelination of the CNS and neurovirulence found to be hallmarks of SFV infection, a number of studies have sought to further characterize the neuropathogenesis of SFV in mice. Using largely histopathological analyses, a number of studies traced the course of infection in the mouse brain after inoculation with SFV (Chew-Lim, 1975; Zlotnik *et al.*, 1972). The pathology of CNS inflammation during infection was analysed in 3–4-week-old mice infected with avirulent SFV. Encephalitis peaked in these mice at day 18 post infection and mice had recovered by day 42. At peak disease, lesions were observed in the medullary substance of the cerebellum including perivascular cuffing of lymphocytes. Similar studies in mice using an avirulent strain of SFV revealed that, in addition to neuronal degeneration and cellular infiltration, focal spongiform lesions were found throughout the CNS during infection (Mackenzie *et al.*, 1978). These lesions were associated with peripheral astrocytic hypertrophy.

To better understand the development of SFV-induced demyelination of the cerebellar white matter, the inflammatory pathology associated with SFV infection was examined using several mouse strains (Suckling *et al.*, 1978b). The virus used was an avirulent strain of SFV derived from the A7 strain. Studies examined the pathogenesis of SFV in 4- to 7-week-old random outbred Swiss/A_2G mice, outbred BSVS, inbred BALB/c, CBA and SJL/J mice inoculated intraperitoneally with 10^4 suckling mouse intracerebral LD_{50} SFV. SJL/J mice were much less prone to developing the neuropathy associated with SFV infection. Although infected mice rarely developed clinical signs of disease a number of mouse strains exhibited a similar incidence of CNS lesions. Typically mild encephalitis and focal microcystic spongiform lesions were found with a loss of myelin within the white matter of the cerebellum, brain stem and spinal cord. By altering parameters of the mouse model Suckling *et al.* observed changes to the type and severity of pathology produced during SFV infection. The age of mice greatly affected disease outcome, with mice older than 10 weeks at inoculation showing fewer and less severe pathological lesions. Also, varying the method of viral passage was found to alter SFV disease pathology in mice. For example, if passaged twice in baby mice and an infected brain suspension then used as inoculum, SFV-induced lesions were characterized by increased perivascular infiltration and a reduction in microcystic vacuolation.

The exact cause of SFV-induced myelin loss was for a number of years a contentious topic, with concurrent studies often reporting conflicting data. In other models of murine encephalomyelitis, virus-induced demyelination was largely accredited to either one of two separate biological pathways. The broad division was made between viral infections where the cytotoxic effect of viral replication was responsible for destruction of the host cell and loss of myelin and immunopathological demyelination in which the mounted immune response to infection induces demyelination. Studies aimed at elucidating the mechanism of SFV-induced demyelination regularly used immune-deficient or immune-compromised mice to analyse the contribution of the immune system to myelin loss. Early studies achieved immunosuppression in mice by whole body irradiation. Compared to controls, SFV infection of irradiated mice resulted in increased SFV titres in the brain and multiple foci of demyelination throughout the brain (Chew-Lim *et al.*, 1977). Equally, in separate studies multiple foci of demyelination were observed in the cerebellum of infected nude (T-lymphocyte-deficient) mice compared to single lesions in infected immune-competent mice (Chew-Lim, 1979). An increased number of demyelinative plaques were also found in the brains of infected nude mice along with an elongated period of replication in brain tissue. Together these data, from studies using immune-compromised mice, suggested that SFV-induced demyelination during infection was due to direct viral cytotoxicity of the CNS rather than an immune-mediated mechanism. In parallel studies however, histopathological changes to the brain of SFV-infected mice were found to occur subsequent to the appearance of infiltrating immune cells (Mackenzie *et al.*, 1978). Furthermore, large bodies of subsequent data have emphasized the fundamental role of immune mediators in SFV-induced demyelination during acute infection. In contrast to previous work, multiple studies observed a complete absence or decreased incidence of demyelination in SFV-infected nude mice (Fazakerley *et al.*, 1983; Gates *et al.*, 1984; Jagelman *et al.*, 1978). More recent studies have also demonstrated the importance of CD8+ T cells, CNS adhesion molecules and the antibody response in the development of demyelination during SFV infection (Smith *et al.*, 2000; Subak-Sharpe *et al.*, 1993). In experimental allergic encephalomyelitis (EAE), a non-viral model of CNS demyelination, demyelination is stimulated by injection of myelin plus adjuvants. It was therefore proposed that release of myelin in SFV-infected oligodendrocytes may occur as a result of cytotoxicity or incorporation of myelin into the budding virus particle. Autoimmunity, driven by host antigen presentation on the viral envelope, may then stimulate demyelination in a manner similar to EAE (Webb and Fazakerley, 1984; Webb *et al.*, 1984). Although further studies are required to elucidate the exact mechanisms and interactions that are responsible for SFV-induced demyelination, it is generally accepted that immune mediators play an important role in the development of inflammation and demyelination in the CNS following SFV infection.

Mouse models of SFV encephalomyelitis have since been used to identify novel therapeutic targets and test effective treatments for viral infections that cause encephalitis. A lethal model of SFV infection in mice was used to analyse the therapeutic value of anti-SFV hyperimmune serum and interferon-inducing poly I:CLC (Coppenhaver *et al.*, 1995). Weanling outbred ICR mice were intracranially inoculated with SFV, causing lethal encephalitis. Survival, viraemia and brain viral titres were measured as outcomes of disease.

Although the hyperimmune serum and poly I:CLC provided no therapeutic protection when administered separately, as a combination therapy hyperimmune serum and poly I:CLC effectively protected mice from alphaviral infection of the CNS. Interestingly, combination therapy was also effective when given after the virus was replicating in the target organ. Mice receiving the combination therapy had a 50% survival rate. No viraemia was detectable following treatment and brain viral load was reduced over the course of infection.

The mouse model established by Fazakerley *et al.* (1993) has the advantage over a number of other models of SFV encephalomyelitis that the strain of SFV used is efficiently neuroinvasive. Using the avirulent A7(74) strain of SFV, although neuroinvasive in neonate mice, A7(74) was unable to disseminate from the original perivascular foci in the brains of 3- to 4-week-old mice. It is suggested that the restriction of A7(74) replication in mouse neuronal cells is likely a reflection of age-related changes in the neurons (Fazakerley *et al.*, 1993). However, efficient neurovirulence of SFV strain A7(74) was achieved by extraneural inoculation of the virus. Infection with neurovirulent A7(74) produces a range of disease signs in mice from clinically unapparent mild encephalitis to fatal panencephalitis. A7(74) is able to cross the blood–brain barrier and initiate perivascular foci of infection in neurons and oligodendrocytes. Moreover, the model of infection produces lesions of inflammatory demyelination (Fazakerley, 2004). A great advantage of the method of inoculation used in this mouse model of SFV disease is that the blood–brain barrier is left undisturbed. This provides a more accurate representation of the route of infection in human disease, which is often overlooked in models that use intracerebral inoculation. Using this model Fragkoudis *et al.* (2008) investigated the role of antibodies in clearance of infectious virus from the brain and generation of lesions of demyelination. For this study, antibody-deficient μMT mice were inoculated intraperitoneally with 5000 PFU of the avirulent A7(74) strain of SFV. Over the course of infection live virus was detectable for 8 weeks in μMT mice brain tissue and systemically. In contrast, from 2 weeks post infection, systemic infectious virus and live virus in the brains of wild-type mice were below the limit of detection. Interestingly, the brains of mice from both strains showed microcystic changes, perivascular mononuclear cell inflammatory infiltrates and lesions of demyelination. Fragkoudis *et al.* (2008) was able to show that, during alphaviral infection, antibodies are required to clear infectious virus from the blood and the CNS. Previous work had established the role of CD8+ T cells in alphavirus-induced myelin loss; however, the role of antibodies in demyelination remained unclear (Subak-Sharpe *et al.*, 1993). Although required for viral clearance, antibodies were found to be non-essential to the formation of lesions of demyelination during SFV infection. Although studies have proposed anti-myelin antibodies also contribute to myelin loss, current evidence suggests that T-cell responses alone are sufficient to induce demyelination (Mokhtarian *et al.*, 2003).

The SFV mouse model developed over a number of years by Mokhtarian and Swoveland (1987) currently uses a scale of disease score to measure the observable clinical manifestations of SFV disease. Female C57BL/6 mice aged 5–6 weeks old are inoculated intraperitoneally with 10^4 PFU A7(74) SFV strain and observed daily for the development of disease signs. Infected mice are scored for disease signs and hind-limb dysfunction on a scale of 0–6, where 6 is the most severe form of disease. Disease score reflects the developing severity of paralysis in the infected mouse based on: 0, no disease signs, up to 6, paralysis of hind-limbs or complete quadriplegia. Peak disease in this model is day 7 post infection, which coincides with high viral titres in brain tissue and CNS inflammation in infected mice. Mice develop a transient encephalomyelitis from which they recover. Demyelination is observed in the cerebellum along with white matter vacuolation and macrophage foci throughout the CNS (Mokhtarian *et al.*, 1994; Mokhtarian and Swoveland, 1987). A number of recent studies have used this model of SFV demyelinating disease together with knockout mouse strains to further dissect the immune mechanisms responsible for the pathogenesis of demyelinating lesions. B-cell-deficient C57BL/6 mice developed moderate to severe clinical signs ~14 days later than control mice, and no white matter vacuolation was observed in the brains of B-cell-deficient mice (Smith-Norowitz *et al.*, 2000). Results suggest that brain-infiltrating B cells and antibodies contribute to SFV encephalomyelitis in mice. Primary demyelination occurs by day 21 post infection and remyelination by day 35 following recovery from SFV encephalomyelitis. Using $\gamma\delta$ T cell KO mice, Safavi *et al.* (2011) used this model to demonstrate the role of antibodies in remyelination and repair of the CNS. In particular serum antibodies to SFV E2 T_h peptide$_2$ enhanced recovery and in KO mice immunized with E2 T_h peptide$_2$, elevated antibody production to this peptide resulted in early remyelination. The potential to use E2 T_h peptide$_2$ for the treatment of other demyelinating diseases, such as MS, has also been examined in the EAE model (Mokhtarian *et al.*, 2012). Histopathological and immunohistochemical studies demonstrated increased remyelination and a higher number of activated oligodendrocytes and astrocytes in treated compared to untreated EAE groups.

Mouse models of Sindbis virus-induced inflammatory disease

Sindbis virus (SINV) is classified as an 'Old World' alphavirus despite grouping as part of the Western equine encephalitis complex. Mosquitoes predominantly of the *Culex* species transmit SINV and the human symptoms of infection are predominantly arthralgia, rash and malaise.

Development of a mouse model to study Sindbis virus

Sindbis virus (SINV) was isolated from mosquitoes in 1952 and takes its name from the village on the Nile Delta, Egypt, where the mosquitoes were collected. SINV was initially described as a 'Coxsackie-like' virus due to both its ability to cause a fatal infection in infant mice but not in adult mice and its myositic disease manifestations. However, with characteristics unfamiliar to the Coxsackie group, SINV was thought to be a virus as yet undescribed. Neonate mice were particularly susceptible to SINV infection, with infection resulting in fatal encephalitis following peripheral inoculation by various routes. Initial investigations therefore focused on the neuroinvasive nature of SINV and regularly used neonate mouse models (Johnson, 1965). Following inoculation virus was found to replicate in skeletal and cardiac muscle and in fibroblast-rich tissues. A high plasma viraemia also developed rapidly, thought to be fuelled by replication in endothelial cells. Viraemia peaked at day 2 post infection and declined prior to death on day 4–5. SINV antigen was found in the endothelium of cerebral vessels before detection in the brain, indicating SINV spread to the CNS from the blood. Once in the CNS, SINV spread rapidly within cerebral parenchyma, infecting both neural and glial cells. As with other mouse models of alphavirus disease, susceptibility to SINV infection is age dependent, with a threshold level of viral replication broken in the brain of newborn verses weanling mice (Reinarz *et al.*, 1971).

Neuroadapted SINV (NSV) was derived from a less virulent strain of SINV, AR339, by serial passage between neonatal and adult BALB/c mice. When infected intranasally, C57BL/6 mice developed fatal encephalitis in all infected adults, although they survived longer as they aged (Thach *et al.*, 2000). Upon infection C57BL/6 mice lost up to 30% of their body mass and died 7–9 days post infection. In contrast, BALB/cBy mice inoculated after 4–5 weeks of age no longer developed fatal disease. Only ruffled fur and mild paralysis was observed in infected BALB/cBy mice. SINV antigen was found in abundance throughout the brain tissue of C57BL/6 mice and to a lesser degree in BALB/cBy mice. Infectious virus increased rapidly in the CNS of C57BL/6 mice and reached higher levels than in BALB/cBy mice. C57BL/6 mice also developed more severe disease signs than BALB/cBy mice, with severe disease typified by kyphoscoliosis, hind-limb paralysis, and muscular atrophy. Interestingly, although infectious virus was cleared faster in C57BL/6 mice, viral RNA was cleared more slowly than in BALB/cBy mice. These data and additional studies demonstrate the differences in mouse strain specific determinants of resistance or susceptibility to fatal encephalitis during SINV infection (Thach *et al.*, 2001). Susceptibility to fatal NSV encephalomyelitis is determined in part by the efficiency of viral replication/RNA load and spread in the central nervous system.

After intranasal inoculation, NSV infected olfactory neurons and spread caudally to the neural pathways of the spinal cord. As with SFV and other viral encephalitides the intranasal route of SINV inoculation is preferable to intracranial infection in which trauma complicates interpretation of these data. Intracranial inoculation of adult animals may also lead to some instances of mortality related to trauma rather than virus infection. Also using the intranasal route of inoculation, lethal mouse models of SINV acute encephalomyelitis have been used to test the efficiency of novel therapeutic strategies (Veckenstedt *et al.*, 1985). With survival as a measure of disease outcome, 4–6–week-old ABD2F1 mice were infected intranasally with 10 LD50 neuroadapted SINV strain AR86. Infection resulted in acute encephalomyelitis with lesions of the CNS, lymphoid organs kidneys and liver and death 3–7 days post infection. Disease signs such as ruffled fur, arched back and paralysis of one or more limbs were also evident prior to death at approximately day 4 post infection. Oral administration of the anticholinergic anti-Parkinson drug Norakin was shown to protect SINV infected animals from death in a dose-dependent manner. The maximum protective effect was achieved when the drug was administered twice daily before, but not after, virus inoculation.

Since the early 1990s, researchers have been using *in vivo* imaging to detect the tissue tropism of microorganisms in living animals. Using recombinant viruses able to express reporter genes it is possible, using *in vivo* imaging, to track viral replication kinetics in real time and identify novel sites of replication that may have otherwise been missed. Studies using this technique with SINV have aimed to identify the determinants of viral virulence and host susceptibility in mice (Cook and Griffin, 2003). Virulent and avirulent strains of SINV together with recombinant chimeric viruses were engineered to express the reporter gene luciferase. *In vivo* imaging was used to monitor the location and extent of virus replication in susceptible and resistant mice. Four-week-old albino C57BL/6-Tyr$^{c\text{-}2J}$ mice and BALB/cAn-NCrlBR mice were inoculated subcutaneously in the hind foot with recombinant virus. Imaging after peripheral inoculation saw significant replication in the lower spinal cord, nose, and brain of mice as well as diffuse light from the abdominal region. Although still unclear, data suggest route of entry into the CNS following peripheral inoculation of SINV by retrograde axonal transport from neurons at the initial site of replication or from the olfactory epithelium after viraemic spread. Replication in either the nose or the lower spinal cord was also shown to precede detection of virus replication in the brain. The amount of light detected by *in vivo* imaging of the brains of susceptible mice was consistently higher than the amount of light detected from the brains of resistant mice. This suggests that the mouse model may be used to phenotype the susceptibilities of different mouse strains to infection. The system could also distinguish between virulent and avirulent strains of virus with the limit of detection approximately 10^3 PFU/g. The benefits of this model over other mouse models of SINV infection include the ability to monitor viral infection over time in a single animal and to link virus replication to SINV disease development in real time.

Mouse models of Eastern equine encephalitis virus disease

Eastern equine encephalitis virus (EEEV) is classified as part of the EEE serological complex and was first described in 1933 in infected horses in Delaware, Maryland and Virginia (Giltner and Shahan, 1933). In humans, a disease similar to that seen in equines develops with headache, fever, seizures and coma (Deresiewicz *et al.*, 1997). *Culiseta melanura* mosquitoes maintain virus spread within the bird population with *Aedes, Coquillettidia,* while *Culex* species are the predominant mosquitoes that carry the virus to humans (Weaver *et al.*, 1991).

Development of a mouse model to study Eastern equine encephalitis virus

The focus of initial mouse studies on EEEV infection was to expand the existing understanding of disease mechanisms and host response. In these experiments a lethal mouse model was used to test various immunization strategies (Casals and Palacios, 1941; King, 1940a,b; Morgan, 1941; Olitsky and Cox, 1936; Olitsky and Harford, 1938).

Initial studies of EEEV histopathology in inbred white mice by King (1940b) showed signs of neuroinflammation and neurodegeneration upon infection by either the intraperitoneal or subcutaneous routes. Hisopathological analysis showed vacuolation and neurodegeneration were also key features in infected animals (King, 1940b). Morgan (1941) examined the effect of age on the susceptibility to EEEV infection using a mouse-adapted strain and Rockefeller albino mice (of Swiss origin) (Morgan, 1941). Animals of 14 days, 22 days, 1 month or 6 months of age were infected either intraperitoneally or intracranially, with results showing that the average survival rates increased with age.

Several decades later, Murphy and Whitfield (1970) inoculated day-old Swiss mice either intraperitoneally or intracranially with EEEV, observing encephalitis as early as 48 hours post infection, with animals moribund by 60 hours. Clinical and histopathological observations in this model included intercellular and perivascular oedema and cytoplasmic swelling of neurons as early as 24 hours post infection, with disease pathogenesis ultimately leading to significant neuronal loss.

More recent studies on EEEV neurovirulence have demonstrated infant NIH Swiss mice to be a good model as the mice usually succumb to intracranial infection of wild-type EEEV between days 3 and 4 post infection (Wang *et al.*, 2007). Alternatively, when the mice are infected intranasally or intraperitoneally with the EEEV strain-FL93–939, they develop clinical signs of encephalitis by day 3–4 post infection and die by days 4–5 (Wang *et al.*, 2007). High viral titres were found in the brains of these adult mice and also in the serum, heart, liver, spleen and kidneys.

To further examine EEEV disease pathogenesis, 5-week-old C57BL/6 mice were inoculated subcutaneously in the footpad with North American strain FL91–4679 (Vogel *et al.*, 2005). Observed clinical signs included lethargy and ruffled fur and by day 4 post infection mice reached humane end points, being hunched and showing signs of tremors. Peak viraemia occurred by 12 hours post infection and was still detectable at day 4 at the experimental end point. EEEV replication was noted in fibroblasts, macrophages and dendritic cells and was also detected in bone tissue of the nasal turbinates, hard palate, and skull. After 24 hours, virus-positive neurons were detected in the CNS. Infected areas included the cerebellum, lateral olfactory tract, and the cerebral and secondary motor cortexes. By day 4 post infection widespread necrosis of neurons was detected with infected glial cells found throughout most regions of the brain (Vogel *et al.*, 2005).

Vaccine candidate efficacy studies on EEEV have used various mouse strains as challenge models. Most recently, a novel pan-alphavirus vaccine was tested for cross-protection against several new world alphaviruses in CD-1 mice. Mice were 100% protected against WEEV challenge and showed significantly greater survival rates with EEEV challenge, irrespective of the route of infection (Phillips *et al.*, 2014).

Mouse models of Western equine encephalitis virus disease

Western equine encephalitis virus (WEEV) is classified as part of the WEEV serological complex and is thought to be a recombinant virus constituting two other alphaviruses, an ancestral Sindbis virus-like virus, and an ancestral Eastern equine encephalitis virus-like virus (Hahn *et al.*, 1988). Mosquitoes of both the *Culex* and *Culiseta* genera transmit WEEV, with *Culex tarsalis* being the most common human vector. Human cases of WEEV encephalitis are rare, with the majority of infections being asymptomatic.

Mouse models to study Western equine encephalitis virus neurovirulence

Initial studies on WEEV disease used young Swiss mice and virus recovered from the brain of a child who died of encephalitis following the 1938 WEEV epidemic in California (Howitt, 1938). The majority of inoculated mice died between 3 and 5 days post infection with signs of CNS disease including tremors and hyperexcitability. Further development of the Swiss mouse (Rockefeller strain) model of WEEV infection was used to identify WEEV associated pathologies in viruses not originally passaged in mouse brains.

Non-neuroadapted WEEV caused mortality and brain lesions in 1- to 2-day-old mice infected subcutaneously, with necrosis and inflammation in connective tissues such as skeletal and smooth muscle, cartilage, and bone marrow. Twenty-one-day-old mice generally exhibited diffuse necrotizing encephalitis, pathologies not seen in younger animals, while the CNS of older mice was largely free of lesions (Aguilar, 1970). Adult Swiss mice infected subcutaneously were used to assess the potential for WEEV to cause myocarditis (Monath *et al.*, 1978). High viral titres were observed in cardiac muscle, and mortality rates reached 40% with all animals developing illness (ruffled fur, hunched posture, fever). Mice that did not recover died between 7 and 14 days post infection (Monath *et al.*, 1978). Both suckling and adult outbred Swiss albino mice were used in a study assessing the virulence markers of several WEEV strains (Bianchi *et al.*, 1993). Results correlated well with the disease severity of these WEEV strains reported in humans and horses, thus supporting the validity of the mouse model and its importance for future studies.

CD-1 and BALB/c mice have also been used to characterize virulence differences of WEEV strains (Logue *et al.*, 2009; Nagata *et al.*, 2006). After intracerebral inoculation of 10^3 PFU in CD-1 mice, the prototype McMillan strain was characterized as neurovirulent, with the mean time to death being 1.9 days. In contrast, the IMP181 strain, which was isolated from a *C. tarsalis* mosquito in California, caused no mortality. Both viruses, however, were shown to be neuroinvasive in this study, with the McMillan strain exhibiting significantly higher titres in the brain (Logue *et al.*, 2009). Logue *et al.* (2009) also examined neurovirulence in CD-1 mice in response to viral challenge with several routes of inoculation including aerosol, intranasal, intravenous, and subcutaneous. Of the six strains examined in this study, only IMP181 was associated with a consistent 0% mortality. Logue *et al.* (2009) examined the course of infection for the low virulence (IMP181) and high virulence (McMillan) isolates after subcutaneous inoculation, finding that in both isolates the brain contained the highest viral titres. The McMillan strain exhibited neural lesions in all infected mice examined that were progressively more severe the closer to mean time of death, with neural necrosis and oedema mostly localized to the forebrain. Interestingly, two-thirds of the mice infected with IMP181 showed unremarkable brain histopathology with the remaining one-third exhibiting a pattern similar to, but much less severe than, McMillan infected animals (Logue *et al.*, 2009). In the study by Nagata *et al.* (2006) BALB/c mice were used in a virulence study of WEEV strains, specifically focused on the intranasal route of infection. In a similar finding, this study reported that WEEV stains could be divided into two groups, high and low virulence, and suggested that variation in infectivity can be attributed to genetic diversity in the structural proteins among the strains (Nagata *et al.*, 2006). Overall, these findings suggest that WEEV has strain-specific neurovirulence determinants and that understanding WEEV strain-specific pathogenicity may identify novel targets for therapeutic intervention and aid vaccine development.

Mouse models of Venezuelan equine encephalitis virus disease

Venezuelan equine encephalitis virus (VEEV) is classified as part of the VEE serological complex, of which there are six different subtypes. VEEV can affect all equine species, including donkeys and zebras. Equine infection can present as a progressive central nervous system disorder, which often results in death. Human infection results in flu-like symptoms; however, in immune-compromised individuals infection can be fatal. The main epizootic mosquito vector of VEEV is *Aedes taeniorhynchus* (Smith *et al.*, 2008) while the enzootic VEEV strains utilize vectors of the *Culex* (*Melanoconion*) subgenus (Weaver *et al.*, 2004).

Development of a mouse model to study Venezuelan equine encephalitis virus

Since the first isolation of VEEV, mice have been used to study VEEV pathogenesis, neurovirulence, neuroinvasion and vaccine development (Beck and Wyckoff, 1938). In these early studies, mice were found to develop paralysis and succumb to infection between 3 and 5 days post infection following intracerebral inoculation (Beck and Wyckoff, 1938). Subsequent mouse models of VEEV disease have been found to develop a very similar course of disease, with high mortality as a common feature. Overall the mouse model of infection results in the onset of a biphasic disease in mice that is similar to that of VEEV-infected horses (Gleiser *et al.*, 1962).

Early studies using Swiss Webster (CFW) suckling mice sought to examine the viraemic spread and tissue tropism of VEEV. Kundin *et al.* (1966) showed that all mice infected subcutaneously or intracranially succumbed to infection after extensive CNS damage. Throughout the course of infection, viral antigen was readily detected in several areas including nerve cells, spinal cord, ganglia, leptomeninges and tooth buds (Kundin *et al.*, 1966).

Similarly, later studies used adult wild-type C57BL/6 mice to further characterize CNS involvement in VEEV pathogenesis (Jackson *et al.*, 1991). In this model, C57BL/6 mice were injected intraperitoneally and developed clinical signs of hunched posture by days 3–4 post infection, progressing to bilateral hind-limb paralysis by day 6 followed by death. VEEV antigen was first detected in neurons of the cerebral cortex and spread to the neurons and glial cells of the hippocampus, thalamus, cerebellum basal ganglia, brainstem and grey matter of the spinal cord (Jackson *et al.*, 1991).

Five-week-old CD-1 mice were also used to investigate the neuroinvasive mechanism of VEEV. Charles *et al.* (1995) identified two distinct routes by which VEEV enters the CNS: the olfactory system and trigeminal nerves. Ryzhikov *et al.* (1995) also used the outbred CD-1 mouse strain to further demonstrate the importance of the olfactory tract in neuroinvasion upon respiratory challenge.

BALB/c mice have been used for a number of different studies on VEEV. Dal Canto and Rabinowitz (1981) infected BALB/c mice, nude mice (homozygotes) and heterozygous nude mice subcutaneously and examined pathological changes in the brain. The BALB/c controls and heterozygous nude mice developed lethargy and hind-limb paralysis, succumbing to infection by day 12 post infection. The nude mice showed a 2-day delay in the development of clinical signs and succumbed between days 12 and 15 post infection. Demyelinating lesions in the spinal cords of the BALB/c mice were reported at day 9 post infection. This was the first report of an arbovirus inducing demyelination (Dal Canto and Rabinowitz, 1981). Steele *et al.* compared the neurovirulence and tissue tropism between wild-type and attenuated strains of VEEV using BALB/c and C3H/HeN mouse strains. Results showed that C3H/HeN are more susceptible than BALB/c to VEEV infection regardless of the strain used. The authors also determined that VEEV tropism is strain specific and that neuroinvasion does not always correlate with neurovirulence (Steele *et al.*, 1998).

C3H/HeN mice are particularly sensitive to VEEV infection, displaying clinical features including lethargy, hunched posture, ruffled fur and weight loss (Hart *et al.*, 1997; Julander *et al.*, 2008; Ludwig *et al.*, 2001). This model also exhibits a similar CNS disease to that seen in other animal models, including the development of meningoencephalitis, gliosis, and multifocal neuropil vacuolation (Julander *et al.*, 2008). Furthermore, this model has been used frequently to test the efficacy of new therapeutic strategies (Kuehne *et al.*, 1977) and has been chosen recently as a model to examine novel anti-VEEV compounds that specifically target the nsP2 protein (Chung *et al.*, 2014).

In VEEV vaccine studies, 4- to 12-week-old BALB/c mice are perhaps the most commonly used mouse model. A diverse range of vaccination strategies have been investigated along with a variety of vaccine candidates including replicons, adenovirus-based vaccine and DNA-based vaccines (Schafer *et al.*, 2009; Tretyakova *et al.*, 2013; Williams *et al.*, 2009). CD-1 mice have also been used, particularly to analyse immune responses to vaccine candidates. Both live-attenuated recombinant vaccine candidates (Charles *et al.*, 1997) and second-generation IRES-based vaccine candidates (Rossi *et al.*, 2013) have been assessed in CD-1 mice. Overall, the mouse model of infection has been crucial in identifying viral determinants of pathogenesis and elucidating the host response to VEEV.

Non-human primate models of alphavirus disease

Non-human primate models of chikungunya virus

Chikungunya virus infects non-human primates in the wild

Shortly after the initial description of CHIKV, it was demonstrated that non-human primates (NHP), specifically African Green Monkeys (*Cercopithecus aethiops*) and Chacma baboons (*Papio ursinus*), in South-East Africa are targeted by the mosquito vector and are hosts of the virus (McIntosh, 1970; McIntosh *et al.*, 1964, 1965). Field studies demonstrated that CHIKV-specific antibodies may persist for more than 4 years within *C. aethiops* (McIntosh, 1970), with a subsequent study showing a temporal relationship between prevalence of CHIKV-specific antibodies in African Green Monkeys and a human outbreak (Kaschula *et al.*, 1978). The high susceptibility of NHP to arboviruses was used by researchers, in parallel to mosquito sampling, to identify transmitted arboviruses in various areas from South-African Republic to Central African Republic, but these studies were not able to provide data on the infectious process or the clinical signs associated with CHIKV infection in NHP (McIntosh, 1970; Saluzzo *et al.*, 1981, 1982). Interestingly, in the early 1960s when chimpanzees (*Pan troglodytes*) were numerous it was found that they, like the human population, are susceptible to CHIKV infection, contributing to the expanse of CHIKV epidemics (Osterrieth and Blanes-Ridaura, 1960; Osterrieth *et al.*, 1960, 1961). Conversely, during an outbreak on the Asian continent in the 1960s only a few Asian monkeys [5.4% (95% confidence interval (CI) 0.5–10%) of tested *Macaca mulatta*] were found positive for CHIKV. A 1967 retrospective study of alphavirus-specific antibodies in six NHP species imported into the USA found that while anti-CHIKV prevalence was high (65% CI 48–83%) within the chimpanzees imported from Africa it was rare (5.5%) in the Asian Rhesus macaque (*Macaca mulatta*) (Harrison *et al.*, 1967). This was the first indication of the difference in epidemic transmission cycles: sylvatic in Africa and anthropophilic in Asia during the 1960s. Finally, after decades of serological studies, it is now understood that NHP are hosts of CHIKV both in Africa and in South-East Asia, but with large variation depending on the NHP species and geographic area studied (Apandi *et al.*, 2010; Inoue *et al.*, 2003; Kirya and Okia, 1977; Marchette *et al.*, 1978; McCrae *et al.*, 1971; Peiris *et al.*, 1993). Additionally, outbreaks have been reported to occur in monkeys, with evidence of viraemia but no pronounced physical manifestations observed in the wild (Binn *et al.*, 1967; Diallo *et al.*, 1999; Jupp *et al.*, 1981). Overall, the role of NHP in the transmission of CHIKV is still a matter of ongoing debate; however, the increasing number of serological studies showing high seroprevalence of CHIKV-specific antibodies indicates that NHP may act as both a reservoir and an amplifying host in an outbreak (Apandi *et al.*, 2010; Weaver *et al.*, 2012).

Early experimental non-human primate models of CHIKV infection

Following the 1965 CHIKV epidemic in India, the first experimental CHIKV infection was performed in rhesus macaques (*Macaca mulatta*) using a locally isolated CHIKV strain (Asian strain). The infected animals showed episodes of fever within a few days of virus inoculation, and the pattern of viral replication was very similar to what had been reported in humans, with plasma viraemia detectable from days 1 to 5 and titres ranging from 3 to 5.3 $\log_{10}$ PFU/ml (Binn *et al.*, 1967). An antibody response, with high neutralizing titre, occurred later and was detectable as early as day 10, reaching a plateau at day 15 post infection (Binn *et al.*, 1967; Paul and Singh, 1968). Using bonnet macaques (*Macaca radiata*), Paul and Singh showed the same response pattern and were able to assess the transmission capacity of mosquitoes from macaque to macaque comparing the infection of *Aedes aegypti* and *Aedes albopictus* (Paul and Singh, 1968). The study found *Ae. albopictus* was more susceptible to CHIKV infection by biting viraemic macaques (Paul and Singh, 1968), which was later confirmed by a similar study by Turell *et al.* (1992). In the studies listed above, no difference was reported comparing adolescent and adult bonnet macaques in both viral production and antibody response. Virus titre in blood reached up to $10^{7.5}$ LD_{50}/ml (suckling mice propagated) or $10^{6.3}$ $TCID_{50}$ (BHK21 cell propagated) at day 1 or 2 post intravenous (i.v.) inoculation or post CHIKV+ mosquito bites. Interestingly, they were able to super-infect six bonnet macaques (*Macaca radiate*) intra-muscularly with 10^5 $TCID_{50}$ of virus (at 3, 14 and 35 weeks post initial infection) and showed anamnestic response in four of the NHP, with no evidence of viral replication at day 2 or 4 post second- inoculation. In our hands, i.v. super-infection of 10 animals (*Macaca fascicularis*) with a similar virus dose at 35 days post-initial infection results in a low but detectable quantity of virus (10^5 RNA equivalent) in peripheral blood at day 3 post second inoculation (Roques *et al.* unpublished data). Barbosa *et al.* (1974) explored the interferon response of the fetal Rhesus monkey against CHIKV 2 days post infection. They showed that during the first trimester (35 days) the fetus was incapable of producing measurable amounts of interferon. During the second and third trimesters substantial amounts of circulating interferon (600/900 units/ml) were detected. Viraemia was present in all fetuses inoculated with CHIKV ranging from $10^{3.5}$ to 10^4 mouse LD_{50}/ml (Barbosa *et al.*, 1974). In these early experiments, limited clinical signs were reported in the macaques, possibly due to technical limitations (Binn *et al.*, 1967; Paul and Singh, 1968).

African green monkeys and macaques were also used as hosts in studies measuring the vector efficacy of different mosquito species (Jupp *et al.*, 1981; Turell *et al.*, 1992). These studies demonstrated the major role of *Aedes* species in CHIKV epidemics but also indicated that CHIKV may use various vectors including *Mansonia africana*.

Recent experimental non-human primate models of CHIKV-infection

Owing to the massive re-emergence of CHIKV, which began in La Réunion and has since spread to new regions of the world, our group has developed a model of CHIKV macaque infection (Table 6.1) using *Macaca fascicularis*. This model was utilized for the rapid implementation of pre-clinical trials following the Réunion epidemic and the risk of extension to the Indian Ocean zone by strains that seemed to be more virulent in humans. We infected cynomolgus macaques (*Macaca fascicularis*) with the CHIKV strain LR-2006-OPY1 from La Réunion harbouring the A226V markers of an *Aedes Albopictus* driven epidemic (Parola *et al.*, 2006). *In vivo* titration of the viral stock was performed by I.V. inoculation with decreasing doses from 10^9 down 10^2 viral copies (Roques *et al.*, 2007b, 2008). Doses as low as 10 PFU were sufficient to infect one macaque and the animal infectious dose 50% was determined to be close to 100 PFU by the method of Spouge (1992). The low viral dose (10 PFU) can infect animals without clinical signs being observed, whereas an intermediate dose (10^2–10^6 PFU) results in

Table 6.1 Histological lesions observed in macaques after chikungunya virus infection. Macaques were euthanized at various time points post infection, from day 2 to 186 (six months), and a large selection of tissues were sampled for histopathology and viral detection. (n = total number, time window during which lesion is present is represented in grey)

		Time post infection (days)							
Tissue	Lesion	2	6	19	32	44	55	82	186
Lymphoid organs	Histiocytosis					*n* = 12/12			
	T cell areas hypertrophy			*n* = 5/6					
	Follicular enlargement						*n* = 7/8		
Liver	Hemosiderosis				*n* = 10/11				
	Hydropic degeneration		2				1		
	Interstitial hepatitis						*n* = 5/5		
Striated muscles	Focal necrosis						1		1
Oral mucosa	Focal erosion	1		1		1			1
Exocrine glands	Interstitial adenitis	1	1			1			

viraemia with fever and rash. A high dose (≥10^7 PFU) results in swelling in the wrist and ankle joints as well as clinical signs of meningoencephalitis (hunching and wobbling, asthenia and ataxia) and/or mortality. The neurological complications and fatalities were similar to the severe forms of disease described in humans during the Réunion Island outbreak (Labadie *et al.*, 2010). In a similar model, subcutaneous infections in the right upper arm of rhesus macaques (*Macaca mulatta*) were performed using an enzootic CHIKV (37997 from West Africa isolated in 1983) or an epidemic strain of CHIKV (DHS-4263 isolated form a US patient returning from India in 2006) (Chen *et al.*, 2010). Messaoudi *et al.* (2013) more recently compared rhesus macaques (*Macaca mulatta*) infected i.v. with either the LR-2006-OPY1 or the West African strain 37997 (10^7–10^9 PFU each). In both rhesus and cynomolgus macaque subspecies, RT-PCR on plasma revealed viral replication kinetics similar to those in humans, together with profound leukopenia consistent with lymphocyte and monocyte recruitment to tissues (Chen *et al.*, 2010; Labadie *et al.*, 2010; Messaoudi *et al.*, 2013).

Innate immune response markers IFN-α and pro-inflammatory cytokines such as IL-6 and MCP-1/CCL-2 were found to be highly elevated in all CHIKV-infected NHP (Chen *et al.*, 2010; Labadie *et al.*, 2010). There is a rapid acute response of MCP-1 (13- to 350-fold increase, correlating with viral input; Fisher's test, $r = 0.831$, $P = 0.039$) but a delayed response of serum levels of IFN-g (≈28–400-fold) and TNF-α (8–46-fold) with concentrations peaking on days 5–10 post infection (Roques *et al.*, 2008). Cytokines IL1Ra, IL-15 and IL-12 were also enhanced, particularly following infection via the subcutaneous route, but to lower levels (Chen *et al.*, 2010; Dupuis-Maguiraga *et al.*, 2012; Labadie *et al.*, 2010). VEGF and IL-13 were observed only by Chen *et al.* (2010) and correlated to the peak of viral replication in rhesus macaques.

In both subspecies, whatever the route of inoculation, acute infection is controlled by the immune response, which renders viral replication undetectable from 10 days post infection (Chen *et al.*, 2010; Labadie *et al.*, 2010; Messaoudi *et al.*, 2013; Roques *et al.*, 2008). Similar to human infection, the acquired immune response is mainly driven by IgG3a (Kam *et al.*, 2012), but the CD8+ specific cytotoxic T-cell (CTL) response was shown to contribute to viral-clearance, as chloroquine treatment impaired the CTL response, resulting in an extended period during which virus could be detected (non-published data and Roques *et al.*, 2007a). Messaoudi *et al.* (2013) characterized CHIKV disease progression and host immune responses in adult (6–13 years of age) and mature rhesus macaques (17–18 years of age) infected with either the Réunion strain (LR-2006-OPY1) or the West African strain (37997). As expected, they observed a reduction in both the virus-specific T-cell and B-cell responses in the mature animals. Their data suggested that the reduced immune responses (together with a lower IFN-α response) in PBMCs of the aged animals promotes long-term virus persistence in LR-2006-OPY1-infected rhesus monkeys (Messaoudi *et al.*, 2013). Interestingly, regardless of age, T-cell and antibody responses were more robust in animals infected with LR-2006-OPY1 than with 37997 CHIKV strain.

CHIKV replicates in several cell types during the acute phase (Labadie *et al.*, 2010), but is thereafter detectable only in macrophages, by immunohistochemical analysis with a monoclonal antibody against the viral E2 glycoprotein. In our cynomolgus macaque (*Macaca fascicularis*) model we detected infected monocytes-macrophages in the blood 6 hours post infection by flow cytometry (Roques *et al.*, 2011) and in most tissues assayed after 24-hours (*in situ* hybridization, immunohistochemistry, RT-PCR and virus isolation). Significant macrophage infiltration was also detected by histological analysis (Labadie *et al.*, 2010) throughout the study and, significantly, after the macaques were no-longer viraemic. This monocyte–macrophage tropism is consistent with human findings from a number of studies showing CHIKV was able to infect primary macrophages *in vitro*, resulting in the production of highly variable amounts of virus, from 10^3 to 10^6 PFU per ml (Gardner *et al.*, 2010; Krejbich-Trotot *et al.*, 2011; Labadie *et al.*, 2010; Solignat *et al.*, 2009; Sourisseau *et al.*, 2007). Furthermore, these results are consistent with human patient studies reporting that macrophages are susceptible to CHIKV infection *in vivo* (Her *et al.*, 2010; Hoarau *et al.*, 2010).

In cynomolgus macaques (*Macaca fascicularis*), CHIKV persists in target tissues after its clearance from the blood, up to 3 months post infection, as demonstrated by viral RNA detection by *in situ* hybridization. Joints, secondary lymphoid organs and, to a lesser extent, muscles are affected. Surprisingly, while we observed viral persistence (but at variable levels) in all animals (from 3–14 years of age) (Dupuis-Maguiraga *et al.*, 2012; Labadie *et al.*, 2010), Messouadi *et al.* (2013) only observed persistence in older animals (15–17 years of age), possibly due to differences in methodology.

In the pregnant rhesus *Macaca mulatta*, regardless of the viral strains, mononuclear cell infiltration was observed, similar to the cynomolgus *Macaca fascicularis* (Chen *et al.*, 2010). Viral RNA was detected at necropsy at 21 days post infection in maternal lymphoid, joint-associated, and spinal cord tissues. Interestingly, the absence of detectable viral RNA and the lack of germinal centre development in fetuses indicated that transplacental transmission did not occur, which reflects human observations (Chen *et al.*, 2010).

Non-human primate models of Ross River virus and other Old World alphaviruses

In the endemic area of RRV there are no native NHP and the natural host is predominantly native marsupials, which have not yet been assessed as potential animal models (Kay *et al.*, 2007). We investigated RRV pathogenesis in a cynomolgus macaque (*Macaca fascicularis)* in order to identify potential key events involved in arthrogenic disease persistence (Dupuis-Maguiraga *et al.*, 2012). Macaques were experimentally infected with the RRV prototype RRV-T48 or RRV QML-1, an Australian clinical strain isolated in 1999, using either intravenous or subcutaneous routes of infection. Both RRV strains induced a very mild acute phase compared to CHIKV infection; however, local inflammation and joint swelling were readily observable near the injection site at the wrist. In addition, plasma levels of IFN-type I and pro-inflammatory cytokines/chemokines indicated a clear and strong innate response followed by a strong humoral response. Specific anti-RRV responses were found from day 7 onwards. Using *in situ* hybridization, RRV was detected in joint and lymphoid tissues up to 52 days post infection as we found for CHIKV. *In vitro* experiments showed that, like CHIKV, RRV was able to infect and persist for weeks in monocyte-derived macrophages from macaques (Roques, unpublished data).

Other arthritogenic alphaviruses, including SINV (McIntosh, 1970; Wolfe *et al.*, 2001) and Mayaro virus (MAYV) in South America (de Thoisy *et al.*, 2003), are able to infect NHP but, to-date, no evaluation of the disease pattern has been explored by experimental infection (Suhrbier *et al.*, 2012).

Non-human primate models of New World alphavirus

Monkeys have been used as a model for human VEEV, WEEV and EEEV infection. Both *Macaca mulatta* (rhesus macaques) as well as *Macaca fascicularis* (cynomolgus macaques) have been used to demonstrate infection by VEEV, WEEV and EEEV through various routes (Danes *et al.*, 1973; Monath *et al.*, 1974; Moreland *et al.*, 1979; Nathanson *et al.*, 1969; Reed *et al.*, 2004, 2005, 2007).

With the exception of direct intracranial inoculation, most macaques infected with VEEV survive infection, whereas macaques infected with WEEV or EEEV display a range of clinical outcomes from mild signs to paralysis, coma and death. Due to large variations in the age of animals, routes of inoculation and strains of viruses used, comparisons between studies are difficult. Common features of VEEV infection in macaques include fever, viraemia, and lymphopenia within 1–3 days of infection and, in some cases, signs of encephalitis a few days later (Victor *et al.*, 1956). The most thorough NHP study of VEEV to date infected 67 rhesus macaques intraperitoneally with various doses of an equine-virulent strain of VEEV. Similar to humans, macaques infected with VEEV developed transient viraemia and a biphasic fever with no other evidence of clinical disease (Gleiser *et al.*, 1962). Wyckoff and Tesar used young Rhesus macaques to demonstrate infection by the subcutaneous, intravenous, intranasal, intracerebral and intralingual routes (Wyckoff and Tesar, 1939). Interestingly, both intranasal and intralingual routes were similarly lethal to intracerebral injection, with similar mortality rates seen for both EEEV and WEEV, although the progression to death was delayed with WEEV. WEEV differed from EEEV in that the peak day of fever occurred later, on day 5–6 as opposed to day 3–4, and was less severe than the fever induced by EEEV. The period in which animals experienced coma at the end of the infection for WEEV also lasted longer than the mean for EEEV. Furthermore, mortality appeared to be age-dependent, with fatal illness observed almost exclusively in young monkeys, while older macaques demonstrated resistance with only a few developing minor neurological signs (Wyckoff and Tesar, 1939). A study of EEEV isolated from a fatal human case was conducted in the 1960s with the virus inoculated into the thalamus of juvenile rhesus macaques (Nathanson *et al.*, 1969). All animals subsequently developed acute encephalitis and were euthanized 4 days post infection with histopathological findings including severe encephalitis with neuronal necrosis. However, the value of this study, due to the chosen route of viral inoculation, remains limited.

Cynomolgus macaques have been primarily used to test infection via the aerosol route, and the vast majority of these have been part of vaccine efficacy studies (Reed *et al.*, 2004, 2005, 2007). In WEEV infection, onset of fever, duration, and peak temperature were similar to what was shown previously with intranasal inoculation of cynomolgus macaques, and the ID50 was found to be $1 \times 10^{6.25}$ PFU. Neurological manifestations of encephalitis after aerosol exposure were found to be more likely with WEEV than VEEV, with duration and severity of tremors more pronounced following WEEV infection (Reed *et al.*, 2005). In a study of 12 adult cynomolgus macaques challenged with aerosolized virulent EEEV, 8 of the 12 macaques developed fever and neurological signs, succumbing to infection by 5–9 days post infection. Histopathology reports showed unremarkable necropsy results in all animals with the major CNS changes characterized as severe meningoencephalomyelitis. In parallel to findings with VEEV, where following aerosol exposure virus enters the CNS through olfactory nerves, WEEV and EEEV have been postulated to enter the CNS in a similar manner. Olfactory entry into the CNS is supported by WEEV and EEEV infection by both aerosol and intranasal inoculation in laboratory animals (Reed *et al.*, 2005; Wyckoff and Tesar, 1939). A review of models of New World alphaviruses extensively compared the pathogenesis and clinical findings of VEEV infection, identifying that neuroinvasion and neurovirulence both occur in a limited fashion in NHPs. This review claims the NHP model to be more similar to human cases of VEEV than the mouse model, particularly as the equivalent VEE viruses are more frequently neuroinvasive and neurovirulent in mice (Steele and Twenhafel, 2010).

Alternative animal models of alphavirus disease

Horse models of New World alphaviruses

Horses experimentally infected intranasally and subcutaneously with VEEV, WEEV and EEEV demonstrated significant pathology in the olfactory bulbs, brainstem, optic thalamus, and hypothalamic region. Histological analysis of horses experimentally infected with VEEV, WEEV and EEEV revealed lesions similar to horses infected naturally during an outbreak. The similarities included a variety of neurological symptoms in horses including loss of motor coordination, fatigue, and differing paralyses depending upon the location of lesions in the CNS (Del Piero *et al.*, 2001; Dietz *et al.*, 1978; Gleiser *et al.*, 1962; Sponseller *et al.*, 1966; Walton *et al.*, 1973). Although horses are clearly important with respect to New World alphaviruses, for several reasons including their size, cost and ethical concerns, they are limited in their use as an experimental model.

Guinea pig and hamster models of New World alphaviruses

Guinea pigs were used as a model for both WEEV and EEV as early as the mid-1930s. Hurst injected WEEV and EEEV intramuscularly and was able to recover virus from blood-free nasal washings from WEEV-infected animals but not EEEV-infected animals. In these early studies it was noticed that following intranasal inoculation, both WEEV and

EEEV could be isolated from the anterior frontal region and olfactory bulbs, and that Guinea pigs with surgically excised olfactory bulbs had delayed neurological symptoms (Hurst, 1936). This model has since been further developed showing that 24 hours post infection the virus is detected in the olfactory neuroepithelium and olfactory bulbs followed by spread to all parts of the brain by day 4 post infection (Roy *et al.*, 2009). Neurons were found to be the main viral target of infection, with brain lesions dominated by neuronal necrosis and the presence of perivascular cuffs with inflammatory infiltrates composed of macrophages and heterophils. Interestingly, despite the aerosol route of infection, only minimal amounts of virus were detected in the lungs of select animals (Roy *et al.*, 2009).

Hamsters experimentally infected with EEEV develop disease with some differences to that observed in Guinea pigs, albeit with a similar trend (Dremov *et al.*, 1978; Paessler *et al.*, 2004). Hamsters inoculated peripherally develop an early visceral phase and viraemic spread followed by neuroinvasion and death due to encephalitis around days 4–6 post infection. Within 24 hours of infection hamsters display signs of fever, vomiting, respiratory signs and coma (Paessler *et al.*, 2004). Early EEEV targets in the hamster are the heart, liver, lungs, kidneys, lymphoid tissues, and skeletal muscle, with EEEV infecting the brain by day 2 post infection (Dremov *et al.*, 1978; Paessler *et al.*, 2004).

Avian models of New World alphaviruses

Various avian species have been used to study the transmission of WEEV from its vector species to avian host. In these experiments WEEV is peripherally inoculated in chickens within the laboratory setting through the bite of *C. tarsalis*. Subcutaneous inoculation resulted in viraemia 24 h post inoculation that was largely undetectable at 48 h and later post infection (Hammon and Reeves, 1943). WEEV isolated from mosquitoes taken from California were compared in their ability to induce both viraemia and antibody responses house finches. This study found age inversely correlated to antibody titres, with younger birds having more WEEV antibodies (Reisen *et al.*, 2000). A study conducted by Reisen *et al.* (2000) found young species of quail (*Callipepla gambelii* and *Callipepla californica*) to be involved in amplification of WEEV whereas older animals were unable to sustain a sufficient viraemia for transmission to mosquitoes. A further study by Reisen *et al.* (2003) inoculated 27 bird species from the Coachella and San Joaquin Valleys of California with isolates of WEEV to determine which species were competent avian hosts. Just under 50% of the species infected were able to support mosquito transmission of WEEV (Reisen *et al.*, 2003). Overall the use of avian species in transmission studies of New World alphaviruses has proven useful in showing that the species range of WEEV during epizootics becomes greatly expanded (Reisen *et al.*, 2003).

Conclusions

Animal models have played a critical role in medical advances for the past century. Studies in animal models of alphavirus-induced disease have proven invaluable for dissecting the intricate nature of the immune response to viral infections and identifying potential targets for the development of antiviral strategies. Alphaviruses are currently expanding into new geographical regions largely driven by expanding vector competence, increased international travel, increasing vector range due to climate change and viral mutation. Alphavirus disease manifestations can be severely debilitating and at times even fatal. With no vaccine currently available and treatment focused on analgesics and non-steroidal anti-inflammatory drugs (NSAIDs) to relieve symptoms, alphavirus animal models provide an ideal tool for examining the exact role each potential drug target could play in the treatment of alphavirus-induced disease.

References

Aaskov, J.G., Ross, P.V., Harper, J.J., and Donaldson, M.D. (1985). Isolation of Ross River virus from epidemic polyarthritis patients in Australia. Aust. J. Exp. Biol. Med. Sci 63 (*Pt 5*), 587–597.

Aguilar, M.J. (1970). Pathological changes in brain and other target organs of infant and weanling mice after infection with non-neuroadapted Western equine encephalitis virus. Infect. Immun. 2, 533–542.

Apandi, Y., Lau, S.K., Izmawati, N., Amal, N.M., Faudzi, Y., Mansor, W., Hani, M.H., and Zainah, S. (2010).

Identification of Chikungunya virus strains circulating in Kelantan, Malaysia in 2009. Southeast Asian J. Trop. Med. Public Health *41*, 1374–1380.

Balluz, I.M., Glasgow, G.M., Killen, H.M., Mabruk, M.J., Sheahan, B.J., and Atkins, G.J. (1993). Virulent and avirulent strains of Semliki Forest virus show similar cell tropism for the murine central nervous system but differ in the severity and rate of induction of cytolytic damage. Neuropathol. Appl. Neurobiol. *19*, 233–239.

Barbosa, L.H., London, W.T., Hamilton, R., and Buckler, C. (1974). Interferon response of the fetal Rhesus monkey after viral infection. Proc. Soc. Exp. Biol. Med. *146*, 398–400.

Beck, C.E., and Wyckoff, R.W. (1938). Venezuelan Equine Encephalomyelitis. Science *88*, 530.

Bianchi, T.I., Aviles, G., Monath, T.P., and Sabattini, M.S. (1993). Western equine encephalomyelitis: virulence markers and their epidemiologic significance. Am. J. Trop. Med. Hyg. *49*, 322–328.

Binn, L.N., Harrison, V.R., and Randall, R. (1967). Patterns of viremia and antibody observed in rhesus monkeys inoculated with chikungunya and other serologically related group A arboviruses. Am. J. Trop. Med. Hyg. *16*, 782–785.

Borgherini, G., Poubeau, P., Staikowsky, F., Lory, M., Le Moullec, N., Becquart, J.P., Wengling, C., Michault, A., and Paganin, F. (2007). Outbreak of chikungunya on Reunion Island: early clinical and laboratory features in 157 adult patients. Clin. Infect. Dis. *44*, 1401–1407.

Borgherini, G., Poubeau, P., Jossaume, A., Gouix, A., Cotte, L., Michault, A., Arvin-Berod, C., and Paganin, F. (2008). Persistent arthralgia associated with chikungunya virus: a study of 88 adult patients on reunion island. Clin. Infect. Dis. *47*, 469–475.

Bradish, C.J., Allner, K., and Maber, H.B. (1971). The virulence of original and derived strains of Semliki forest virus for mice, guinea-pigs and rabbits. J. Gen. Virol. *12*, 141–160.

Casals, J., and Palacios, R. (1941). the complement fixation test in the diagnosis of virus infections of the central nervous system. J. Exp. Med. *74*, 409–426.

Chandak, N.H., Kashyap, R.S., Kabra, D., Karandikar, P., Saha, S.S., Morey, S.H., Purohit, H.J., Taori, G.M., and Daginawala, H.F. (2009). Neurological complications of Chikungunya virus infection. Neurol. India *57*, 177–180.

Charles, P.C., Walters, E., Margolis, F., and Johnston, R.E. (1995). Mechanism of neuroinvasion of Venezuelan equine encephalitis virus in the mouse. Virology *208*, 662–671.

Charles, P.C., Brown, K.W., Davis, N.L., Hart, M.K., and Johnston, R.E. (1997). Mucosal immunity induced by parenteral immunization with a live attenuated Venezuelan equine encephalitis virus vaccine candidate. Virology *228*, 153–160.

Chen, C.I., Clark, D.C., Pesavento, P., Lerche, N.W., Luciw, P.A., Reisen, W.K., and Brault, A.C. (2010). Comparative pathogenesis of epidemic and enzootic Chikungunya viruses in a pregnant Rhesus macaque model. Am. J. Trop. Med. Hyg. *83*, 1249–1258.

Chen, W., Foo, S.S., Rulli, N.E., Taylor, A., Sheng, K.C., Herrero, L.J., Herring, B.L., Lidbury, B.A., Li, R.W., Walsh, N.C., *et al.* (2014). Arthritogenic alphaviral infection perturbs osteoblast function and triggers pathologic bone loss. Proc. Natl. Acad. Sci. U.S.A. *111*, 6040–6045.

Chew-Lim, M. (1975). Mouse encephalitis induced by avirulent Semliki forest virus. Vet. Pathol. *12*, 387–393.

Chew-Lim, M. (1979). Brain viral persistence and myelin damage in nude mice. Can. J. Comp. Med. *43*, 39–43.

Chew-Lim, M., Webb, H.E., and Jagelman, S. (1977). The effect of irradiation on demyelination induced by avirulent Semliki Forest virus. Br. J. Exp. Pathol. *58*, 459–464.

Chung, D.H., Jonsson, C.B., Tower, N.A., Chu, Y.K., Sahin, E., Golden, J.E., Noah, J.W., Schroeder, C.E., Sotsky, J.B., Sosa, M.I., *et al.* (2014). Discovery of a novel compound with anti-venezuelan equine encephalitis virus activity that targets the nonstructural protein 2. PLoS Pathog. *10*, e1004213.

Clarris, B.J., Doherty, R.L., Fraser, J.R., French, E.L., and Muirden, K.D. (1975). Epidemic polyarthritis: a cytological, virological and immunochemical study. Aust. N. Z. J. Med. *5*, 450–457.

Cook, S.H., and Griffin, D.E. (2003). Luciferase imaging of a neurotropic viral infection in intact animals. J. Virol. *77*, 5333–5338.

Coppenhaver, D.H., Singh, I.P., Sarzotti, M., Levy, H.B., and Baron, S. (1995). Treatment of intracranial alphavirus infections in mice by a combination of specific antibodies and an interferon inducer. Am. J. Trop. Med. Hyg. *52*, 34–40.

Couderc, T., Chretien, F., Schilte, C., Disson, O., Brigitte, M., Guivel-Benhassine, F., Touret, Y., Barau, G., Cayet, N., Schuffenecker, I., *et al.* (2008). A mouse model for Chikungunya: young age and inefficient type-I interferon signaling are risk factors for severe disease. PLoS Pathog. *4*, e29.

Dal Canto, M.C., and Rabinowitz, S.G. (1981). Central nervous system demyelination in Venezuelan equine encephalomyelitis infection. J. Neurol. Sci. *49*, 397–418.

Danes, L., Rychterova, V., Kufner, J., and Hruskova, J. (1973). The role of the olfactory route on infection of the respiratory tract with Venezuelan equine encephalomyelitis virus in normal and operated Macaca rhesus monkeys. II. Results of histological examination. Acta Virol. *17*, 57–60.

Del Piero, F., Wilkins, P.A., Dubovi, E.J., Biolatti, B., and Cantile, C. (2001). Clinical, pathologic, immunohistochemical, and virologic findings of eastern equine encephalomyelitis in two horses. Vet. Pathol. *38*, 451–456.

Deresiewicz, R.L., Thaler, S.J., Hsu, L., and Zamani, A.A. (1997). Clinical and neuroradiographic manifestations of eastern equine encephalitis. New Eng. J. Med. *336*, 1867–1874.

Dhanwani, R., Khan, M., Alam, S.I., Rao, P.V., and Parida, M. (2011). Differential proteome analysis of Chikungunya virus-infected new-born mice tissues reveal implication of stress, inflammatory and apoptotic pathways in disease pathogenesis. Proteomics *11*, 1936–1951.

Dhanwani, R., Khan, M., Lomash, V., Rao, P.V., Ly, H., and Parida, M. (2014). Characterization of chikungunya virus induced host response in a mouse model of viral myositis. PLOS ONE 9, e92813.

Diallo, M., Thonnon, J., Traore-Lamizana, M., and Fontenille, D. (1999). Vectors of Chikungunya virus in Senegal: current data and transmission cycles. Am. J. Trop. Med. Hyg. *60*, 281–286.

Dietz, W.H., Jr., Alvarez, O., Jr., Martin, D.H., Walton, T.E., Ackerman, L.J., and Johnson, K.M. (1978). Enzootic and epizootic Venezuelan equine encephalomyelitis virus in horses infected by peripheral and intrathecal routes. J. Infect. Dis. *137*, 227–237.

Doherty, R.L., H., W.R., M., G.B., and K., O.G.A. (1963). The isolation of a third group A arbovirus in Australia with preliminary observations on its relationship to epidemic polyarthritis. Aust. J. Sci. *26*, 183–184.

Dremov, D.P., Solyanik, R.G., Miryutova, T.L., and Laptakova, L.M. (1978). Attenuated variants of eastern equine encephalomyelitis virus: pathomorphological, immunofluorescence and virological studies of infection in Syrian hamsters. Acta Virol. *22*, 139–145.

Dupuis-Maguiraga, L., Larcher, T., Vezmar, K., Dereuddre-Bosquet, N., Babarit, C., Delache, B., Le Grand, R., Mahalingam, S., Aaskov, J., Suhrbier, A., *et al.* (2012). Arthrogenic alphaviruses: comparison between Chikungunya and Ross River virus in non-human primates. Paper presented at: 11th ArboVirus Research in Australia and the 10th Mosquito Control Association of Australia Symposium (Surfer Paradise, Gold Coast, Queensland, Australia).

Economopoulou, A., Dominguez, M., Helynck, B., Sissoko, D., Wichmann, O., Quenel, P., Germonneau, P., and Quatresous, I. (2009). Atypical Chikungunya virus infections: clinical manifestations, mortality and risk factors for severe disease during the 2005–2006 outbreak on Reunion. Epidemiol. Infect. *137*, 534–541.

Fazakerley, J.K. (2004). Semliki forest virus infection of laboratory mice: a model to study the pathogenesis of viral encephalitis. Arch. Virol. Suppl. *18*, 179–190.

Fazakerley, J.K., Amor, S., and Webb, H.E. (1983). Reconstitution of Semliki forest virus infected mice, induces immune mediated pathological changes in the CNS. Clin. Exp. Immunol. *52*, 115–120.

Fazakerley, J.K., Pathak, S., Scallan, M., Amor, S., and Dyson, H. (1993). Replication of the A7(74) strain of Semliki Forest virus is restricted in neurons. Virology *195*, 627–637.

Fleming, S., Ikeda, H., and Pessoa, V.F. (1982). Pattern and extent of demyelination in the optic-nerve of the mouse infected with Semliki Forest virus. J. Physiol. *327*, P28–P28.

Fragkoudis, R., Ballany, C.M., Boyd, A., and Fazakerley, J.K. (2008). In Semliki Forest virus encephalitis, antibody rapidly clears infectious virus and is required to eliminate viral material from the brain, but is not required to generate lesions of demyelination. J. Gen. Virol. *89*, 2565–2568.

Fraser, J.R. (1986). Epidemic polyarthritis and Ross River virus disease. Clin. Rheum. Dis. *12*, 369–388.

Fraser, J.R., Cunningham, A.L., Clarris, B.J., Aaskov, J.G., and Leach, R. (1981). Cytology of synovial effusions in epidemic polyarthritis. Aust. N. Z. J. Med. *11*, 168–173.

Fraser, J.R., Ratnamohan, V.M., Dowling, J.P., Becker, G.J., and Varigos, G.A. (1983). The exanthem of Ross River virus infection: histology, location of virus antigen and nature of inflammatory infiltrate. J. Clin. Pathol. *36*, 1256–1263.

Gardner, C.L., Burke, C.W., Higgs, S.T., Klimstra, W.B., and Ryman, K.D. (2012). Interferon-alpha/beta deficiency greatly exacerbates arthritogenic disease in mice infected with wild-type chikungunya virus but not with the cell culture-adapted live-attenuated 181/25 vaccine candidate. Virology *425*, 103–112.

Gardner, J., Anraku, I., Le, T.T., Larcher, T., Major, L., Roques, P., Schroder, W.A., Higgs, S., and Suhrbier, A. (2010). Chikungunya virus arthritis in adult wild-type mice. J. Virol. *84*, 8021–8032.

Gates, M.C., Sheahan, B.J., and Atkins, G.J. (1984). The pathogenicity of the M9 mutant of Semliki Forest virus in immune-compromised mice. J. Gen. Virol. *65 (Pt 1)*, 73–80.

Gates, M.C., Sheahan, B.J., O'Sullivan, M.A., and Atkins, G.J. (1985). The pathogenicity of the A7, M9 and L10 strains of Semliki Forest virus for weanling mice and primary mouse brain cell cultures. J. Gen. Virol. *66*(Pt 11), 2365–2373.

Giltner, L.T., and Shahan, M.S. (1933). The immunological relationship of Eastern and Western strains of equine encephalomyelitis virus. Science *78*, 587–588.

Gleiser, C.A., Gochenour, W.S., Jr., Berge, T.O., and Tigertt, W.D. (1962). The comparative pathology of experimental Venezuelan equine encephalomyelitis infection in different animal hosts. J. Infect. Dis. *110*, 80–97.

Grieder, F.B., and Vogel, S.N. (1999). Role of interferon and interferon regulatory factors in early protection against Venezuelan equine encephalitis virus infection. Virology 257, 106–118.

Gunn, B.M., Morrison, T.E., Whitmore, A.C., Blevins, L.K., Hueston, L., Fraser, R.J., Herrero, L.J., Ramirez, R., Smith, P.N., Mahalingam, S., *et al.* (2012). Mannose binding lectin is required for alphavirus-induced arthritis/myositis. PLoS Pathog. *8*, e1002586.

Hahn, C.S., Lustig, S., Strauss, E.G., and Strauss, J.H. (1988). Western equine encephalitis virus is a recombinant virus. Proc. Natl. Acad. Sci. U.S.A. *85*, 5997–6001.

Hammon, W.M., and Reeves, W.C. (1943). Laboratory transmission of Western equine encephalomyelitis virus by mosquitoes of the genera Culex and Culiseta. J. Exp. Med. *78*, 425–434.

Harrison, V.R., Marshall, J.D., and Guilloud, N.B. (1967). The presence of antibody to Chikungunya and other serologically related viruses in the sera of sub-human primate imports to the United States. J. Immunol. *98*, 979–981.

Hart, M.K., Pratt, W., Panelo, F., Tammariello, R., and Dertzbaugh, M. (1997). Venezuelan equine encephalitis virus vaccines induce mucosal IgA responses and protection from airborne infection in BALB/c, but not C3H/HeN mice. Vaccine *15*, 363–369.

Hazelton, R.A., Hughes, C., and Aaskov, J.G. (1985). The inflammatory response in the synovium of a patient with Ross River arbovirus infection. Aust. N. Z. J. Med. *15*, 336–339.

Her, Z., Malleret, B., Chan, M., Ong, E.K., Wong, S.C., Kwek, D.J., Tolou, H., Lin, R.T., Tambyah, P.A., Renia, L., *et al.* (2010). Active infection of human blood monocytes by Chikungunya virus triggers an innate immune response. J. Immunol. *184*, 5903–5913.

Hoarau, J.J., Jaffar Bandjee, M.C., Trotot, P.K., Das, T., Li-Pat-Yuen, G., Dassa, B., Denizot, M., Guichard, E., Ribera, A., Henni, T., *et al.* (2010). Persistent chronic inflammation and infection by Chikungunya arthritogenic alphavirus in spite of a robust host immune response. J. Immunol. *184*, 5914–5927.

Howitt, B. (1938). Recovery of the virus of equine encephalomyelitis from the brain of a child. Science *88*, 455–456.

Illavia, S.J., Webb, H.E., and Pathak, S. (1982). Demyelination induced in mice by avirulent Semliki Forest virus. 1. Virology and effects on optic-nerve. Neuropath. Appl. Neuro. *8*, 35–42.

Inoue, S., Morita, K., Matias, R.R., Tuplano, J.V., Resuello, R.R., Candelario, J.R., Cruz, D.J., Mapua, C.A., Hasebe, F., Igarashi, A., *et al.* (2003). Distribution of three arbovirus antibodies among monkeys (*Macaca fascicularis*) in the Philippines. J. Med. Primatol. *32*, 89–94.

Jackson, A.C., SenGupta, S.K., and Smith, J.F. (1991). Pathogenesis of Venezuelan equine encephalitis virus infection in mice and hamsters. Vet. Pathol. *28*, 410–418.

Jagelman, S., Suckling, A.J., Webb, H.E., and Bowen, F.T. (1978). The pathogenesis of avirulent Semliki Forest virus infections in athymic nude mice. J. Gen. Virol. *41*, 599–607.

Johnson, R.T. (1965). Virus Invasion of the Central Nervous System: A study of Sindbis virus infection in the mouse using fluorescent antibody. Am. J. Pathol. *46*, 929–943.

Julander, J.G., Skirpstunas, R., Siddharthan, V., Shafer, K., Hoopes, J.D., Smee, D.F., and Morrey, J.D. (2008). C3H/HeN mouse model for the evaluation of antiviral agents for the treatment of Venezuelan equine encephalitis virus infection. Antiviral Res. *78*, 230–241.

Jupp, P.G., McIntosh, B.M., Dos Santos, I., and DeMoor, P. (1981). Laboratory vector studies on six mosquito and one tick species with chikungunya virus. Trans. R. Soc. Trop. Med. Hyg. 75, 15–19.

Kam, Y.W., Lum, F.M., Teo, T.H., Lee, W.W., Simarmata, D., Harjanto, S., Chua, C.L., Chan, Y.F., Wee, J.K., Chow, A., *et al.* (2012). Early neutralizing IgG response to Chikungunya virus in infected patients targets a dominant linear epitope on the E2 glycoprotein. EMBO Mol. Med. *4*, 330–343.

Kaschula, V.R., Van Dellen, A.F., and de Vos, V. (1978). Some infectious diseases of wild vervet monkeys (Cercopithecus aethiops pygerythrus) in South Africa. J. S. Afr. Vet. Assoc. *49*, 223–227.

Kay, B.H., Boyd, A.M., Ryan, P.A., and Hall, R.A. (2007). Mosquito feeding patterns and natural infection of vertebrates with Ross River and Barmah Forest viruses in Brisbane, Australia. Am. J. Trop. Med. Hyg. *76*, 417–423.

King, L.S. (1940a). Studies on Eastern equine encephalomyelitis: Iv. Infection in the mouse with fresh and fixed virus. J. Exp. Med. *71*, 95–106.

King, L.S. (1940b). Studies on Eastern equine encephalomyelitis: V. Histopathology in the mouse. J. Exp. Med. *71*, 107–112.

Kirya, B.G., and Okia, N.O. (1977). A yellow fever epizootic in Zika Forest, Uganda, during 1972: Part 2: Monkey serology. Trans. R. Soc. Trop. Med. Hyg. *71*, 300–303.

Krejbich-Trotot, P., Denizot, M., Hoarau, J.J., Jaffar-Bandjee, M.C., Das, T., and Gasque, P. (2011). Chikungunya virus mobilizes the apoptotic machinery to invade host cell defenses. FASEB J. *25*, 314–325.

Kuehne, R.W., Pannier, W.L., and Stephen, E.L. (1977). Indirect mouse model for the evaluation of potential antiviral compounds: results with Venezuelan equine encephalomyelitis virus. Antimicrob. Agents Chemother. *11*, 683–687.

Kundin, W.D., Liu, C., and Rodina, P. (1966). Pathogenesis of Venezuelan equine encephalomyelitis virus. I. Infection in suckling mice. J. Immunol. *96*, 39–48.

Labadie, K., Larcher, T., Joubert, C., Mannioui, A., Delache, B., Brochard, P., Guigand, L., Dubreil, L., Lebon, P., Verrier, B., *et al.* (2010). Chikungunya disease in nonhuman primates involves long-term viral persistence in macrophages. J. Clin. Invest. *120*, 894–906.

Levy, O. (2007). Innate immunity of the newborn: basic mechanisms and clinical correlates. Nat. Rev. Immunol. 7, 379–390.

Lidbury, B.A., Simeonovic, C., Maxwell, G.E., Marshall, I.D., and Hapel, A.J. (2000). Macrophage-induced muscle pathology results in morbidity and mortality for Ross River virus-infected mice. J. Infect. Dis. *181*, 27–34.

Lidbury, B.A., Rulli, N.E., Suhrbier, A., Smith, P.N., McColl, S.R., Cunningham, A.L., Tarkowski, A., van Rooijen, N., Fraser, R.J., and Mahalingam, S. (2008). Macrophage-derived proinflammatory factors contribute to the development of arthritis and myositis after infection with an arthrogenic alphavirus. J. Infect. Dis. *197*, 1585–1593.

Linn, M.L., Aaskov, J.G., and Suhrbier, A. (1996). Antibody-dependent enhancement and persistence in macrophages of an arbovirus associated with arthritis. J. Gen. Virol. 77(Pt 3), 407–411.

Logue, C.H., Bosio, C.F., Welte, T., Keene, K.M., Ledermann, J.P., Phillips, A., Sheahan, B.J., Pierro, D.J., Marlenee, N., Brault, A.C., *et al.* (2009). Virulence variation among isolates of western equine encephalitis virus in an outbred mouse model. J. Gen. Virol. *90*, 1848–1858.

Ludwig, G.V., Turell, M.J., Vogel, P., Kondig, J.P., Kell, W.K., Smith, J.F., and Pratt, W.D. (2001). Comparative neurovirulence of attenuated and non-attenuated strains of Venezuelan equine encephalitis virus in mice. Am. J. Trop. Med. Hyg. *64*, 49–55.

McCrae, A.W., Henderson, B.E., Kirya, B.G., and Sempala, S.D. (1971). Chikungunya virus in the Entebbe area of Uganda: isolations and epidemiology. Trans. R. Soc. Trop. Med. Hyg. *65*, 152–168.

McIntosh, B.M. (1970). Antibody against Chikungunya virus in wild primates in Southern Africa. S. Afr. J. Med. Sci. *35*, 65–74.

McIntosh, B.M., Worth, C.B., and Kokernot, R.H. (1961). Isolation of Semliki Forest virus from Aedes (Aedimorphus) argenteopunctatus (Theobald) collected in Portuguese East Africa. Trans. R. Soc. Trop. Med. Hyg. *55*, 192–198.

McIntosh, B.M., Paterson, H.E., McGillivray, G., and Desousa, J. (1964). Further studies on the Chikungunya outbreak in Southern Rhodesia in 1962. I. Mosquitoes, wild primates and birds in relation to the epidemic. Ann. Trop. Med. Parasitol. *58*, 45–51.

McIntosh, B.M., Sweetnam, J., McGillivray, G.M., and De Sousa, J. (1965). Laboratory transmission of chikungunya virus by Mansonia (Mansonioides) africana (Theobald). Ann. Trop. Med. Parasitol. *59*, 390–392.

Mackenzie, A., Suckling, A.J., Jagelman, S., and Wilson, A.M. (1978). Histopathological and enzyme histochemical changes in experimental Semliki Forest virus infection in mice and their relevance to scrapie. J. Comp. Pathol. *88*, 335–344.

Marchette, N.J., Rudnick, A., Garcia, R., and MacVean, D.W. (1978). Alphaviruses in Peninusular Malaysia: I. Virus isolations and animal serology. Southeast Asian J. Trop. Med Public Health *9*, 317–329.

Mateo, L., La Linn, M., McColl, S.R., Cross, S., Gardner, J., and Suhrbier, A. (2000). An arthrogenic alphavirus induces monocyte chemoattractant protein-1 and interleukin-8. Intervirology *43*, 55–60.

Messaoudi, I., Vomaske, J., Totonchy, T., Kreklywich, C.N., Haberthur, K., Springgay, L., Brien, J.D., Diamond, M.S., Defilippis, V.R., and Streblow, D.N. (2013). Chikungunya virus infection results in higher and persistent viral replication in aged rhesus macaques due to defects in antiviral immunity. PLoS Negl. Trop. Dis. *7*, e2343.

Mims, C.A., Murphy, F.A., Taylor, W.P., and Marshall, I.D. (1973). Pathogenesis of Ross River virus infection in mice. I. Ependymal infection, cortical thinning, and hydrocephalus. J. Infect. Dis. *127*, 121–128.

Mokhtarian, F., and Swoveland, P. (1987). Predisposition to EAE induction in resistant mice by prior infection with Semliki Forest virus. J. Immunol. *138*, 3264–3268.

Mokhtarian, F., Shi, Y., Zhu, P.F., and Grob, D. (1994). Immune responses, and autoimmune outcome, during virus infection of the central nervous system. Cell. Immunol. *157*, 195–210.

Mokhtarian, F., Huan, C.M., Roman, C., and Raine, C.S. (2003). Semliki Forest virus-induced demyelination and remyelination--involvement of B cells and anti-myelin antibodies. J. Neuroimmunol. *137*, 19–31.

Mokhtarian, F., Safavi, F., and Sarafraz-Yazdi, E. (2012). Immunization with a peptide of Semliki Forest virus promotes remyelination in experimental autoimmune encephalomyelitis. Brain Res. *1488*, 92–103.

Monath, T.P., Calisher, C.H., Davis, M., Bowen, G.S., and White, J. (1974). Experimental studies of rhesus monkeys infected with epizootic and enzootic subtypes of Venezuelan equine encephalitis virus. J. Infect. Dis. *129*, 194–200.

Monath, T.P., Kemp, G.E., Cropp, C.B., and Chandler, F.W. (1978). Necrotizing myocarditis in mice infected with Western equine encephalitis virus: clinical, electrocardiographic, and histopathologic correlations. J. Infect. Dis. *138*, 59–66.

Moreland, A.F., Schimpff, R.D., and Gaskin, J.M. (1979). Fetal mortality and malformations associated with experimental infections of western equine encephalomyelitis vaccine virus in rhesus monkeys (Macaca mulatta). Teratology *20*, 65–74.

Morgan, I.M. (1941). Influence of age on susceptibility and on immune response of mice to Eastern equine encephalomyelitis virus. J. Exp. Med. *74*, 115–132.

Morrison, T.E., Whitmore, A.C., Shabman, R.S., Lidbury, B.A., Mahalingam, S., and Heise, M.T. (2006). Characterization of Ross River virus tropism and virus-induced inflammation in a mouse model of viral arthritis and myositis. J. Virol. *80*, 737–749.

Morrison, T.E., Fraser, R.J., Smith, P.N., Mahalingam, S., and Heise, M.T. (2007). Complement contributes to inflammatory tissue destruction in a mouse model of Ross River virus-induced disease. J. Virol. *81*, 5132–5143.

Morrison, T.E., Simmons, J.D., and Heise, M.T. (2008). Complement receptor 3 promotes severe ross river virus-induced disease. J. Virol. *82*, 11263–11272.

Morrison, T.E., Oko, L., Montgomery, S.A., Whitmore, A.C., Lotstein, A.R., Gunn, B.M., Elmore, S.A., and Heise, M.T. (2011). A mouse model of chikungunya virus-induced musculoskeletal inflammatory disease: evidence of arthritis, tenosynovitis, myositis, and persistence. Am. J. Pathol. *178*, 32–40.

Murphy, F.A., and Whitfield, S.G. (1970). Eastern equine encephalitis virus infection: electron microscopic studies of mouse central nervous system. Exp. Mol. Pathol. *13*, 131–146.

Murphy, F.A., Taylor, W.P., Mims, C.A., and Marshall, I.D. (1973). Pathogenesis of Ross River virus infection in mice. II. Muscle, heart, and brown fat lesions. J. Infect. Dis. *127*, 129–138.

Nagata, L.P., Hu, W.G., Parker, M., Chau, D., Rayner, G.A., Schmaltz, F.L., and Wong, J.P. (2006). Infectivity variation and genetic diversity among strains of Western equine encephalitis virus. J. Gen. Virol. *87*, 2353–2361.

Nakaya, H.I., Gardner, J., Poo, Y.S., Major, L., Pulendran, B., and Suhrbier, A. (2012). Gene profiling of Chikungunya virus arthritis in a mouse model reveals significant overlap with rheumatoid arthritis. Arthritis Rheum. *64*, 3553–3563.

Nathanson, N., Stolley, P.D., and Boolukos, P.J. (1969). Eastern equine encephalitis. Distribution of central nervous system lesions in man and Rhesus monkey. J. Comp. Pathol. *79*, 109–115.

Nimmannitya, S., Halstead, S.B., Cohen, S.N., and Margiotta, M.R. (1969). Dengue and chikungunya virus infection in man in Thailand, 1962–1964. I. Observations on hospitalized patients with hemorrhagic fever. Am. J. Trop. Med. Hyg. *18*, 954–971.

Olitsky, P.K., and Cox, H.R. (1936). Active immunication of guinea pigs with the virus of equine encephalomyelitis: I. Quantitative experiments with various preparations of active virus. J. Exp. Med. *63*, 311–324.

Olitsky, P.K., and Harford, C.G. (1938). Intraperitoneal and intracerebral routes in serum protection tests with the virus of equine encephalomyelitis: I. A comparison of the two routes in protection tests. J. Exp. Med. *68*, 173–189.

Osterrieth, P., and Blanes-Ridaura, G. (1960). [Research on the Chikungunya virus in the Belgian Congo. I. Isolation of the virus in upper Uele.]. Ann. Soc. Belg. Med. Trop. *40*, 199–203.

Osterrieth, P., Deleplanque-Liegeois, P., and Renoirte, R. (1960). [Research on the Chikungunya virus in the Belgian Congo. II. Serological investigation.]. Ann. Soc. Belg. Med. Trop. *40*, 205–213.

Osterrieth, P., Rathe, E., and Deleplanque-Liegeois, P. (1961). [Simultaneous isolation of the yellow fever and Chikungunya viruses at Bili (Haut-Uele, Belgian Congo).]. Ann. Soc. Belg. Med. Trop. *41*, 207–212.

Ozden, S., Huerre, M., Riviere, J.P., Coffey, L.L., Afonso, P.V., Mouly, V., de Monredon, J., Roger, J.C., El Amrani, M., Yvin, J.L., *et al.* (2007). Human muscle satellite cells as targets of Chikungunya virus infection. PLOS ONE 2, e527.

Paessler, S., Aguilar, P., Anishchenko, M., Wang, H.Q., Aronson, J., Campbell, G., Cararra, A.S., and Weaver, S.C. (2004). The hamster as an animal model for eastern equine encephalitis--and its use in studies of virus entrance into the brain. J. Infect. Dis. *189*, 2072–2076.

Parola, P., de Lamballerie, X., Jourdan, J., Rovery, C., Vaillant, V., Minodier, P., Brouqui, P., Flahault, A., Raoult, D., and Charrel, R.N. (2006). Novel chikungunya virus variant in travelers returning from Indian Ocean islands. Emerg. Infect. Dis. *12*, 1493–1499.

Paul, S.D., and Singh, K.R. (1968). Experimental infection of Macaca radiata with Chikungunya virus and transmission of virus by mosquitoes. Indian J. Med. Res. *56*, 802–811.

Peiris, J.S., Dittus, W.P., and Ratnayake, C.B. (1993). Seroepidemiology of dengue and other arboviruses in a natural population of toque macaques (Macaca sinica) at Polonnaruwa, Sri Lanka. J. Med. Primatol. *22*, 240–245.

Phillips, A.T., Schountz, T., Toth, A.M., Rico, A.B., Jarvis, D.L., Powers, A.M., and Olson, K.E. (2014). Liposome-antigen-nucleic acid complexes protect mice from lethal challenge with western and eastern equine encephalitis viruses. J. Virol. *88*, 1771–1780.

Powers, A.M., and Logue, C.H. (2007). Changing patterns of chikungunya virus: re-emergence of a zoonotic arbovirus. J. Gen. Virol. *88*, 2363–2377.

Pusztai, R., Gould, E.A., and Smith, H. (1971). Infection patterns in mice of an avirulent and virulent strain of Semliki Forest virus. Br. J. Exp. Pathol. *52*, 669–677.

Reed, D.S., Lind, C.M., Sullivan, L.J., Pratt, W.D., and Parker, M.D. (2004). Aerosol infection of cynomolgus macaques with enzootic strains of venezuelan equine encephalitis viruses. J. Infect. Dis. *189*, 1013–1017.

Reed, D.S., Larsen, T., Sullivan, L.J., Lind, C.M., Lackemeyer, M.G., Pratt, W.D., and Parker, M.D. (2005). Aerosol exposure to western equine encephalitis virus causes fever and encephalitis in cynomolgus macaques. J. Infect. Dis. *192*, 1173–1182.

Reed, D.S., Lackemeyer, M.G., Garza, N.L., Norris, S., Gamble, S., Sullivan, L.J., Lind, C.M., and Raymond, J.L. (2007). Severe encephalitis in cynomolgus macaques exposed to aerosolized Eastern equine encephalitis virus. J. Infect. Dis. *196*, 441–450.

Reinarz, A.B., Broome, M.G., and Sagik, B.P. (1971). Age-dependent resistance of mice to sindbis virus infection: viral replication as a function of host age. Infect. Immun. *3*, 268–273.

Reisen, W.K., Kramer, L.D., Chiles, R.E., Martinez, V.M., and Eldridge, B.F. (2000). Response of house finches to infection with sympatric and allopatric strains of western equine encephalomyelitis and St. Louis encephalitis viruses from California. J. Med. Entomol. *37*, 259–264.

Reisen, W.K., Chiles, R.E., Martinez, V.M., Fang, Y., and Green, E.N. (2003). Experimental infection of California birds with western equine encephalomyelitis and St. Louis encephalitis viruses. J. Med. Entomol. *40*, 968–982.

Robin, S., Ramful, D., Le Seach, F., Jaffar-Bandjee, M.C., Rigou, G., and Alessandri, J.L. (2008). Neurologic manifestations of pediatric chikungunya infection. J. Child Neurol. *23*, 1028–1035.

Roques, P., Delache, B., Brochard, P., Joubert, C., de Lamballerie, X., Cherel, Y., Aumont, G., Lotto, V., and Le Grand, R. (2007a). Chloroquine prevention of Chikungunya infection in experimental Macaques model. Paper presented at: Colloque international – Chikungunya et Autres Arboviroses Émergentes en Milieu Tropical (Saint-Pierre, Ile de La Reunion).

Roques, P., Delache, B., Brochard, P., Joubert, C., de Lamballerie, X., Cherel, Y., Aumont, G., Lotto, V., and Le Grand, R. (2007b). Pathogenesis of Chikungunya virus infection in macaques. Paper presented at: Colloque international – Chikungunya et Autres Arboviroses Émergentes en Milieu Tropical (Saint-Pierre, Ile de La Reunion).

Roques, P., Larcher, T., Labadie, K., Joubert, C., Malleret, B., Lebon, P., de Lamballerie, X., Cherel, Y., and Le Grand, R. (2008). Chikungunya outbreak in Reunion: experimental infection of macaques and physiopathology of chikungunya infection. Paper presented at: ArboVirus Ressearch in Coffs Harbour, Australia.

Roques, P., Gras, G., Labadie, K., Larcher, T., Cherel, Y., Suhrbier, A., and Le Grand, R. (2011). Chikungunya virus infection involved monocytes and during chronic phase of the disease persisted in tissue macrophages. 45th Annual Scientific Meeting of the European

Society for Clinical Investigation. Crete, Greece, 13–16 April 2011. Eur J. Clin. Invest. *41*, 23–52.

Ross, R.W. (1956). The Newala epidemic. III. The virus: isolation, pathogenic properties and relationship to the epidemic. J. Hyg. (Lond.) *54*, 177–191.

Rossi, S.L., Guerbois, M., Gorchakov, R., Plante, K.S., Forrester, N.L., and Weaver, S.C. (2013). IRES-based Venezuelan equine encephalitis vaccine candidate elicits protective immunity in mice. Virology *437*, 81–88.

Roy, C.J., Reed, D.S., Wilhelmsen, C.L., Hartings, J., Norris, S., and Steele, K.E. (2009). Pathogenesis of aerosolized Eastern Equine Encephalitis virus infection in guinea pigs. Virol. J. *6*, 170.

Rudd, P.A., Wilson, J., Gardner, J., Larcher, T., Babarit, C., Le, T.T., Anraku, I., Kumagai, Y., Loo, Y.M., Gale, M. Jr, *et al.* (2012). Interferon response factors 3 and 7 protect against Chikungunya virus hemorrhagic fever and shock. J. Virol. *86*, 9888–9898.

Rulli, N.E., Guglielmotti, A., Mangano, G., Rolph, M.S., Apicella, C., Zaid, A., Suhrbier, A., and Mahalingam, S. (2009). Amelioration of alphavirus-induced arthritis and myositis in a mouse model by treatment with bindarit, an inhibitor of monocyte chemotactic proteins. Arthritis Rheum. *60*, 2513–2523.

Rulli, N.E., Rolph, M.S., Srikiatkhachorn, A., Anantapreecha, S., Guglielmotti, A., and Mahalingam, S. (2011). Protection from arthritis and myositis in a mouse model of acute chikungunya virus disease by bindarit, an inhibitor of monocyte chemotactic protein-1 synthesis. J. Infect. Dis. *204*, 1026–1030.

Ryman, K.D., Klimstra, W.B., Nguyen, K.B., Biron, C.A., and Johnston, R.E. (2000). Alpha/beta interferon protects adult mice from fatal Sindbis virus infection and is an important determinant of cell and tissue tropism. J. Virol. *74*, 3366–3378.

Ryzhikov, A.B., Ryabchikova, E.I., Sergeev, A.N., and Tkacheva, N.V. (1995). Spread of Venezuelan equine encephalitis virus in mice olfactory tract. Arch. Virol. *140*, 2243–2254.

Safavi, F., Feliberti, J.P., Raine, C.S., and Mokhtarian, F. (2011). Role of gammadelta T cells in antibody production and recovery from SFV demyelinating disease. J. Neuroimmunol. *235*, 18–26.

Saluzzo, J.F., Gonzalez, J.P., Herve, J.P., and Georges, A.J. (1981). [Serological survey for the prevalence of certain arboviruses in the human population of the south-east area of Central African Republic (author's transl)]. Bull. Soc. Pathol. Exot. Filiales *74*, 490–499.

Saluzzo, J.F., Ivanoff, B., Languillat, G., and Georges, A.J. (1982). [Serological survey for arbovirus antibodies in the human and simian populations of the South-East of Gabon (author's transl)]. Bull. Soc. Pathol. Exot. Filiales *75*, 262–266.

Schafer, A., Whitmore, A.C., Konopka, J.L., and Johnston, R.E. (2009). Replicon particles of Venezuelan equine encephalitis virus as a reductionist murine model for encephalitis. J. Virol. *83*, 4275–4286.

Schilte, C., Couderc, T., Chretien, F., Sourisseau, M., Gangneux, N., Guivel-Benhassine, F., Kraxner, A., Tschopp, J., Higgs, S., Michault, A., *et al.* (2010). Type I IFN controls chikungunya virus via its action on nonhematopoietic cells. J. Exp. Med. *207*, 429–442.

Seamer, J., Randles, W.J., and Fitzgeorge, R. (1967). The course of Semliki Forest virus infection in mice. Br. J. Exp. Pathol. *48*, 395–402.

Seay, A.R., and Wolinsky, J.S. (1982). Ross River virus-induced demyelination: I. Pathogenesis and histopathology. Ann. Neurol. *12*, 380–389.

Seay, A.R., and Wolinsky, J.S. (1983). Ross River virus--induced demyelination: II. Ultrastructural studies. Ann. Neurol. *14*, 559–568.

Seay, A.R., Griffin, D.E., and Johnson, R.T. (1981). Experimental viral polymyositis: age dependency and immune responses to Ross River virus infection in mice. Neurology *31*, 656–660.

Sissoko, D., Malvy, D., Ezzedine, K., Renault, P., Moscetti, F., Ledrans, M., and Pierre, V. (2009). Post-epidemic Chikungunya disease on Reunion Island: course of rheumatic manifestations and associated factors over a 15-month period. PLoS Negl. Trop. Dis. *3*, e389.

Smith, D.R., Adams, A.P., Kenney, J.L., Wang, E., and Weaver, S.C. (2008). Venezuelan equine encephalitis virus in the mosquito vector Aedes taeniorhynchus: infection initiated by a small number of susceptible epithelial cells and a population bottleneck. Virology *372*, 176–186.

Smith, J.P., Morris-Downes, M., Brennan, F.R., Wallace, G.J., and Amor, S. (2000). A role for alpha4-integrin in the pathology following Semliki Forest virus infection. J. Neuroimmunol. *106*, 60–68.

Smith-Norowitz, T.A., Sobel, R.A., and Mokhtarian, F. (2000). B cells and antibodies in the pathogenesis of myelin injury in Semliki Forest Virus encephalomyelitis. Cell. Immunol. *200*, 27–35.

Smithburn, K.C., and Haddow, A.J. (1944). Semliki Forest virus. I. Isolation and pathogenic properties. J. Immunol. *49*, 141–157.

Soden, M., Vasudevan, H., Roberts, B., Coelen, R., Hamlin, G., Vasudevan, S., and La Brooy, J. (2000). Detection of viral ribonucleic acid and histologic analysis of inflamed synovium in Ross River virus infection. Arthritis Rheum. *43*, 365–369.

Solignat, M., Gay, B., Higgs, S., Briant, L., and Devaux, C. (2009). Replication cycle of chikungunya: a re-emerging arbovirus. Virology *393*, 183–197.

Sourisseau, M., Schilte, C., Casartelli, N., Trouillet, C., Guivel-Benhassine, F., Rudnicka, D., Sol-Foulon, N., Le Roux, K., Prevost, M.C., Fsihi, H., *et al.* (2007). Characterization of reemerging Chikungunya virus. PLoS Pathog. *3*, e89.

Sponseller, M.L., Binn, L.N., Wooding, W.L., and Yager, R.H. (1966). Field strains of western encephalitis virus in ponies: virologic, clinical, and pathologic observations. Am. J. Vet. Res. *27*, 1591–1598.

Spouge, J.L. (1992). Statistical analysis of sparse infection data and its implications for retroviral treatment trials in primates. Proc. Natl. Acad. Sci. U.S.A. *89*, 7581–7585.

Steele, K.E., and Twenhafel, N.A. (2010). Review paper: pathology of animal models of Alphavirus encephalitis. Vet. Pathol. *47*, 790–805.

Steele, K.E., Davis, K.J., Stephan, K., Kell, W., Vogel, P., and Hart, M.K. (1998). Comparative neurovirulence and tissue tropism of wild-type and attenuated strains of Venezuelan equine encephalitis virus administered by aerosol in C3H/HeN and BALB/c mice. Vet. Pathol. *35*, 386–397.

Subak-Sharpe, I., Dyson, H., and Fazakerley, J. (1993). In vivo depletion of CD8+ T cells prevents lesions of demyelination in Semliki Forest virus infection. J. Virol. *67*, 7629–7633.

Suckling, A.J., Jagelman, S., and Webb, H.E. (1978a). A comparison of brain lysosomal enzyme activities in four experimental togavirus encephalitides. J. Neurol. Sci. *35*, 355–364.

Suckling, A.J., Pathak, S., Jagelman, S., and Webb, H.E. (1978b). Virus-associated demyelination. A model using avirulent Semliki Forest virus infection of mice. J. Neurol. Sci. *39*, 147–154.

Suhrbier, A., Jaffar-Bandjee, M.C., and Gasque, P. (2012). Arthritogenic alphaviruses--an overview. Nat. Rev. Rheumatol. *8*, 420–429.

Taylor, W.P., and Marshall, I.D. (1975). Adaptation studies with Ross River virus: laboratory mice and cell cultures. J. Gen. Virol. *28*, 59–72.

Tesh, R.B., McLean, R.G., Shroyer, D.A., Calisher, C.H., and Rosen, L. (1981). Ross River virus (Togaviridae: Alphavirus) infection (epidemic polyarthritis) in American Samoa. Trans. R. Soc. Trop. Med. Hyg. *75*, 426–431.

Thach, D.C., Kimura, T., and Griffin, D.E. (2000). Differences between C57BL/6 and BALB/cBy mice in mortality and virus replication after intranasal infection with neuroadapted Sindbis virus. J. Virol. *74*, 6156–6161.

Thach, D.C., Kleeberger, S.R., Tucker, P.C., and Griffin, D.E. (2001). Genetic control of neuroadapted sindbis virus replication in female mice maps to chromosome 2 and associates with paralysis and mortality. J. Virol. *75*, 8674–8680.

Thangamani, S., Higgs, S., Ziegler, S., Vanlandingham, D., Tesh, R., and Wikel, S. (2010). Host immune response to mosquito-transmitted chikungunya virus differs from that elicited by needle inoculated virus. PLOS ONE *5*, e12137.

de Thoisy, B., Gardon, J., Salas, R.A., Morvan, J., and Kazanji, M. (2003). Mayaro virus in wild mammals, French Guiana. Emerg. Infect. Dis. *9*, 1326–1329.

Tournebize, P., Charlin, C., and Lagrange, M. (2009). [Neurological manifestations in Chikungunya: about 23 cases collected in Reunion Island]. Rev. Neurol. (Paris) *165*, 48–51.

Tremain, K.E., and Ikeda, H. (1983). Physiological deficits in the visual-system of mice infected with Semliki Forest virus and their correlation with those seen in patients with demyelinating disease. Brain *106*, 879–895.

Tretyakova, I., Lukashevich, I.S., Glass, P., Wang, E., Weaver, S., and Pushko, P. (2013). Novel vaccine against Venezuelan equine encephalitis combines advantages of DNA immunization and a live attenuated vaccine. Vaccine *31*, 1019–1025.

Turell, M.J., Beaman, J.R., and Tammariello, R.F. (1992). Susceptibility of selected strains of *Aedes aegypti* and *Aedes albopictus* (Diptera: Culicidae) to chikungunya virus. J. Med. Entomol. *29*, 49–53.

Veckenstedt, A., Guttner, J., and Schroeder, C. (1985). Inhibition by Norakin (triperiden) of Sindbis virus infection in mice. Acta Virol. *29*, 209–215.

Victor, J., Smith, D.G., and Pollack, A.D. (1956). The comparative pathology of Venezuelan equine encephalomyelitis. J. Infect. Dis. *98*, 55–66.

Vogel, P., Kell, W.M., Fritz, D.L., Parker, M.D., and Schoepp, R.J. (2005). Early events in the pathogenesis of eastern equine encephalitis virus in mice. Am. J. Pathol. *166*, 159–171.

Walton, T.E., Alvarez, O., Jr., Buckwalter, R.M., and Johnson, K.M. (1973). Experimental infection of horses with enzootic and epizootic strains of Venezuelan equine encephalomyelitis virus. J. Infect. Dis. *128*, 271–282.

Wang, E., Petrakova, O., Adams, A.P., Aguilar, P.V., Kang, W., Paessler, S., Volk, S.M., Frolov, I., and Weaver, S.C. (2007). Chimeric Sindbis/eastern equine encephalitis vaccine candidates are highly attenuated and immunogenic in mice. Vaccine *25*, 7573–7581.

Wang, E., Volkova, E., Adams, A.P., Forrester, N., Xiao, S.Y., Frolov, I., and Weaver, S.C. (2008). Chimeric alphavirus vaccine candidates for chikungunya. Vaccine *26*, 5030–5039.

Weaver, S.C., Scott, T.W., Lorenz, L.H., and Repik, P.M. (1991). Detection of eastern equine encephalomyelitis virus deposition in Culiseta melanura following ingestion of radiolabeled virus in blood meals. Am. J. Trop. Med. Hyg. *44*, 250–259.

Weaver, S.C., Ferro, C., Barrera, R., Boshell, J., and Navarro, J.C. (2004). Venezuelan equine encephalitis. Annu. Rev. Entomol. *49*, 141–174.

Weaver, S.C., Osorio, J.E., Livengood, J.A., Chen, R., and Stinchcomb, D.T. (2012). Chikungunya virus and prospects for a vaccine. Expert Rev. Vaccines *11*, 1087–1101.

Webb, H.E., and Fazakerley, J.K. (1984). Can viral envelope glycolipids produce auto-immunity, with reference to the CNS and multiple sclerosis? Neuropathol. Appl. Neurobiol. *10*, 1–10.

Webb, H.E., Mehta, S., Gregson, N.A., and Leibowitz, S. (1984). Immunological reaction of the demyelinating Semliki Forest virus with immune serum to glycolipids and its possible importance to central nervous system viral auto-immune disease. Neuropathol. Appl. Neurobiol. *10*, 77–84.

Werneke, S.W., Schilte, C., Rohatgi, A., Monte, K.J., Michault, A., Arenzana-Seisdedos, F., Vanlandingham, D.L., Higgs, S., Fontanet, A., Albert, M.L., *et al.* (2011). ISG15 is critical in the control of Chikungunya virus infection independent of UbE1L mediated conjugation. PLoS Pathog. *7*, e1002322.

Williams, A.J., O'Brien, L.M., Phillpotts, R.J., and Perkins, S.D. (2009). Improved efficacy of a gene optimised adenovirus-based vaccine for venezuelan equine encephalitis virus. Virol. J. *6*, 118.

Wolfe, N.D., Kilbourn, A.M., Karesh, W.B., Rahman, H.A., Bosi, E.J., Cropp, B.C., Andau, M., Spielman,

A., and Gubler, D.J. (2001). Sylvatic transmission of arboviruses among Bornean orangutans. Am. J. Trop. Med. Hyg. *64*, 310–316.

Wyckoff, R.W.G., and Tesar, W.C. (1939). Equine Encephalomyelitis in Monkeys. J. Immunol. *37*, 329–343.

Ziegler, S.A., Lu, L., da Rosa, A.P., Xiao, S.Y., and Tesh, R.B. (2008). An animal model for studying the pathogenesis of chikungunya virus infection. Am. J. Trop. Med. Hyg. *79*, 133–139.

Ziegler, S.A., Nuckols, J., McGee, C.E., Huang, Y.J., Vanlandingham, D.L., Tesh, R.B., and Higgs, S. (2011). *In vivo* imaging of chikungunya virus in mice and Aedes mosquitoes using a Renilla luciferase clone. Vector Borne Zoonotic Dis. *11*, 1471–1477.

Zisman, B., Hirsch, M.S., and Allison, A.C. (1970). Selective effects of anti-macrophage serum, silica and anti-lymphocyte serum on pathogenesis of herpes virus infection of young adult mice. J. Immunol. *104*, 1155–1159.

Zlotnik, I., Grant, D.P., and Batter-Hatton, D. (1972). Encephalopathy in mice following inapparent Semliki Forest Virus (S.F.V.) infection. Br. J. Exp. Pathol. *53*, 125–129.

Clinical Manifestations of Arthritogenic Alphaviruses

7

Lara J. Herrero, Adam Taylor, Suan Sin Foo, Lynden Roberts, Natkunam Ketheesan and Suresh Mahalingam

Abstract

Increasing geographical range and the ability of alphaviruses to cause major disease outbreaks highlight their growing importance in clinical practice. Alphaviruses are now the most common cause of infectious arthritis worldwide. A significant number of symptomatic infections result in arthralgia that can incapacitate for weeks. Acute manifestations are often severe and highly debilitating, affecting multiple joints. Furthermore, rheumatic symptoms may persist for months. Chronic arthritis has also been described; however, symptoms beyond 3–6 months can usually be attributed to an alternative disease. Fatigue and malaise often accompany the infection-related arthritis, causing high morbidity, which has an important economic impact. In this chapter, the clinical manifestations associated with some of the more important arthritogenic alphaviruses are described.

Introduction

Arthropod-borne alphaviruses are globally widespread, infecting a large variety of vertebrates and causing considerable human morbidity and mortality. Alphaviruses can cause a range of diseases in humans, which in turn can present with a variety of clinical symptoms. Disease manifestations vary from fever or rashes to significant inflammatory pathologies including encephalitis and severe arthritis. Infection may also resolve asymptomatically. Infections with Old World alphaviruses such as Ross River virus (RRV), Barmah Forest virus (BFV), O'nyong-nyong virus (ONNV), Mayaro virus, chikungunya virus (CHIKV), Sindbis virus (SINV), and Semliki Forest virus (SFV) are often associated with acute febrile illnesses, malaise, maculopapular rashes and severe arthralgias (Kiwanuka *et al.*, 1999; Laine *et al.*, 2004; Munoz and Navarro, 2012; Paquet *et al.*, 2006; Phillips *et al.*, 1990; Rulli *et al.*, 2007; Suhrbier and La Linn, 2004). See Fig. 7.1 for geographic distribution of Old World alphaviruses. In contrast, New World alphaviruses such as Venezuelan equine encephalitis virus (VEEV), Western equine encephalitis virus (WEEV) and Eastern equine encephalitis virus (EEEV) are neuroinvasive, regularly causing potentially fatal encephalitis in infected individuals (Zacks and Paessler, 2010) (see Chapter 8 for details of encephalitic alphaviruses). However, exceptions exist to this method of symptomatic classification. The Old World viruses RRV, CHIKV and SINV are also known to cause neurological disease in infected hosts (Schuffenecker *et al.*, 2006).

Clinical spectrum of infections caused by arthritogenic alphaviruses

Alphaviruses are the most common cause of infectious arthritis. Males and females are equally at risk of alphavirus infection. Based on recent advances on alphaviral tropism, a pathway of infection has been postulated (Fig. 7.2). Following transmission through the bite of an alphavirus-infected mosquito, the virus is able to replicate in the primary site of infection: the epithelial layer and fibroblasts of the skin. Replication of virus attracts a vast number of immune regulators to the virus inoculation site. Viraemia rapidly occurs

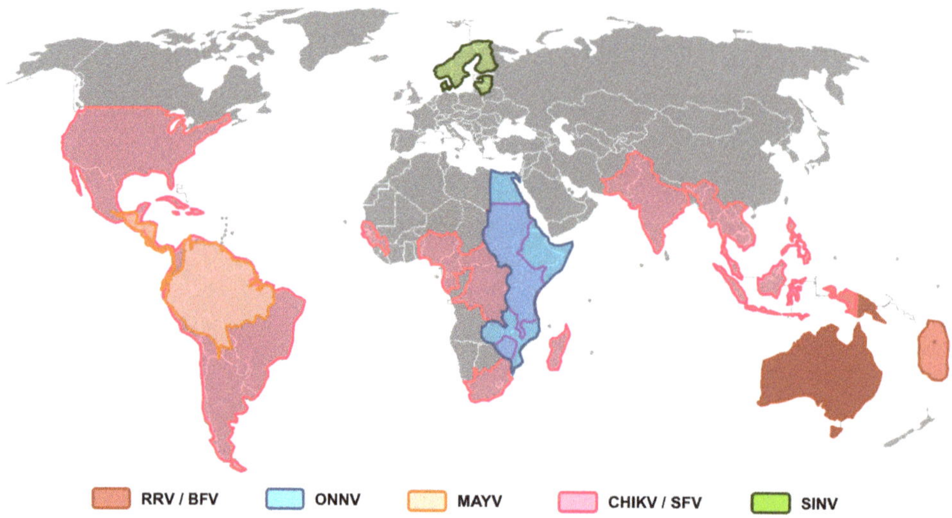

Figure 7.1 Predominant distribution of arthritogenic alphaviruses. Ross River virus (RRV) outbreaks have been reported mainly in Australia (Kelly-Hope *et al*., 2004), Papua New Guinea (Scrimgeour *et al*., 1987) and several islands in the Oceania region (Faragher *et al*., 1985; Fauran *et al*., 1984; Rosen *et al*., 1981). Regions of Barmah Forest virus (BFV) outbreaks overlap with RRV, affecting only Australia (Flexman *et al*., 1998; Passmore *et al*., 2002). O'Nyong Nyong virus (ONNV) is currently endemic to Eastern Africa (Lanciotti *et al*., 1998), while Mayaro virus (MAYV) outbreaks are contained in Central and Southern America (Neumayr *et al*., 2012; Pinheiro *et al*., 1981). Chikungunya virus (CHIKV) outbreaks have been reported in Africa, India, many regions of Asia and now the Americas (Weaver and Lecuit, 2015). Semliki Forest virus (SFV) outbreaks overlap with CHIKV, endemic only in central Africa (Mathiot *et al*., 1990). Sindbis virus (SINV) outbreak is currently endemic in northern Europe (Laine *et al*., 2004).

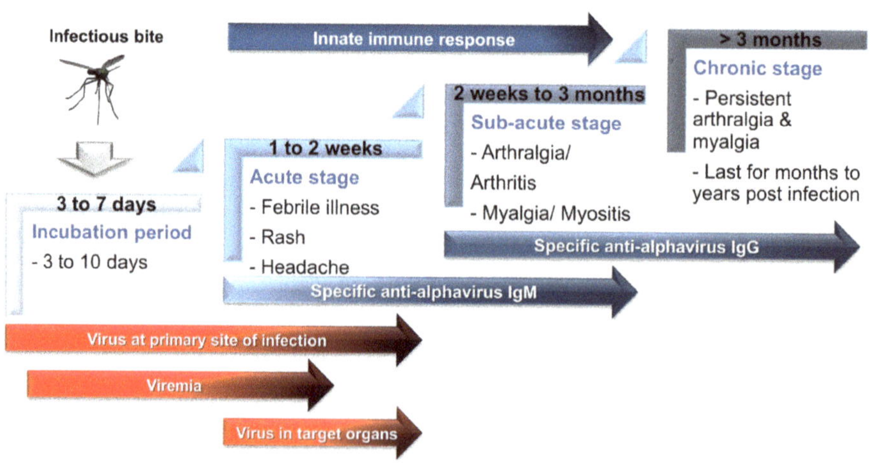

Figure 7.2 Clinical pathogenesis of a typical arthritogenic alphaviral infection. Following an infectious bite, virus replication commences at the primary site of infection and viraemia rapidly occurs, with an incubation period of 3 to 7 days (Harley *et al*., 2001; Schwartz and Albert, 2010). This is followed by the onset of acute clinical symptoms, such as febrile illness, rash and headache 1 to 2 weeks post infection (Carey, 1971; Russell, 2002), and is accompanied by the activation on innate immunity. The innate immune response elicited gives rise to the sub-acute stage of infection, displaying myalgia/myositis and arthralgia/arthritis (Morrison *et al*., 2006). In severe cases, persistent arthralgia or myalgia may be experienced which can last years. Specific anti-alphavirus IgM can be detected around 3 to 7 days post illness onset and persists up to three to four months. Specific anti-alphavirus IgG can be detected around 1 week post illness onset and persists for more than six months.

and infected immune cells then migrate into the lymph node, allowing spread of virus to target organs (Kam *et al.*, 2009). Typically, most patients experience full recovery within 2–3 months.

Disease following infection with arthritogenic alphaviruses is usually self-limiting and characterized by fever, rash, muscle pain, and severe joint pain commonly affecting distal large joints, such as the ankle, knee, wrist and elbow, as well as the small joints in the fingers and toes, but generally sparing the distal interphalangeal joints (Figs 7.3 and 7.4). Patterns of

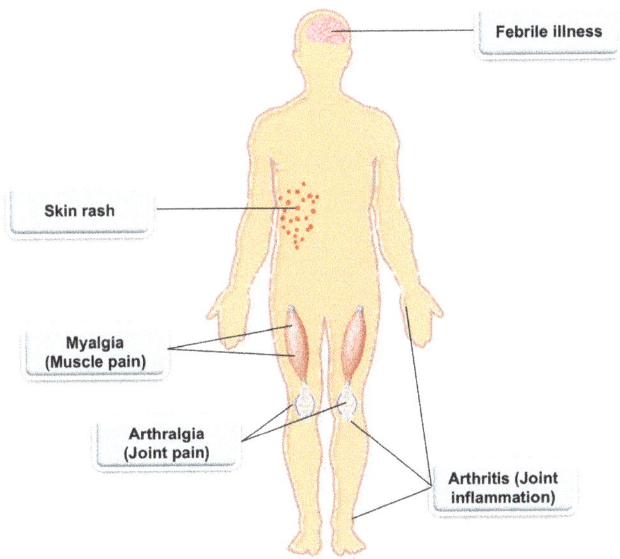

Figure 7.3 Anatomical illustration of alphaviral disease manifestation. Typically, arthritogenic alphaviral disease manifestations include febrile illness, skin rash, arthritis, arthralgia and myalgia.

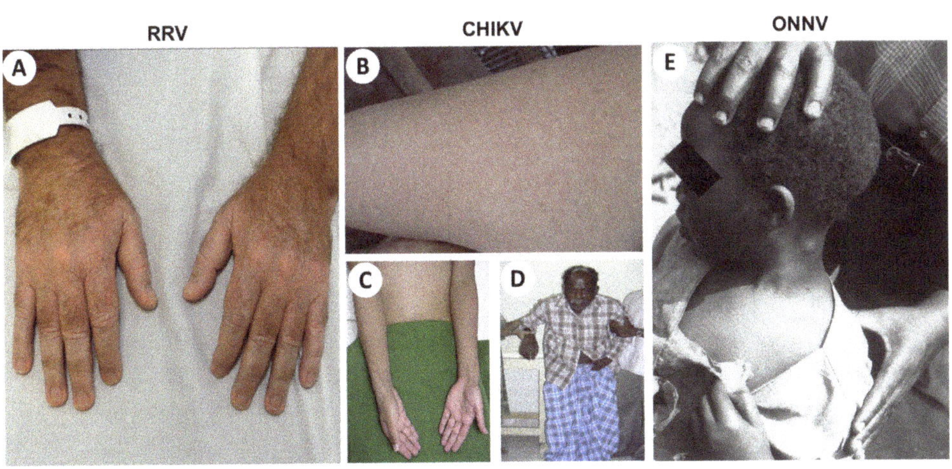

Figure 7.4 Clinical presentations of arthritogenic alphaviral diseases. (A) RRV patients presenting subtle evidence of metacarpophalangeal joint swelling in both hands with some loss of the usual skin depressions between the distal metacarpal heads. The metacarpal joints of both thumbs and second right proximal interphalangeal joint are also swollen. CHIKV patients presenting (B) skin rash on limb, (C) hands and (D) stooped posture due to CHIKV-elicited incapacitating pain. Figure courtesy of Dr Sarma Rachakonda, India. (E) ONNV patient presenting posterior cervical lymphadenopathy (Kiwanuka *et al.*, 1999).

monoarthritis, oligoarthritis and polyarthritis have been described. Resolution of disease symptoms generally occurs over several weeks; however, many patients endure prolonged rheumatic pain lasting six months or longer (Table 7.1). The chronic symptoms and ongoing health issues associated with alphavirus infection and the ability of this group of viruses to cause large-scale epidemics makes alphavirus-associated disease an illness of major socioeconomic concern (Suhrbier and Mahalingam, 2009).

Ross River virus

RRV is endemic to Australia and Papua New Guinea and is associated with outbreaks of polyarthritis and arthralgia in humans. In Australia, up to 8000 cases of RRV infections are recorded annually, making it the most common arboviral disease in Australia. The symptoms of RRV infection were initially described in 1928, and recurring Australian outbreaks of epidemic polyarthritis have continued to be reported since. In 1960, an outbreak of polyarthritis occurred in South Australia affecting more than 1000 individuals. The causative agent was not identified at the time, with paired sera failing to react in haemagglutinin-inhibiting antibody tests to the known group A arboviruses of the time (which included CHIKV, SFV, MAYV, Middelburg virus, VEEV, WEEV, EEEV and SINV) (Shope and Anderson, 1960). Shortly after, in 1963, an alphavirus was isolated from an *Aedes vigilax* mosquito that had been obtained near the Ross River in northern Queensland. Using serological techniques, this new virus was confirmed as the causative agent of epidemic polyarthritis and was named Ross River virus (Doherty *et al.*, 1963). It was not until 1981 that infectious RRV was first isolated from a human patient presenting with epidemic polyarthritis (Aaskov *et al.*, 1981; Nimmo, 1928).

The disease is debilitating at onset and has been compared in severity to that experienced by patients with osteoarthritis, rheumatoid arthritis, or the arthritis of Lyme disease (Mylonas *et al.*, 2002). Generally, the symptoms of RRV infection resolve within days to weeks, and in almost all patients the disease resolves within three to six months. Subclinical infection is also common and presumably confers immunity, although studies have not been done to test this. Some patients experience severe arthralgia, myalgia, and lethargy beyond six months. In addition, recurrent or persistent arthralgia has also been reported in patients after resolution of acute RRV disease symptoms (Proll *et al.*, 1999). The incubation period for RRV disease ranges from 3–21 days and symptoms can vary greatly from person to person. Classically, patients present with arthritis, malaise, fever, headache and a maculopapular erythematous rash commonly affecting the limbs and trunk. The rash, which is not usually itchy, may persist for 5–10 days before spontaneous resolution. Acute-onset arthritis is the predominant symptom in most patients with RRV infection, causing joint swelling, heat and tenderness (Jacups *et al.*, 2008). Monoarthritis (one joint), oligoarthritis (up to six joints), and polyarthritis (more than six joints) patterns of joint involvement have been described. Symmetrical joint pain is typical and commonly involved joints include the wrists, metacarpophalangeal, proximal interphalangeal finger joints, ankles and knees. Less common symptoms of acute infection include nausea, sore eyes, lymphadenopathy and diarrhoea (Pollard and Benz, 1998). Glomerulonephritis has also been reported in one case of acute phase RRV disease (Fraser *et al.*, 1988). Encephalitis has also been reported as a possible complication (Lucas and Qiao, 1999; Scrimgeour, 1999).

Barmah Forest virus

RRV causes 90% of the total cases of epidemic polyarthritis recorded annually in Australia. BFV is responsible for the remaining 10%, making it the second most common arboviral disease in Australia (Rulli *et al.*, 2007). Like RRV, epidemics of BFV are also frequently described, with 1895 cases notified in Australia during the 2005/2006-summer period (Jacups *et al.*, 2008). BFV is found only in mainland Australia and, like RRV disease, arthritis is the predominant symptom of infection. The most common symptoms of BFV infection are acute arthritis or arthralgia, myalgia, fever, rash and headache. These symptoms can occur individually or in combination. The disease affects males and females equally, and has an incubation period of about 7–9 days (Mackenzie and

Table 7.1 Overview of clinical features observed during arthritogenic alphaviral diseases

Features	RRV	BFV	ONNV	MAYV	CHIKV	SINV	SFV
Incubation period	8 days (range of 3 to 21 days)		Not yet determined	7 to 12 days	3 to 7 days	»4 days (range: 2–18 days)	Not yet determined
Asymptomatic population (%)	»50	Majority asymptomatic, exact % not determined	Not yet determined	»8	»10 to 15	Majority asymptomatic, exact % not determined	Majority asymptomatic, exact % not determined
Typical clinical features	Polyarthritis, athralgia, maculopapular rash, fever, fatigue and myalgia	Polyarthritis, arthralgia, rash (particularly on palm), low-grade fever, fatigue, myalgia	Arthralgia, maculopapular rash, fever, pruritus, myalgia and lymphadenopathy (particularly posterior cervical region)	Arthralgias, fever, headache, maculopapular rash, myalgias, retro-orbital pain and articular oedema	Athralgia, petechial or maculopapular rash, fever, chills, headache, nausea, photophobia and vomiting	Polyarthritis, fever, rash, fatigue, myalgia, malaise, headache and nausea	Fever, severe persistent headaches, myalgia, and arthralgia
Other clinical symptoms (less common)	Lymphadenopathy, sore throat, coryza, headache, neck stiffness, and photophobia	Headache and stiffness	Headache, red eyes and bleeding gums	Pruritus, nausea, cough, sore throat, anorexia, abdominal pain, nasal congestion, swollen lymph nodes, vomiting and bleeding gums	Neurological, haemorrhagic and ocular manifestations	Retro-orbital pain	Not reported
Complications	Prolong arthralgia, myalgia and fatigue	Prolong arthralgia, myalgia and fatigue experienced by »10% patients	Prolong athralgia	Prolong athralgia	Prolong joint pain experienced by 30 to 40% patients	Prolong athralgia experienced by >50% patients.	Meningoencephal omyand mortality (rare cases)
Diagnosis	Detection of >4-fold rise in RRV IgG antibody; positive RRV IgM	Detection of >4-fold rise in BFV IgG antibody; positive BFV IgM	ONNV-specific real-time polymerase chain reaction (RT-PCR) assay; detection of >4-fold rise in ONNV IgG antibody; positive ONNV IgM	Detection of >4-fold rise in MAYV IgG antibody; positive MAYV IgM	CHIKV-specific real-time polymerase chain reaction (RT-PCR) assay; detection of >4-fold rise in CHIKV IgG antibody; positive CHIKV IgM	Detection of >4-fold rise in SINV IgG antibody; positive SINV virus IgM	Detection of SFV IgG and IgM antibodies
Treatment	No specific treatment. Symptomatic treatment with Non-steroidal anti-inflammatory drugs (NSAIDs)	No specific treatment. Symptomatic treatment with NSAIDs(NSAIDs)	No specific treatment. Symptomatic treatment with NSAIDs	No specific treatment. Symptomatic treatment with NSAIDs	No specific treatment. Symptomatic treatment with NSAIDs)	No specific treatment. Symptomatic treatment with NSAIDs	No specific treatment. Symptomatic treatment with NSAIDs
References	Curran *et al*. (1997), Harley *et al*. (2001)	Jacups *et al*. (2008)	Kiwanuka *et al*. (1999), Lutwama *et al*. (1999), Posey *et al*. (2005), Williams et *al*. (1965)	Azevedo *et al*. (2009), Taylor *et al*. (2005), Tesh (1982), Tesh *et al*. (1999), Torres *et al*. (2004)	Lemant *et al*. (2008), Morrison (1979), Mourya and Mishra (2006), Schwartz and Albert (2010), Yazdani and Kaushik (2007)	Kurkela *et al*. (2005), Sane *et al*. (2011), Turunen *et al*. (1998)	Fazakerley (2002), Mathiot *et al*. (1990), Willems *et al*. (1979)

Smith, 1996). Less common symptoms of BFV infection include kidney inflammation, glomerulonephritis and Guillain-Barre syndrome (Katz *et al.*, 1997; Phan *et al.*, 1998). In general, arthritis is a more prominent symptom of RRV infection, while rash is often more common and florid in BFV-infected patients (Flexman *et al.*, 1998). Moreover, patients with RRV infection appear to be more likely to suffer arthralgia and myalgia for longer, up to 6 months, whereas its been reported that only 10% of patients with BFV suffer for this extended duration (Jacups *et al.*, 2008). Due to the similarity in the clinical symptoms presented by infection with RFV or RRV, formal diagnosis is often based on rising antibody titres observed by serological testing (Flexman *et al.*, 1998). Importantly, neither RRV nor BFV arthritic disease symptoms are associated with long-term damage to the affected joints.

O'nyong-nyong virus

ONNV and CHIKV are antigenically related, and both were isolated in the 1950s in Africa. Specific symptoms of ONNV infection were first recognized among the Acholi people of northwest Uganda in 1959 (Haddow *et al.*, 1960). The virus was subsequently isolated from both human blood and *Anopheles* mosquitoes in Gulu, Uganda, and given the designation O'nyong nyong, or 'weakening of the joints', as the local term to describe the disease. The virus, which continues to circulate endemically in East Africa, produces a similar disease presentation to CHIKV infection, with an acute febrile illness that is associated with severe arthralgic symptoms. Arthralgia is often prolonged, incapacitating patients for up to 2 weeks (Kiwanuka *et al.*, 1999). Patients infected with ONNV regularly experience migratory polyarthralgias or arthritis in the small joints of the hands, feet, wrists, and ankles. Many patients present with an itchy maculopapular rash and a number of cases have experienced paraesthesias (Posey *et al.*, 2005). Joint swelling, diffuse myalgia and back or shoulder pain have also been described (Laine *et al.*, 2004). ONNV infection also causes morning stiffness, and epistaxis has been reported in a number of cases (Naides, 2003; Shore, 1961). In some circumstances, eye pain and reddening with no corresponding discharge, occasional chest pain, and a general malaise have been reported (Rwaguma *et al.*, 1997). Lymphadenopathy of the posterior cervical lymph node is a commonly reported clinical manifestation of O'nyong nyong fever and is considered by some to be the principal clinical manifestation that differentiates ONNV infection from CHIKV infection in patients with high fever (Jupp and McIntosh, 1988). However, a number of patients infected with CHIKV have presented with enlarged cervical lymph nodes during recent outbreaks of CHIKV (Retuya *et al.*, 1998).

Between 1959 and 1962 there was a large outbreak of ONNV in East Africa involving an estimated 2 million people. The outbreak began in Uganda and spread through Kenya and Tanzania (Posey *et al.*, 2005; Rwaguma *et al.*, 1997; Williams *et al.*, 1965). Although there are limited clinical reports of this outbreak, the predominant clinical features were low-grade fever, symmetrical polyarthralgia, generalized maculopapular exanthem and cervical lymphadenopathy (Shore, 1961).

In 1996–1997 ONNV re-emerged in south-central Uganda (Kiwanuka *et al.*, 1999). During this outbreak all patients presented with arthralgia and fever. The majority also reported headache and rash, and almost half the cases reported signs of lymphadenopathy and red eyes (Kiwanuka *et al.*, 1999). The median duration of arthralgia was 6–7 days, with a range of 1–90 days, while the median period of joint immobilization was 4 days, with a range of 1–14 days (Kiwanuka *et al.*, 1999).

The most recent outbreak of ONNV occurred in 2003 in western Côte d'Ivoire. Patients presented with fever and rash, which was originally thought to be due to measles, but was later attributed to ONNV by the Centers for Disease Control and Prevention (CDC). During this outbreak, the major clinical findings were reported as fever (90%), rash (84%), pruritus (87%), arthralgia (58%), and myalgia (71%) (Posey *et al.*, 2005).

It is worth noting that a related virus, Igbo Ora virus, was first isolated in western Nigeria in 1967. Originally thought to be a separate species of the *Alphavirus* genus with clinical manifestations akin to ONNV, this virus has since been shown to be an antigenic variant of ONNV (Lanciotti *et al.*, 1998; Powers *et al.*, 2000).

Mayaro virus

Mayaro virus (MAYV) was first isolated in Trinidad in 1954 from patients presenting with acute febrile illness and is now endemic to tropical South America (Anderson *et al.*, 1957). MAYV infections have an incubation period of 8–12 days and are classically associated with persons who live or work in tropical forests or rural communities of the Amazon Basin (Causey and Maroja, 1957; Pinheiro *et al.*, 1981). Typical manifestations of MAYV disease last up to 5 days and consist of fever, chills, vomiting, retro-orbital pain, generalized myalgia, arthralgia, diarrhoea, frontal headaches, and finish with the appearance of a non-pruritic maculopapular rash (Munoz and Navarro, 2012). The primary characteristic of this non-fatal disease is severe arthralgia immobilizing patients for up to two months (Taylor *et al.*, 2005). Joint stiffness after inactivity is a major complaint. Although MAYV disease is largely self-limiting, an exceptional case of recurrent arthralgias has been reported in a patient known to have had a previous MAYV infection. The individual described similar recurrences of arthralgias, on average every six months since the resolution of the original disease symptoms (Taylor *et al.*, 2005). Less frequently reported symptoms include sore throat, abdominal pain, nasal congestion and haemorrhagic manifestations including bleeding gums (Munoz and Navarro, 2012).

Cases of MAYV infection and outbreaks have been documented in Trinidad, Surinam, Brazil, Bolivia, French Guinea and Peru (Tesh *et al.*, 1999). Recognized outbreaks of MAYV have been relatively small, more than likely due to the association with rural communities with limited populations (Causey and Maroja, 1957; Munoz and Navarro, 2012). However, inadequate surveillance and diagnostic laboratories in MAYV endemic areas suggests that the incidence of MAYV infection is potentially heavily under reported.

Interestingly, imported cases of MAYV infection have been recently reported in three European countries, demonstrating a potential for this virus to spread to non-endemic areas (Hassing *et al.*, 2010; Neumayr *et al.*, 2012; Receveur *et al.*, 2010). In 2008 a Dutch couple returning from Surinam developed symptoms of fever, rash and arthralgia. MAYV diagnosis was confirmed 7 weeks after initial presentation of illness. Despite receiving symptomatic treatment, the couple continued to suffer with persistent arthralgia for more than 12 months after acute infection (Hassing *et al.*, 2010). In 2010, a male traveller returning to France after time in the Amazon Basin of Brazil presented with ongoing and incapacitating arthralgia that was diagnosed as acute MAYV infection (Receveur *et al.*, 2010). This patient was diagnosed with acute IgM+ MAYV infection despite symptoms of fever and polyarthritis first presenting while in Brazil 2 months earlier. In 2011, a Swiss traveller returned home after a 3.5-week trip to the Amazon Basin of northern Peru. During the second week of his travels the man developed chills, malaise, frontal headache, cervical and inguinal lymphadenopathy, myalgia, non-pruritic maculopapular rash and pronounced polyarthalgic pain. Initially diagnosed with dengue, the patient's symptoms improved within a week, with the exception of the persisting polyarthritis and arthralgia, which continued for several months. Upon returning to Switzerland he was positively diagnosed with MAYV infection with no evidence of dengue infection (Neumayr *et al.*, 2012)

Chikungunya virus

Chikungunya virus (CHIKV) was first isolated from a febrile patient during an outbreak of dengue-like disease in southern Tanzania in 1952–53 (Robinson, 1955). Owing to the severity and rapid onset of arthralgia, the disease became known as 'chikungunya' in the local Makonde language, or 'that which bends up'. The term refers to the bent or stooping posture adopted by patients who experience debilitating arthralgic clinical manifestations during infection (Pialoux *et al.*, 2007).

The average incubation period ranges from 3 to 7 days and is followed by the sudden onset of chikungunya fever (CHIKF) (Burt *et al.*, 2012; Schwartz and Albert, 2010), although asymptomatic infection occurs in approximately 15% of infected individuals (Lemant *et al.*, 2008). Two stages of CHIKV disease have been described: a severe febrile and eruptive polyarthritis, followed by a disabling peripheral rheumatism (Simon *et al.*, 2007). The most commonly observed symptoms associated with the acute phase of CHIKV infection in patients include a high fever (up to 40°C) (99.6%), headache (84.1%), myalgia (97.7%) (Paquet *et al.*, 2006), a transient maculopapular rash (75%) found on the trunk, and joint pain and generalized pruritus (15%) (Taubitz *et al.*, 2007). Almost all patients (99.2%) experience migratory polyarthralgia in multiple joints (Paquet *et al.*, 2006). The arthralgia is often symmetrical and commonly affects peripheral joints such as the fingers, wrists, ankles, elbows, toes, and knees. Joints can also show signs of soft tissue swelling (Borgherini *et al.*, 2007). Patients regularly experience acute difficulty, over a period of 2–3 days, in performing basic daily activities such as walking, lying down and standing straight (Sudeep and Parashar, 2008).

CHIKF is a self-limiting disease, and acute clinical manifestations of CHIKV infection usually resolve within 7–10 days. Febrile illness usually subsides within 3–7 days, whereas arthralgia typically persists for around two months (Taubitz *et al.*, 2007). Although most patients recover completely following the acute stage of the disease, approximately 40% of infected individuals experienced rheumatic symptoms that persist for a number of months or even years (Burt *et al.*, 2012; Schwartz and Albert, 2010). Persistence of rheumatic symptoms in 43–75% of CHIKF patients was reported two years post onset of illness (Gerardin *et al.*, 2012; Marimoutou *et al.*, 2012). A number of studies have reported on the chronic arthritic stages of CHIKV disease. Findings suggest that CHIKV-induced arthritis in the chronic stage mimics the clinical presentation of rheumatoid arthritic syndrome (RA) (Manimunda *et al.*, 2010). Furthermore, a study on the severity of diagnosed RA, in a follow-up of 21 individuals known to have previously been infected with CHIKV, revealed most patients showed signs of erosive arthritis and joint space narrowing in the hands and feet (Bouquillard and Combe, 2009). Patients with chronic CHIKV disease may also develop hypertrophic tenosynovitis, commonly involving the extensors and flexors of wrists and fingers (Simon *et al.*, 2007). Clinical evidence of erosive arthritis, joint effusion, marrow oedema, synovial thickening, tendinitis and tenosynovitis in chronic cases of CHIKF was revealed with the use of magnetic resonance imaging (MRI) (Manimunda *et al.*, 2010; Mizuno *et al.*, 2011).

More atypical clinical manifestations of CHIKV infection, including pre-renal failure, pneumonia and cardiovascular and neurological disorders, are often associated with underlying medical conditions (Economopoulou *et al.*, 2009). The severity of these atypical symptoms was first described during the 2005–2006 La Réunion outbreak of CHIKV where significant mortality was observed (Renault *et al.*, 2007). More recently, biological abnormalities such as elevated liver enzyme and creatine levels, anaemia and lymphocytopenia were also reported in severe cases of CHIKF with high viral load ($\geq$ 100,000 DNA copies) (Nkoghe *et al.*, 2012). Neutropenia (counts $< 1 \times 10^3$ cells/mm^3) and thrombocytopenia (platelets < 150,000 cells/mm^3) were also observed in 66.7% and 89% of CHIKF patients, respectively, in the 2008 Malaysia outbreak (Chew and Chua, 2009). The elderly, infants and those with concomitant diseases such as diabetes, ischaemic heart disease, alcoholic liver disease or renal failure were most at risk of developing severe forms of CHIKV disease. A clinical study conducted on 97 CHIKF patients from the 2008 Singapore outbreak suggested females with lower creatine level were to be more likely to develop persistent athralgia (Win *et al.*, 2010). Common complications in CHIKV infection included abrupt onset high-grade fever, meningoencephalitis (Paquet *et al.*, 2006), acute hepatitis (Lemant *et al.*, 2008) and severe skin rashes (Pialoux *et al.*, 2007). Vesiculobullous lesions were also observed in 12.3% of patients who were admitted to the hospital in Kerala, India, in 2009, during the widespread CHIKV epidemic (Pakran *et al.*, 2011)

Although CHIKV is classically grouped as an arthritogenic alphavirus, reports since the 1960s have also described neurological manifestations associated with CHIKV infection (Nimmannitya *et al.*, 1969). The neurological disorders associated with severe cases of atypical CHIKV infection during the Réunion outbreak included encephalitis, malaise, meningoencephalitis, syndrome of meningeal irritation and cerebellar syndrome (Economopoulou *et al.*, 2009). Again, a number of these patients had underlying disorders such as diabetes mellitus, epilepsy, hypertension, and stroke. A recent case report of a 55-year-old male CHIKF patient also revealed clinical evidence of reversible demyelinating encephalitis (Gauri *et al.*, 2012). Obesity has also been suggested as one of the risk factors in contributing towards the development of chronic arthritis, likely due to the higher production of inflammatory cytokines in the adipose tissues due to chronic disruption of metabolic homeostasis (Padmakumar *et al.*, 2010).

Sindbis virus

SINV is found world-wide and is the most widely distributed of all known alphaviruses (Tesh, 1982). It was first identified in 1952, being isolated from *Culex* mosquitoes collected in the village of Sindbis near Cairo (Taylor *et al.*, 1955). The first description of clinical symptoms of SINV infection came from a single case in Uganda in 1961 and included fever, malaise, pains in the joints, muscles and tendons, and rash (Malherbe *et al.*, 1963).

Currently SINV circulates endemically in Europe, Africa, Asia and Oceania (Tesh, 1982). The majority of clinical cases of SINV are reported in Europe, particularly in Sweden, Finland and Russia, where it is the causative agent of Ockelbo disease, Pogosta disease and Karelian fever, respectively. These syndromes are named according to the regions where they circulate, despite being similar in clinical presentations and causing similar morbidity. The major symptoms of SINV infection are arthritis, arthralgia and myalgia, in addition to fever, fatigue, headache and rash (Laine *et al.*, 2004). Along with the acute form of infection, SINV frequently leads to chronic symptoms. The musculoskeletal symptoms of SINV-associated disease have been documented as lasting up to 2.5 years after onset (Laine *et al.*, 2000), similar to what has been observed in CHIKV patients.

SINV outbreaks occur in Europe in common cycles. In 1981 Sweden reported 54 cases (Ockelbo), Russia 200 cases (Karelian) and Finland 300 cases (Pogosta) of clinically and serologically diagnosed SINV infection (Brummer-Korvenkontio *et al.*, 2002; Espmark and Niklasson, 1984; L'Vov *et al.*, 1982). Outbreaks of Pogosta disease have thus far emerged every seven years (for reasons that remain currently unknown) since the first outbreak was noted in 1974, including a large outbreak in 2002 (Sane *et al.*, 2010).

Semliki Forest virus

SFV was first isolated in 1942 in Uganda from *Aedes abnormalis* mosquitoes in the Semliki Forest (Smithburn and Haddow, 1944). Shortly after its initial isolation, SFV was found to be neurotropic in mice (Smithburn *et al.*, 1946). This early finding has led to SFV being one of the most extensively studied members of the alphavirus genus, despite rarely having been documented as a cause of disease in humans. The few reported cases of human disease have involved headache, arthralgia and myalgia, presentations that are commonly associated with other alphavirus disease manifestations. The only recorded outbreak of SFV to date occurred in 1987 in the Central African Republic (Mathiot *et al.*, 1990). The clinical features reported in this outbreak included headache, fever, arthralgia and myalgia in all cases. Less common symptoms included abdominal pain, diarrhoea and conjunctivitis. The headaches were reported as being severe and prolonged, lasting for up to 15 days. The acute phase of infection lasted between 2 and 4 days, with a lengthy convalescence and notable aesthenia (Mathiot *et al.*, 1990). Interestingly, it was reported that the SFV syndrome in this outbreak was less severe than that frequently reported for other related alphaviruses, including RRV, CHIKV and ONNV. In particular, there were no signs of rash or lymphadenopathy (Mathiot *et al.*, 1990).

It is worth noting that there has been a single case of neurological disease, resulting in death, attributed to a laboratory-acquired case of SFV infection (Willems *et al.*, 1979). This unusual case involved a research scientist working with a particularly virulent strain of SFV who became ill with viral meningoencephalomyelitis and succumbed after respiratory and cardiac arrest. The route of infection remains unknown. Clinicians reporting on the case concluded that there were potential extenuating circumstances, including the patient being previously immunocompromised, the virulent SFV strain, and an unknown route and dosage which may have contributed to the extraordinary course of infection (Willems *et al.*, 1979).

Treatment of alphavirus infection

Most primary therapies aimed at alleviating the arthritic symptoms of alphaviral disease rely on non-steroidal anti-inflammatory drugs; however, these often provide only partial relief. No specific treatments or preventative vaccines are currently available against any of the old world alphaviruses. The increasing incidence and severity of alphaviral outbreaks highlights the pressing need to develop cost-effective and efficacious antiviral strategies, including vaccines and drugs, to alleviate the arthritogenic manifestations caused by this group of viruses. See Chapter 10 for a case study on chikungunya virus that describes approaches on vaccines and therapeutics.

Conclusion

Arthritogenic alphaviruses are endemic to a number of countries worldwide and have the potential to cause explosive epidemics. Alphavirus-induced arthritis is a self-limiting disease often associated with febrile illness, rash and myalgia in symptomatic infections. With no vaccine currently available, treatment with analgesics and non-steroidal anti-inflammatory drugs (NSAIDs) aims to relieve arthritic symptoms. Arthritic disease manifestations can be severely debilitating, and infection in the elderly, infants and patients with underlying illnesses can prove fatal. This, together with the expanding geographical range and vector competence of this group of viruses, probably as a result of increased international travel, climate change and viral mutation, makes alphavirus-induced disease an area of intense study.

References

Aaskov, J.G., Mataika, J.U., Lawrence, G.W., Rabukawaqa, V., Tucker, M.M., Miles, J.A.R., and Dalglish, D.A. (1981). An epidemic of ross river virus-infection in Fiji, 1979. Am. J. Trop. Med. Hyg. *30*, 1053–1059.

Anderson, C.R., Downs, W.G., Wattley, G.H., Ahin, N.W., and Reese, A.A. (1957). Mayaro virus: a new human disease agent. II. Isolation from blood of patients in Trinidad, B.W.I. Am. J. Trop. Med. Hyg. *6*, 1012–1016.

Borgherini, G., Poubeau, P., Staikowsky, F., Lory, M., Le Moullec, N., Becquart, J.P., Wengling, C., Michault, A., and Paganin, F. (2007). Outbreak of chikungunya on Reunion Island: early clinical and laboratory features in 157 adult patients. Clin. Infect. Dis. *44*, 1401–1407.

Bouquillard, E., and Combe, B. (2009). Rheumatoid arthritis after Chikungunya fever: a prospective follow-up study of 21 cases. Ann. Rheum. Dis. *68*, 1505–1506.

Brummer-Korvenkontio, M., Vapalahti, O., Kuusisto, P., Saikku, P., Manni, T., Koskela, P., Nygren, T., Brummer-Korvenkontio, H., and Vaheri, A. (2002). Epidemiology of Sindbis virus infections in Finland 1981–96: possible factors explaining a peculiar disease pattern. Epidemiol. Infect. *129*, 335–345.

Burt, F.J., Rolph, M.S., Rulli, N.E., Mahalingam, S., and Heise, M.T. (2012). Chikungunya: a re-emerging virus. Lancet *379*, 662–671.

Carey, D.E. (1971). Chikungunya and dengue: a case of mistaken identity? J. Hist. Med. Allied Sci. *26*, 243–262.

Causey, O.R., and Maroja, O.M. (1957). Mayaro virus: a new human disease agent. III. Investigation of an epidemic of acute febrile illness on the river Guama in Para, Brazil, and isolation of Mayaro virus as causative agent. Am. J. Trop. Med. Hyg. *6*, 1017–1023.

Chew, L.P., and Chua, H.H. (2009). Outbreak of chikungunya in Johor Bahru, Malaysia: clinical and laboratory features of hospitalized patients. Med. J. Malaysia *64*, 220–222.

Doherty, R.L., H., W.R., M., G.B., and K., O.G.A. (1963). The isolation of a third group A arbovirus in Australia with preliminary observations on its relationship to epidemic polyarthritis. Aust. J. Sci. *26*, 183–184.

Economopoulou, A., Dominguez, M., Helynck, B., Sissoko, D., Wichmann, O., Quenel, P., Germonneau, P., and Quatresous, I. (2009). Atypical Chikungunya virus infections: clinical manifestations, mortality and risk factors for severe disease during the 2005–2006 outbreak on Reunion. Epidemiol. Infect. *137*, 534–541.

Espmark, A., and Niklasson, B. (1984). Ockelbo disease in Sweden: epidemiological, clinical, and virological data from the 1982 outbreak. Am. J. Trop. Med. Hyg. *33*, 1203–1211.

Faragher, S.G., Marshall, I.D., and Dalgarno, L. (1985). Ross River virus genetic variants in Australia and the Pacific Islands. Aust. J. Exp. Biol. Med. Sci. *63 (Pt 4)*, 473–488.

Fauran, P., Donaldson, M., Harper, J., Oseni, R.A., and Aaskov, J.G. (1984). Characterization of Ross River viruses isolated from patients with polyarthritis in New Caledonia and Wallis and Futuna Islands. Am. J. Trop. Med. Hyg. 33, 1228–1231.

Flexman, J.P., Smith, D.W., Mackenzie, J.S., Fraser, J.R., Bass, S.P., Hueston, L., Lindsay, M.D., and Cunningham, A.L. (1998). A comparison of the diseases caused by Ross River virus and Barmah Forest virus. Med. J. Aust. *169*, 159–163.

Fraser, J.R., Cunningham, A.L., Muller, H.K., Sinclair, R.A., and Standish, H.G. (1988). Glomerulonephritis in the acute phase of Ross River virus disease (epidemic polyarthritis). Clin. Nephrol. *29*, 149–152.

Gauri, L.A., Ranwa, B.L., Nagar, K., Vyas, A., and Fatima, Q. (2012). Post chikungunya brain stem encephalitis. J. Assoc. Physicians India *60*, 68–70.

Gerardin, P., Fianu, A., Malvy, D., Mussard, C., Boussaid, K., Rollot, O., Michault, A., Gauzere, B.A., Breart, G., and Favier, F. (2012). [Perceived morbidity and community burden of chikungunya in La Reunion]. Med. Trop. (Mars) *72*, 76–82.

Haddow, A.J., Davies, C.W., and Walker, A.J. (1960). O'nyong-nyong fever: an epidemic virus disease in east Africa. Trans. R. Soc. Trop. Med. Hyg. *54*, 517–522.

Harley, D., Sleigh, A., and Ritchie, S. (2001). Ross River virus transmission, infection, and disease: a cross-disciplinary review. Clin. Microbiol. Rev. *14*, 909–932, table of contents.

Hassing, R.J., Leparc-Goffart, I., Blank, S.N., Thevarayan, S., Tolou, H., van Doornum, G., and van Genderen, P.J. (2010). Imported Mayaro virus infection in the Netherlands. J. Infect. *61*, 343–345.

Jacups, S.P., Whelan, P.I., and Currie, B.J. (2008). Ross River virus and Barmah Forest virus infections: a review of history, ecology, and predictive models, with implications for tropical northern Australia. Vector Borne Zoonotic Dis. *8*, 283–297.

Jupp, P.G., and McIntosh, B.M. (1988). Chikungunya virus disease. In The Arboviruses: Epidemiology And Ecology, T.P. Monath, ed. (CRC Press, Boca Raton, FL, USA), pp. 137–157.

Kam, Y.W., Ong, E.K., Renia, L., Tong, J.C., and Ng, L.F. (2009). Immuno-biology of Chikungunya and implications for disease intervention. Microbes Infect. *11*, 1186–1196.

Katz, I.A., Hale, G.E., Hudson, B.J., Ibels, L.S., Eckstein, R.P., and Dermott, P.L. (1997). Glomerulonephritis secondary to Barmah Forest virus infection. Med. J. Aust. *167*, 21–23.

Kelly-Hope, L.A., Purdie, D.M., and Kay, B.H. (2004). Differences in climatic factors between Ross River virus disease outbreak and nonoutbreak years. J. Med. Entomol. *41*, 1116–1122.

Kiwanuka, N., Sanders, E.J., Rwaguma, E.B., Kawamata, J., Ssengooba, F.P., Najjemba, R., Were, W.A., Lamunu, M., Bagambisa, G., Burkot, T.R., *et al.* (1999). O'nyong-nyong fever in south-central Uganda, 1996–1997: clinical features and validation of a clinical case definition for surveillance purposes. Clin. Infect. Dis. *29*, 1243–1250.

L'Vov, D.K., Skvortsova, T.M., Kondrashina, N.G., Vershinskii, B.V., and Lesnikov, A.L. (1982). [Etiology of Karelia fever--a new arbovirus infection]. Vopr. Virusol. *27*, 690–692.

Laine, M., Luukkainen, R., Jalava, J., Ilonen, J., Kuusisto, P., and Toivanen, A. (2000). Prolonged arthritis associated with sindbis-related (Pogosta) virus infection. Rheumatology (Oxford) *39*, 1272–1274.

Laine, M., Luukkainen, R., and Toivanen, A. (2004). Sindbis viruses and other alphaviruses as cause of human arthritic disease. J. Int. Med. *256*, 457–471.

Lanciotti, R.S., Ludwig, M.L., Rwaguma, E.B., Lutwama, J.J., Kram, T.M., Karabatsos, N., Cropp, B.C., and Miller, B.R. (1998). Emergence of epidemic O'nyong-nyong fever in Uganda after a 35-year absence: genetic characterization of the virus. Virology *252*, 258–268.

Lemant, J., Boisson, V., Winer, A., Thibault, L., Andre, H., Tixier, F., Lemercier, M., Antok, E., Cresta, M.P., Grivard, P., *et al.* (2008). Serious acute chikungunya virus infection requiring intensive care during the Reunion Island outbreak in 2005–2006. Crit. Care Med. *36*, 2536–2541.

Lucas, R.E., and Qiao, M. (1999). A case of encephalitis in central Australia due to Ross River virus? Aust. N. Z. J. Med. *29*, 268–270.

Mackenzie, J.S., and Smith, D.W. (1996). Mosquito-borne viruses and epidemic polyarthritis. Med. J. Aust. *164*, 90–93.

Malherbe, H., Strickland-Cholmley, M., and Jackson, A.L. (1963). Sindbis virus infection in man. Report of a case with recovery of virus from skin lesions. S. Afr. Med. J. *37*, 547–552.

Manimunda, S.P., Vijayachari, P., Uppoor, R., Sugunan, A.P., Singh, S.S., Rai, S.K., Sudeep, A.B., Muruganandam, N., Chaitanya, I.K., and Guruprasad, D.R. (2010). Clinical progression of chikungunya fever during acute and chronic arthritic stages and the changes in joint morphology as revealed by imaging. Trans. R. Soc. Trop. Med. Hyg. *104*, 392–399.

Marimoutou, C., Vivier, E., Oliver, M., Boutin, J.P., and Simon, F. (2012). Morbidity and impaired quality of life 30 months after chikungunya infection: comparative cohort of infected and uninfected French military policemen in Reunion Island. Medicine *91*, 212–219.

Mathiot, C.C., Grimaud, G., Garry, P., Bouquety, J.C., Mada, A., Daguisy, A.M., and Georges, A.J. (1990). An outbreak of human Semliki Forest virus infections in Central African Republic. Am. J. Trop. Med. Hyg. *42*, 386–393.

Mizuno, Y., Kato, Y., Takeshita, N., Ujiie, M., Kobayashi, T., Kanagawa, S., Kudo, K., Lim, C.K., and Takasaki, T. (2011). Clinical and radiological features of imported chikungunya fever in Japan: a study of six cases at the National Center for Global Health and Medicine. J. Infect. Chemother. *17*, 419–423.

Morrison, T.E., Whitmore, A.C., Shabman, R.S., Lidbury, B.A., Mahalingam, S., and Heise, M.T. (2006). Characterization of Ross River virus tropism and virus-induced inflammation in a mouse model of viral arthritis and myositis. J. Virol. *80*, 737–749.

Munoz, M., and Navarro, J.C. (2012). [Mayaro: A re-emerging Arbovirus in Venezuela and Latin America]. Biomedica *32*, 286–302.

Mylonas, A.D., Brown, A.M., Carthew, T.L., McGrath, B., Purdie, D.M., Pandeya, N., Vecchio, P.C., Collins, L.G., Gardner, I.D., de Looze, F.J., *et al.* (2002). Natural history of Ross River virus-induced epidemic polyarthritis. Med. J. Aust. *177*, 356–360.

Naides, S.J. (2003). Viral arthritis. In Rheumatology, M.C. Hochberg, A.J. Silman, J.S. Smolen, M.E. Weinblatt, and M.H. Weisman, eds. (Mosby, London, UK), pp. 1105–1113.

Neumayr, A., Gabriel, M., Fritz, J., Gunther, S., Hatz, C., Schmidt-Chanasit, J., and Blum, J. (2012). Mayaro virus infection in traveler returning from Amazon Basin, northern Peru. Emerg. Infect. Dis. *18*, 695–696.

Nimmannitya, S., Halstead, S.B., Cohen, S.N., and Margiotta, M.R. (1969). Dengue and chikungunya virus infection in man in Thailand, 1962–1964. I. Observations on hospitalized patients with hemorrhagic fever. Am. J. Trop. Med. Hyg. *18*, 954–971.

Nimmo, J.R. (1928). An unusual epidemic. Med. J. Aust. *1*, 549–550.

Nkoghe, D., Kassa, R.F., Caron, M., Grard, G., Mombo, I., Bikie, B., Paupy, C., Becquart, P., Bisvigou, U., and Leroy, E.M. (2012). Clinical forms of chikungunya in Gabon, 2010. PLoS Negl. Trop. Dis. *6*, e1517.

Padmakumar, B., Jayan, J.B., Menon, R., and Kottarathara, A.J. (2010). Clinical profile of chikungunya sequelae, association with obesity and rest during acute phase. Southeast Asian J. Trop. Med. Public Health *41*, 85–91.

Pakran, J., George, M., Riyaz, N., Arakkal, R., George, S., Rajan, U., Khader, A., Thomas, S., Abdurahman, R., Sasidharanpillai, S., *et al.* (2011). Purpuric macules with vesiculobullous lesions: a novel manifestation of Chikungunya. Int. J. Dermatol. *50*, 61–69.

Paquet, C., Quatresous, I., Solet, J.L., Sissoko, D., Renault, P., Pierre, V., Cordel, H., Lassalle, C., Thiria, J., Zeller, H., *et al.* (2006). Chikungunya outbreak in Reunion: epidemiology and surveillance, 2005 to early January 2006. Eurosurveillance *11*, E060202.

Passmore, J., O'Grady, K.A., Moran, R., and Wishart, E. (2002). An outbreak of Barmah Forest virus disease in Victoria. Commun. Dis. Intell. Q Rep. *26*, 600–604.

Phan, T.G., Somerville, E., Hueston, L., and Dwyer, D.E. (1998). Probable association between Barmah Forest virus infection and Guillain-Barre syndrome. Med. J. Aust. *168*, 422–423.

Phillips, D.A., Murray, J.R., Aaskov, J.G., and Wiemers, M.A. (1990). Clinical and subclinical Barmah Forest virus infection in Queensland. Med. J. Aust. *152*, 463–466.

Pialoux, G., Gauzere, B.A., Jaureguiberry, S., and Strobel, M. (2007). Chikungunya, an epidemic arbovirosis. Lancet Infect. Dis. *7*, 319–327.

Pinheiro, F.P., Freitas, R.B., Travassos da Rosa, J.F., Gabbay, Y.B., Mello, W.A., and LeDuc, J.W. (1981). An outbreak of Mayaro virus disease in Belterra, Brazil. I. Clinical and virological findings. Am. J. Trop. Med. Hyg. *30*, 674–681.

Pollard, H., and Benz, G. (1998). Manifestations of the ross river and barmah forest arboviruses: a clinical challenge for chiropractors. Australas. Chiropr. Osteopathy *7*, 116–119.

Posey, D.L., O'Rourke, T., Roehrig, J.T., Lanciotti, R.S., Weinberg, M., and Maloney, S. (2005). O'Nyong-nyong fever in West Africa. Am. J. Trop. Med. Hyg. *73*, 32.

Powers, A.M., and Logue, C.H. (2007). Changing patterns of chikungunya virus: re-emergence of a zoonotic arbovirus. J. Gen. Virol. *88*, 2363–2377.

Powers, A.M., Brault, A.C., Tesh, R.B., and Weaver, S.C. (2000). Re-emergence of Chikungunya and O'nyong-nyong viruses: evidence for distinct geographical lineages and distant evolutionary relationships. J. Gen. Virol. *81*, 471–479.

Proll, S., Dobler, G., Pfeffer, M., Jelinek, T., Nothdurft, H.D., and Loscher, T. (1999). [Persistent arthralgias in Ross-River-Virus disease after travel to the South Pacific]. Dtsch. Med. Wochenschr. *124*, 759–762.

Receveur, M., Ezzedine, K., Pistone, T., and Malvy, D. (2010). Chikungunya infection in a French traveller returning from the Maldives, October, 2009. Eurosurveillance *15*, 19494.

Renault, P., Solet, J.L., Sissoko, D., Balleydier, E., Larrieu, S., Filleul, L., Lassalle, C., Thiria, J., Rachou, E., de Valk, H., *et al.* (2007). A major epidemic of chikungunya virus infection on Reunion Island, France, 2005–2006. Am. J. Trop. Med. Hyg. *77*, 727–731.

Retuya, T.J.A., Ting, D.L., and Dacula, B.D. (1998). Chikungunya fever outbreak in an agricultural village in Indang, Cavite, Philippines. Philippine J. Med. Infect. Dis. *27*, 93–96.

Robinson, M.C. (1955). An epidemic of virus disease in Southern Province, Tanganyika Territory, in 1952–53. I. Clinical features. Trans. R. Soc. Trop. Med. Hyg. *49*, 28–32.

Rosen, L., Gubler, D.J., and Bennett, P.H. (1981). Epidemic polyarthritis (Ross River) virus infection in the Cook Islands. Am. J. Trop. Med. Hyg. *30*, 1294–1302.

Rulli, N.E., Melton, J., Wilmes, A., Ewart, G., and Mahalingam, S. (2007). The molecular and cellular aspects of arthritis due to alphavirus infections: lesson learned from Ross River virus. Ann. N. Y. Acad. Sci. *1102*, 96–108.

Russell, R.C. (2002). Ross River virus: ecology and distribution. Annu. Rev. Entomol. *47*, 1–31.

Rwaguma, E.B., Lutwama, J.J., Sempala, S.D., Kiwanuka, N., Kamugisha, J., Okware, S., Bagambisa, G., Lanciotti, R., Roehrig, J.T., and Gubler, D.J. (1997). Emergence of epidemic O'nyong-nyong fever in southwestern Uganda, after an absence of 35 years. Emerg. Infect. Dis. *3*, 77.

Sane, J., Guedes, S., Kurkela, S., Lyytikainen, O., and Vapalahti, O. (2010). Epidemiological analysis of mosquito-borne Pogosta disease in Finland, 2009. Eurosurveillance *15*, pii=19462.

Schuffenecker, I., Iteman, I., Michault, A., Murri, S., Frangeul, L., Vaney, M.C., Lavenir, R., Pardigon, N., Reynes, J.M., Pettinelli, F., *et al.* (2006). Genome microevolution of Chikungunya Viruses causing the Indian Ocean outbreak. PLoS Med. *3*, 1058–1070.

Schwartz, O., and Albert, M.L. (2010). Biology and pathogenesis of chikungunya virus. Nat. Rev. Microbiol. *8*, 491–500.

Scrimgeour, E.M. (1999). Suspected Ross River virus encephalitis in Papua New Guinea. Aust. N. Z. J. Med. *29*, 559.

Scrimgeour, E.M., Aaskov, J.G., and Matz, L.R. (1987). Ross River virus arthritis in Papua New Guinea. Trans. R. Soc. Trop. Med. Hyg. *81*, 833–834.

Shope, R.E., and Anderson, S.G. (1960). The virus aetiology of epidemic exanthem and polyarthritis. Med. J. Aust. 47(1), 156–158.

Shore, H. (1961). O'nyong-nyong fever: an epidemic virus disease in East Africa. III. Some clinical and epidemiological observations in the Northern Province of Uganda. Trans. R. Soc. Trop. Med. Hyg. *55*, 361–373.

Simon, F., Parola, P., Grandadam, M., Fourcade, S., Oliver, M., Brouqui, P., Hance, P., Kraemer, P., Ali Mohamed, A., de Lamballerie, X., *et al.* (2007). Chikungunya infection: an emerging rheumatism among travelers returned from Indian Ocean islands. Report of 47 cases. Medicine (Baltimore) *86*, 123–137.

Smithburn, K.C., and Haddow, A.J. (1944). Semliki Forest virus I. Isolation and pathogenic properties. J. Immunol. *49*, 141–157.

Smithburn, K.C., Haddow, A.J., and Mahaffy, A.F. (1946). a neurotropic virus isolated from Aedes mosquitoes caught in the Semliki Forest. Am. J. Trop. Med. *26*, 189–208.

Sudeep, A.B., and Parashar, D. (2008). Chikungunya: an overview. J. Biosci. 33, 443–449.

Suhrbier, A., and La Linn, M. (2004). Clinical and pathologic aspects of arthritis due to Ross River virus and other alphaviruses. Curr. Opin. Rheumatol. *16*, 374–379.

Suhrbier, A., and Mahalingam, S. (2009). The immunobiology of viral arthritides. Pharmacol. Ther. *124*, 301–308.

Taubitz, W., Cramer, J.P., Kapaun, A., Pfeffer, M., Drosten, C., Dobler, G., Burchard, G.D., and Loscher, T. (2007). Chikungunya fever in travelers: clinical presentation and course. Clin. Infect. Dis. *45*, e1–4.

Taylor, R.M., Hurlbut, H.S., Work, T.H., Kingston, J.R., and Frothingham, T.E. (1955). Sindbis virus: a newly recognized arthropodtransmitted virus. Am. J. Trop. Med. Hyg. *4*, 844–862.

Taylor, S.F., Patel, P.R., and Herold, T.J. (2005). Recurrent arthralgias in a patient with previous Mayaro fever infection. South. Med. J *98*, 484–485.

Tesh, R.B. (1982). Arthritides caused by mosquito-borne viruses. Annu. Rev. Med. *33*, 31–40.

Tesh, R.B., Watts, D.M., Russell, K.L., Damodaran, C., Calampa, C., Cabezas, C., Ramirez, G., Vasquez, B., Hayes, C.G., Rossi, C.A., *et al.* (1999). Mayaro virus disease: an emerging mosquito-borne zoonosis in tropical South America. Clin. Infect. Dis. *28*, 67–73.

Weaver, S.C., and Lecuit, M. (2015). Chikungunya virus and the global spread of a mosquito-borne disease. N. Engl. J. Med. *372*, 1231–1239.

Willems, W.R., Kaluza, G., Boschek, C.B., Bauer, H., Hager, H., Schutz, H.J., and Feistner, H. (1979). Semliki forest virus: cause of a fatal case of human encephalitis. Science *203*, 1127–1129.

Williams, M.C., Woodall, J.P., Corbet, P.S., and Gillett, J.D. (1965). O'nyong-nyong Fever:An epidemic virus disease in East Africa. 8. Virus isolations from Anopheles mosquitoes. Trans. R. Soc. Trop. Med. Hyg. *59*, 300–306.

Win, M.K., Chow, A., Dimatatac, F., Go, C.J., and Leo, Y.S. (2010). Chikungunya fever in Singapore: acute clinical and laboratory features, and factors associated with persistent arthralgia. J. Clin. Virol. *49*, 111–114.

Zacks, M.A., and Paessler, S. (2010). Encephalitic alphaviruses. Vet. Microbiol. *140*, 281–286.

Encephalitic Alphaviruses

8

Clive S. McKimmie and John K. Fazakerley

Abstract

Alphaviruses are a group of medically important viruses spread by arthropods. Recent years has witnessed an increase in their geographic range and the number of infections. Following inoculation by infected arthropods, alphaviruses rapidly replicate in peripheral tissues, giving rise to a high titre plasma viraemia. In the majority of cases this results is an otherwise asymptomatic infection, but in some can progress to life-threatening encephalitis. Well-characterized animal models have demonstrated that virus spreads to the brain, primarily infecting neurons, often as perivascular foci of infection. Virus infection of neural tissue triggers an influx of inflammatory mediators that, depending on a combination of virus and host factors, results in either death or clearance of infectious virus by neutralizing high affinity immunoglobulins. Following clearance of brain virus, lesions of T-cell-driven demyelinating inflammation can result. In the absence of efficacious vaccines or specific medical treatments for alphavirus infection, a better understanding of its pathogenesis will aid the design of novel therapies.

Introduction

Increasing migration, globalization and changing countryside management are allowing more opportunities for the spread of infections (Johnston and Peters, 1996). In addition, a warming planet is enlarging the geographic range of endemic viruses and their vectors; including viruses spread by arthropod vectors.

Old World alphaviruses have a wide geographic distribution from Africa, through Asia to Australia, and include; Ross River virus (RRV), chikungunya virus (CHIKV), O'nyong-nyong virus and Semliki Forest virus (SFV) (Johnston and Peters, 1996; Powers *et al.*, 2001). These four viruses are known as the Semliki Forest group. As with the related Sindbis virus, infection of humans typically causes rash, headache, fever, arthritis/arthralgia and myositis/myalgia. Viraemia is rapid but short-lived, although subsequent arthralgia can persist for months to years (Burt *et al.*, 2012; Rulli *et al.*, 2007). Clinically apparent invasion of neural tissue by naturally circulating Old World alphaviruses is rare, but has been reported, particularly in children (Arpino *et al.*, 2009; Burt *et al.*, 2012; Sebastian *et al.*, 2009).

New World alphaviruses are geographically restricted to the Americas and include several important human pathogens; Venezuelan equine encephalitis virus (VEEV), Eastern equine encephalitis virus (EEEV) and Western equine encephalitis virus (WEEV). As suggested by their names, these viruses cause occasional large outbreaks of debilitating encephalitis and constitute a significant risk to the health of both humans and economically important livestock, particularly horses (Steele and Twenhafel, 2010). Like other neuroinvasive viruses, such as West Nile virus (Gould and Solomon, 2008), most infections in humans are either clinically unapparent or cause a mild febrile illness (Johnston and Peters, 1996). In a significant minority of cases, infection causes debilitating encephalitis with high mortality. In such cases the course of infection has three phases; an initial peripheral phase in which virus replicates at the inoculation site and disseminates; a secondary phase in which virus invades neural tissue; and a CNS phase in which virus disseminates within the CNS. Despite high morbidity and mortality, no specific medications are available, nor are there widely available vaccines (Reed *et al.*, 2005b; Steele and Twenhafel, 2010). This review will focus on the pathogenesis of alphavirus encephalitis.

Old World alphaviruses

Semliki Forest virus

Semliki Forest virus (SFV) occurs naturally in Central Africa where it cycles between *Aedes* spp. mosquitos and vertebrate hosts. The latter include horses, humans and putatively rodents (Mathiot *et al.*, 1990). SFV infection in humans can result in fever, persistent headaches, myalgia and arthralgia; more serious encephalitic complications have not been reported following natural infection (Mathiot *et al.*, 1990). Clinically, SFV has most in common with the other SFV serogroup viruses; RRV and CHIKV, which primarily cause arthritis and myalgia and only occasionally encephalitis (Arpino *et al.*, 2009). Thus, natural SFV infection is not a significant health problem. Indeed, antibodies to SFV have been found in healthy people in endemic areas, whilst subclinical seroconversion of laboratory workers who come into contact with SFV is not uncommon. An exception to this is a case of fatal meningoencephalomyelitis in a laboratory worker who was accidently inoculated with the Osterrieth strain (Mathiot *et al.*, 1990; Willems *et al.*, 1979).

Following isolation from pools of mosquitoes, a number of SFV strains have been adapted for use in the laboratory. In rodents, some of these strains produce encephalitis. The most commonly used neuroinvasive strains include A7(74), which is virulent in suckling mice but avirulent in adult mice, and L10, which is highly virulent in both adult and suckling mice (Bradish *et al.*, 1971). The prototype strain was the first to be transformed into an 'infectious' cDNA clone; virus derived from this is known as SFV4. Molecular clones are now also available for the A7(74) and L10 strains of the SFV. Transcription of SFV, as with other alphavirus genomes from cDNA constructs, gives rise to RNA, that when generated within, or transferred into, cells, can be translated and replicated to generate fully infectious virions. Using standard molecular biological tools, numerous virus mutants have been generated, which when combined with the ability to study this virus under widely available biological containment level 2 facilities, has greatly aided understanding of alphavirus encephalitis (Atkins *et al.*, 1999). In particular, SFV4 has been widely used to generate SFV mutants and expression vectors (Berglund *et al.*, 1998).

Sindbis virus

Sindbis virus (SINV) is the prototype alphavirus and the most widely distributed arbovirus (Johnston and Peters, 1996). Based on sequence analysis it is most similar to WEEV (Hahn *et al.*, 1988) and Mayaro virus, an emerging cause of rash and arthritis in South America (Tesh *et al.*, 1999). Although studied as a model for virus encephalitis, it was for many years considered 'a virus without a disease' (Laine *et al.*, 2004). However, it is now known to be the causative agent of Karelian fever, Pogosta disease and Ockelbo disease, which occur in Finland and Sweden (Tesh, 1982). Clinical features include rash, fever and arthritis/arthralgia (Laine *et al.*, 2004). SINV strains passaged in the laboratory are generally neurovirulent for rodents; the most commonly used is AR339 (Johnson *et al.*, 1972). In adult mice, AR339 is avirulent regardless of dose or inoculation route, although like SFV, suckling mice succumb to infection (Griffin, 1976; Johnson *et al.*, 1972). The TR339 strain has a genome sequence more similar than AR339 to original non-passaged virus isolates. This virus is more virulent in mice, infecting a wide range of tissues including muscle, joints and neural tissue, although adult mice still do not succumb to infection (Klimstra *et al.*, 1998; Ryman and Klimstra, 2008). SINV has been widely and successfully used to study many aspects of alphavirus virology and pathogenesis. As with SFV, SINV has been much used as a eukaryotic protein expression system (Schlesinger, 2001).

Chikungunya virus

Chikungunya virus (CHIKV) is an important human pathogen that has recently re-emerged to initiate epidemics that have infected millions of people. In humans, CHIKV occasionally infects neural tissue, but it is predominantly an arthritogenic virus (Arpino *et al.*, 2009). The nature of the disease caused by CHIKV is discussed in Chapters 7 and 10.

New World alphaviruses

VEEV, EEEV and WEEV are New World alphaviruses that cause encephalitic disease in equids and humans. In horses, these viruses can cause devastating epizootics that spill over into surrounding human populations, with children being particularly susceptible to neurological complications. Between epizootic outbreaks, enzootic strains of these viruses cycle between vertebrate hosts and mosquitoes (Calisher, 1994; Johnston and Peters, 1996; Pratt *et al.*, 2006; Steele and Twenhafel, 2010). High infectivity, ease of synthesis, physical stability, potential for aerosolization, the susceptibility of humans to aerosol infection, and the rapid neuroinvasive ability of these viruses, has generated research and concern for their misuse (Hanson *et al.*, 1967; Slepushkin, 1959; Steele and Twenhafel, 2010). This is particularly true for VEEV, which causes overt disease in humans and can initiate natural epidemics in humans (Bowen *et al.*, 1976; Johnston and Peters, 1996). The threat of natural epidemics and bioterrorism, combined with the absence of specific therapies or vaccines, drives research on the pathogenesis of this group of viruses.

Venezuelan equine encephalitis

VEEV is arguably the most well studied New World alphavirus. It has caused a number of epidemics in North, Central and South America. VEEV infection of humans most often presents as a non-lethal but incapacitating illness typified by fever, malaise, vomiting and acute lymphopenia. After a few days a small percentage of cases progress to neurological disease (Steele and Twenhafel, 2010). Mortality rates of these cases can be as high as 35% in children and 10% in adults. Survivors can exhibit long-term neurological sequelae (Bowen *et al.*, 1976; la Monte *et al.*, 1985; Weaver *et al.*, 1993). CNS pathology includes cerebral oedema, meningitis, vasculitis and leucocyte infiltrates (Johnston and Peters, 1996; la Monte *et al.*, 1985; Powers *et al.*, 2001). Intriguingly, outwith the CNS, VEEV infection results in substantial lymphopenia and acute necrosis of lymphoid tissue (Burt *et al.*, 2012; Johnson *et al.*, 1968; Rulli *et al.*, 2007), perhaps as a result of an ability to infect leucocytes (Arpino *et al.*, 2009; Burt *et al.*, 2012; Gardner *et al.*, 2008; Levitt *et al.*, 1979; Sebastian *et al.*, 2009).

Eastern equine encephalitis

EEEV has a wide geographic distribution in the eastern Americas, from Canada to Argentina (Calisher, 1994; Steele and Twenhafel, 2010). Similar to VEEV, infections in humans are mostly asymptomatic, although encephalitic disease when it occurs tends to be more severe, especially in children. Mortality rates have been estimated to be as high as 75% for those presenting with CNS disease (Ayres and Feemster, 1949; Dent, 1955; Deresiewicz *et al.*, 1997; Gould and Solomon, 2008; Hauser, 1948). The early phase of flu-like illness is somewhat more protracted than VEEV, lasting 5 days in humans following which neurological manifestations such as paralysis and seizures occur. Histological findings are dominated by severe neuronal damage along with meningeal, perivascular and parenchymal leucocyte infiltration. Gliosis, demyelination, vasculitis, thrombosis and oedema are also evident, and lesions are found across the brain (Calisher, 1994; Dent, 1955; Deresiewicz *et al.*, 1997; Hauser, 1948; Johnston and Peters, 1996). Interestingly, unlike VEEV, EEEV does not appear to infect leucocytes, nevertheless it causes widespread lymphocyte apoptosis in lymphoid tissue (Gardner *et al.*, 2008; Levitt *et al.*, 1979; Reed *et al.*, 2005b; Steele and Twenhafel, 2010).

Western equine encephalitis

WEEV shares many common features with EEEV, including a wide geographic distribution in the Americas, similar clinical and histological features and frequent neurological involvement (Hahn *et al.*, 1988; Mathiot *et al.*, 1990; Steele and Twenhafel, 2010). Up to 90% of infected children under 1 year of age present with overt neurological disease (Calisher, 1994; Mathiot *et al.*, 1990). Mortality rates are as high as 15% in natural infections and as high as 40% in the few cases resulting from aerosolized infection in the laboratory (Arpino *et al.*, 2009; Hanson *et al.*,

1967). For reasons that are yet to be fully determined, in recent decades there has been a steady decline in the number of WEE cases in both equids and humans. Interestingly, studies have shown that there is no evidence for a decline in virus virulence *per se*; ecological factors that affect both human and equine exposure are the most likely explanation (Calisher, 1994; Forrester *et al.*, 2008; Mathiot *et al.*, 1990; Steele and Twenhafel, 2010; Willems *et al.*, 1979).

Peripheral virus replication and innate immunity *in vivo*

The pathogenesis of encephalitic alphaviruses in humans remains unclear but animal models, particularly rodents infected with SINV and SFV, have been studied in detail. Susceptibility to a particular alphavirus depends on a number of factors including; the species, strain and age of the animal, virus dose, route of inoculation and the strain of virus (Fazakerley and Buchmeier, 1993; Fazakerley, 2002; Gorchakov *et al.*, 2004; Ryman and Klimstra, 2008; Sawicki *et al.*, 2003; Steele and Twenhafel, 2010). Alphavirus infections of laboratory mice have provided excellent models for studying virus neuroinvasion, blood–brain barrier events, virus neurotropism, CNS immune responses, the pathogenesis of encephalitis, demyelinating disease and CNS virus persistence.

Complementing rodent studies are those that have been undertaken in non-human primates (NHPs), including the macaques. NHP studies are closer to events in humans but come with considerable ethical issues, require expensive specialist facilities, are time-consuming and lack the power of host genetic modification (Steele and Twenhafel, 2010). Compared to mice, like their human counterparts, NHPs are generally less susceptible to infection with New World alphaviruses (Brown and Officer, 1975; Logue *et al.*, 2009; Pratt *et al.*, 2006; Steele and Twenhafel, 2010). For example, neuroinvasion of VEEV in humans and NHPs occurs much less frequently than in mice. Following infection with VEEV, NHPs exhibit viraemia but often lack overt neurological infection and even less frequently demonstrate obvious clinical disease. Infection of NHPs with EEEV, and to a lesser extent WEEV, results more often in fatal panencephalitis (Reed *et al.*, 2005a, 2007). Those NHPs that do succumb to alphavirus infection exhibit many of the features of human disease such as vasculitis that are often absent in mouse models. Neuronal infection and neuronal death is a common and dominant feature of both NHP and rodent models of alphavirus encephalitis.

Dissemination of virus from the site of inoculation to the CNS

The early events of infection are important for host survival (Ryman and Klimstra, 2008). There is a close relationship between early peripheral virus load and subsequent mortality. Following natural mosquito bite, these early events include spread from the inoculation site in the skin to initiate a viraemia. This important stage of the virus life cycle is poorly understood (Kam *et al.*, 2009; Ryman and Klimstra, 2008). Alphaviruses must replicate to high levels in the blood of at least one host species, generally a vertebrate, such that the probability of take up by a feeding arthropod is sufficiently high for animal-to-animal spread (Ryman and Klimstra, 2008). The high viraemia is associated with a debilitating febrile illness. Following natural skin inoculation, prior to the establishment of viraemia, virus first spreads to draining lymph nodes (LN) where virus could either be amplified or removed by macrophages (Junt *et al.*, 2007; MacDonald and Johnston, 2000; Ryman and Klimstra, 2008; Styer *et al.*, 2011; 2007). Alphaviruses most efficiently infect a variety of non-haematopoietic cells that include fibroblasts, skeletal muscle, cardiac myocytes, osteocytes and keratinocytes (Fragkoudis *et al.*, 2007; Klimstra *et al.*, 1998; Steele and Twenhafel, 2010; Vogel *et al.*, 2005), although some alphaviruses can also infect leucocytes (Gardner *et al.*, 2008; Ryman and Klimstra, 2008; Shabman *et al.*, 2006).

Experimentally, the outcome of alphavirus infection is greatly affected by the route of inoculation. A good example is provided by the TC-83 strain of VEEV, which when administered subcutaneously (SC) in C3H/HeN mice is almost completely avirulent, but when administered as an aerosol, is highly virulent (Hart *et al.*, 2000). The V3014 strain of VEEV, all strains of SINV and the SFV4 strain of SFV all fail to initiate a

CNS infection when inoculated intraperitoneally or subcutaneously but are fatal when inoculated intracerebrally (IC) (Grieder *et al.*, 1995). Likewise, in NHPs, IC inoculation of VEEV results in fulminant lethal encephalitis, whilst inoculation by peripheral routes, including by aerosol, causes mild encephalitis (Reed *et al.*, 2005a,b). A study using a virulent strain of WEEV showed that mean survival times were decreased from 6 to 2 days when virus was delivered IC compared with IV (Logue *et al.*, 2009). In a study of EEEV in mice, the LD50 was 1250 pfu for SC administration, 500 pfu for aerosol, 400 pfu for intraperitoneal inoculation (IP) and just 1 pfu for IC inoculation (Brown and Officer, 1975; Pratt *et al.*, 2006).

CNS pathology and tropism of encephalitic alphaviruses

Neuroinvasion via the olfactory tract

Aerosolized alphaviruses can 'naturally' infect the CNS via the olfactory nerve. This has been demonstrated by laboratory acquired infections and is of interest to those concerned with biowarfare and bioterrorism. The olfactory nerve provides the only single cell route, olfactory neurons, from the environment to the brain. Infection by this route leads to a rapid onset of encephalitis (Steele and Twenhafel, 2010).

Experimentally following intranasal inoculation, VEEV, EEEV, WEEV, SFV and SINV can all enter the CNS olfactory bulb by infection along the olfactory nerve (Oliver and Fazakerley, 1998; Reed *et al.*, 2005a; Sammin *et al.*, 1999; Steele and Twenhafel, 2010; Thach *et al.*, 2000). VEEV has been studied the most extensively. Unlike most other arboviral encephalitides, direct invasion of neural tissue from the blood through the blood brain barrier is not a prominent feature of VEEV. Instead, a substantial body of work has demonstrated that infection of the brain occurs via the olfactory system (Charles *et al.*, 1995; Grieder *et al.*, 1995; Jackson *et al.*, 1991; Pratt *et al.*, 2006; Ryzhikov *et al.*, 1995; 1991; Steele *et al.*, 1998; Vogel *et al.*, 1996). Virions exit the bloodstream through fenestrated capillaries that lie beneath the olfactory mucosa of the nasal tract. This tract possesses a large number of unmyelinated axons susceptible to VEEV infection and their infection transmits virus along olfactory axons to the olfactory bulb (Charles *et al.*, 1995; Steele and Twenhafel, 2010). Interestingly, virus tropism for olfactory neuroepithelium is specific, since neither squamous nor respiratory epithelia of the nasal tract are susceptible (Pratt *et al.*, 2006). Once within the olfactory bulb, VEEV rapidly disseminates throughout the brain, infecting regions from the pyriform cortex to the cerebellum. These findings, obtained mostly in mice, are also observed in NHPs (Gleiser *et al.*, 1962; Steele and Twenhafel, 2010).

That the olfactory nerve is the route of CNS infection with VEEV explains why disease onset is accelerated by aerosol infection (Ryzhikov *et al.*, 1991; Steele *et al.*, 1998; Vogel *et al.*, 1996). Following inoculation by aerosol, VEEV can be detected in the CNS within 16 hours (Ryzhikov *et al.*, 1991). VEEV neutralizing IgG antibody in the serum does not afford protection from aerosol infection, whereas IgA production within the nasal mucosa does (Hart *et al.*, 1997, 2000). This is an important consideration for vaccines to combat putative weaponized VEEV (Steele and Twenhafel, 2010). Infection of NHPs by the intranasal route results in brain lesions within 48 hours, compared with 6 days for those inoculated IP (Danes *et al.*, 1973). CNS infection via the olfactory route may explain why mice succumb more readily to VEEV than NHPs and humans; relative to their size, mice possess a large and highly developed olfactory system. In the case of the A7(74) strain of SFV, which can infect the rodent brain via the olfactory nerve, the extent of brain infection is age related and determined by the maturity of the olfactory system (Oliver and Fazakerley, 1998).

Neuroinvasion via the blood brain barrier

Most alphaviruses enter the brain by passage across the blood–brain barrier (BBB). In the absence of infection or inflammation, the BBB is a complex structure highly impermeable to molecules and cells in the blood. It is composed of vascular endothelial and smooth muscle cells, with scattered pericytes, enveloped on the brain side by a layer of astrocytic foot processes (Abbott *et al.*, 2006). Unlike endothelia in extraneural tissues, BBB endothelial cells are connected by tight junctions, which together with associated glia, regulate the entry of cells and macromolecules into the brain. Most Old World alphaviruses, including SFV and SINV and the New World virus EEEV, enter the brain by infecting or traversing cerebral endothelial cells (Fazakerley, 2002; Ryman and Klimstra, 2008; Soilu-Hanninen *et al.*, 1994; Steele and Twenhafel, 2010; Vogel *et al.*, 2005).

For SFV, brain infection is initiated as perivascular foci. For the A7(74) strain, at least in the mature adult mouse brain, these foci enlarge little with time (Fazakerley, 2002). In contrast, the more virulent neuroinvasive L10 strain of SFV rapidly spreads outward from initial perivascular foci to produce a fatal panencephalitis (Fazakerley, 2004; Pathak and Webb, 1974; Soilu-Hanninen *et al.*, 1994). Entry of virus into the brain from the blood may be enhanced by the presence of systemic inflammatory mediators. Viraemia generally correlates with high levels of cytokines including IFNα/β, TNF-α and IFN-γ (Grieder *et al.*, 1997; Klimstra *et al.*, 1999; Ryman and Klimstra, 2008). TNF-α and IFN-γ permeabilize the BBB allowing plasma components, including virus if present, to enter the CNS (Lustig *et al.*, 1992; Wang *et al.*, 2004). Once infection has become established within the brain, localized pro-inflammatory cytokine production further facilitates BBB leakage (Fazakerley, 2002; Ryman and Klimstra, 2008). Measurement of the integrity of the BBB during SFV infection, as indicated by the IgG index, indicates that it is leaky to plasma proteins between 4 and 10 days after SFV A7(74) infection (Parsons and Webb, 1982a).

Cellular targets of alphaviruses in neural tissues

A characteristic feature of all encephalitic alphavirus infections of the brain is the infection of neurons. The extent of neuronal infection is age-related for many virus strains. In adult mice, alphaviruses infect a variety of neuronal subtypes that include pyramidal, granular and Purkinje neurons (Fazakerley, 2002; Jackson *et al.*, 1991; Steele *et al.*, 1998; Vogel *et al.*, 1996). In comparison, astrocytes, oligodendrocytes, microglia, ependymal, meningeal and choroid plexus cells are rarely observed to be infected *in vivo* (McKimmie and Fazakerley, 2005; Pratt *et al.*, 2006; Schoneboom *et al.*, 1999; Steele and Twenhafel, 2010; Vogel *et al.*, 1996). There are clear exceptions; VEEV can infect macrophage-like cells (Vogel *et al.*, 1996) and SFV readily infects oligodendrocytes (Fazakerley *et al.*, 1993). As demonstrated by infection of IFNα/β-R null mice, at least in the case of SFV, many of these cell types are protected from infection by the IFN response as they can be infected in its absence (Fragkoudis *et al.*, 2007).

Host age and genetics, virus strain and route of infection are all determinants of the brain areas infected. VEEV infects neurons of the olfactory bulb, before proceeding to infect pyramidal neurons of the neocortex, granular neurons of the hippocampus and Purkinje neurons of the cerebellum. EEEV also primarily targets neurons and occasional glial cells, with the caudate nucleus, thalamus and pons most affected (Steele and Twenhafel, 2010; Vogel *et al.*, 2005). In comparison, WEEV in NHPs infects microglia cells along with Purkinje neurons of the cerebellum and motor neurons of the spinal cord (Reed *et al.*, 2005a).

SFV provides a good example of the difference between virus strains. In the adult mouse CNS, infection can be either productive, with infection spreading from neuron to neuron, or highly restricted (Fazakerley *et al.*, 1993). In the adult mouse brain, the virulent L10 strain rapidly spreads throughout the brain parenchyma, while the avirulent SFV A7(74) strain is highly restricted to perivascular foci (Fazakerley, 2004; Fazakerley and Webb, 1987a). Electron microscopy has demonstrated that L10 infection of mature neurons is productive with cytopathic replication complexes and budding mature virions readily observed. In comparison, infection of mature neurons with SFV A7(74) results in dense fibrous aggregates of viral capsid protein, but cytopathic vacuole replication complexes and budding viruses are rarely seen and there are only low levels of infectious virus production (Pathak and Webb, 1978; Pathak *et al.*, 1976). The genetic locus responsible for this phenotypic difference maps to the carboxy-terminus of the nsP3 gene, the function of this protein domain remains unknown (Tuittila *et al.*, 2000).

Age-related virulence

Infection of mice with encephalitic alphaviruses generally demonstrates a clear age-dependent outcome. With virtually all strains of alphaviruses, young mice, that is neonatal and suckling mice up to around 12 days of age, succumb within days of infection to fulminant encephalitis that invariably leads to death. In comparison, adult mice are able to clear many strains of alphavirus without obvious clinical signs. An analogous situation is observed with human infections; children present with more severe neurological disease (Deresiewicz *et al.*, 1997; Ehrenkranz and Ventura, 1974; Zacks and Paessler, 2010). A well-studied example is infection of mice with the A7(74) strain of SFV (Fazakerley, 2002). In young mice, irrespective of inoculation route, SFV A7(74) is neuroinvasive and spreads throughout the brain resulting in death within 48 hours. In contrast in adult mice, A7(74) CNS infection is restricted and there is a mild encephalitis with no or only mild clinical signs (Fazakerley and Webb, 1987a). Strikingly, mice infected at 12 days of age or younger die, whereas those infected at 14 days of age or older survive (Fleming, 1977). With increasing age at infection, 0–12 days, the virus infection becomes increasingly restricted to fewer brain areas (Oliver *et al.*, 1997). This age-related virulence is not a function of a maturing immune system, since it is still observed in adult mice lacking either T cells (Amor *et al.*, 1996) or a functional type I IFN system (Fragkoudis *et al.*, 2007).

Age-related virulence is related to neuronal maturation (Allsopp *et al.*, 1998; Fazakerley *et al.*, 1993). In the first 2 weeks after birth substantial re-modelling of the CNS occurs including axonogenesis, synapatogenesis, gliogenesis and myelination (Fazakerley and Allsopp, 2001). These processes, or more specifically the molecular processes associated with them, facilitate productive viral replication (Fazakerley, 2001). Following completion of axonogenesis and neuronal connectivity it seems likely that changes in neurons restrict virus replication; virus replication structures and mature virions are not readily observed. Associated with this seems to be a change in the propensity of neurons to undergo apoptosis and this is considered in greater detail in the next section. An interesting observation is that treatment of adult mice with gold compounds converts SFV A7(74) infection of 'mature' neurons from restricted to productive (Fazakerley, 2002; Pathak and Webb, 1983; Scallan and Fazakerley, 1999). Heavy metals have been associated with neuronal dedifferentiation. It should be noted that even in the adult mouse neuronal populations vary in their maturity. Neurons in the cerebellum, the dentate gyrus of the hippocampus and the olfactory bulb may retain some capacity for renewal and those in the subventricular zone and the rostral migratory stream which migrate from this region to the olfactory bulb do remain capable of division.

As with SFV A7(74), strains of SINV demonstrate an age-related virulence in mice. SINV AR339 exhibit highly restricted perivascular foci of infection in mature CNS tissue (Griffin, 2005; Ryman and Klimstra, 2008). In comparison, if infected during the first two weeks of life, mice rapidly succumb to infection (Griffin, 1976; Johnson *et al.*, 1972). Like SFV, this age-related susceptibility is unlikely to be a consequence of the maturity of immune responses (Griffin, 1976; Levine *et al.*, 1991; Ryman *et al.*, 2000; Ryman and Klimstra, 2008). There is again a considerable body of evidence that the restricted infection is related to post-natal CNS development, in which infected mature neurons are less able to support productive viral replication (Griffin, 2005). As with SFV, increased neuronal maturity correlates with reduced productive infection and a reduced propensity to undergo apoptotic cell death upon infection (Burdeinick-Kerr and Griffin, 2005; Havert *et al.*, 2000; Levine *et al.*, 1991; Lewis *et al.*, 1996).

Neuronal cell death

CNS development involves significant neuronal apoptosis. Neurons survive by making synaptic connections that provide pro-survival cues. Neurons that make successful connections stop dividing and become more resistant to apoptosis by up-regulating the expression of anti-apoptotic genes, such as Bcl-2 (Scallan *et al.*, 1997). There are many studies and considerable evidence that alphavirus infection of cell lines in continuous culture, or alphavirus infection of primary epithelial or fibroblast cell cultures, triggers apoptotic death, usually within 48 hours (Glasgow *et al.*, 1997; Levine *et al.*, 1993; Scallan *et al.*, 1997). Similar observations have been made with primary neuronal cultures and neuronal cell lines, which by their nature are respectively, immature, generally embryonic and not post-mitotic. Studies with mature, differentiated, post-mitotic neurons are at best difficult *in vitro*, since post-mitotic cells cannot be maintained as continuous cell lines. Differentiation of neuronal-like cell lines in culture with growth factors, or maturation in culture of primary embryonic neurons, reduces the propensity of these cells to undergo apoptosis upon subsequent alphavirus infection (Allsopp *et al.*, 1998; Vernon and Griffin, 2005). Studies *in vivo* confirm that, at least with SFV and SINV, neuronal apoptosis upon infection is related to neuronal differentiation state (Fazakerley, 2001). An interesting example is infection of the rostral migratory stream of immature neurons, which shows apoptosis upon infection with SFV even in the adult rat brain (Sammin *et al.*, 1999). Mature neurons *in vivo* are relatively resistant to alphavirus induced apoptosis and, at least in the absence of immune responses, SFV and SINV can persist in these cells for many months (Allsopp *et al.*, 1998; Amor *et al.*, 1996; Fazakerley *et al.*, 1993). This has led to an appreciation that whilst virus-induced apoptosis in renewable cell populations in most tissues is an evolved altruistic strategy to limit virus spread, mature neurons are a specialized case in which selective pressures have led to constitutive protection from cell death (Allsopp and Fazakerley, 2000). Resistance to apoptosis of mature neurons and other CNS cell types may underlie the persistence of many viruses in the mature brain.

Whereas, with avirulent strains of SFV and SINV, virus replication and neuronal survival are age-related, with neurovirulent strains of these viruses, virus replication in mature neurons is productive and destructive. This includes infected cells that are positive by TUNEL staining (Allsopp and Fazakerley, 2000; Lewis *et al.*, 1996). This could be indicative of apoptosis, although TUNEL positivity has also been associated with non-apoptotic cell death. Studies using markers of activated caspases are needed to resolve whether this is, or is not, apoptotic neuronal death. During infection with VEEV, neurons exhibit the characteristic cellular features of both necrosis and apoptosis (Jackson and Rossiter, 1997; Schoneboom *et al.*, 2000a; Steele *et al.*, 1998, 2006), depending of the type of neuron infected (Steele *et al.*, 2006). Larger neurons more frequently become necrotic, such as those in the pyramidal layer of the pyriform cortex. This is in contrast with the granule-type neurons of the hippocampus, which show morphological changes typical of apoptosis (Steele and Twenhafel, 2010).

Although not strictly a form of cell death, autophagy is a related, highly conserved mechanism for recycling cellular contents by delivering cytoplasmic material to the lysosome for degradation. Since neurons are post-mitotic and long-lived, protein quality control and organelle turnover are important in maintaining cellular health. Accordingly, it is now appreciated that constitutive low levels of autophagy are neuroprotective (Orvedahl and Levine, 2008). Furthermore, it has been suggested that autophagy may defend against virus infection by targeting viral RNA, including SINV RNA, for sequestration and elimination (Levine and Deretic, 2007; Liang *et al.*, 1998). For example, Beclin 1 is an autophagy protein that when overexpressed in neurons protects mice against SINV infection (Liang *et al.*, 1998).

Pathological features in the adult mouse brain

The fully developed brain is an immunospecialized tissue that lacks many immunological features present in extraneural tissues. Inflammatory responses when they do occur are tightly controlled (Ransohoff and Engelhardt, 2012). Swelling associated with CNS inflammation is highly dangerous and as most neurons are post-mitotic, neuronal nuclei and circuits are generally not replaced if damaged or lost. As a consequence, in the resting brain lymphocytes, antibodies, complement and MHC-I expression are mostly absent. Occasionally patrolling T cells can be found in the cerebrospinal fluid (Reboldi *et al.*, 2009). This immunologically quiescent state is maintained by constitutive interactions between neurons and glial cells, the integrity of the BBB and high levels of immunosuppressive TGFβ (Griffin, 2010; Ransohoff and Engelhardt, 2012).

Nevertheless, florid CNS inflammatory responses can develop (Griffin, 2003).

In the adult mouse brain, during acute CNS infection with SFV A7(74), inflammatory cells are observed in the perivascular foci of infection. These perivascular cuffs are most prominent between days 7 and 10-post infection. They contain large numbers of MHC-I+ cells, some of which migrate into the parenchyma and develop dendritic-like morphology (Morris *et al.*, 1997). MHC-II positive pericytes can be found adjacent to blood vessel in areas of infection, as can MHC-II+ astrocytes and microglia. Infiltrating leucocytes include CD3+ and CD8+ T cells, with occasional CD45+/B220+ B cells within lesions (Morris *et al.*, 1997). Significant F4/80 staining is evident, which may represent activated microglia or the presence of infiltrating myeloid leucocytes. These inflammatory lesions are positive for IL-1b, IL-2, IL-6, type I IFNs and IFN-γ (Brown *et al.*, 2005; Fragkoudis *et al.*, 2007; McKimmie *et al.*, 2005; Morris *et al.*, 1997). Likewise, leucocytes are present in cerebrospinal fluid from day 4 onwards, peaking at day 7. Once virus is cleared, leucocyte numbers fall but are maintained at an elevated level for at least several months (Parsons and Webb, 1982b). These cells are predominantly of the B-lymphocyte lineage (Parsons and Webb, 1989).

Mice infected with VEEV demonstrate many of the pathological and clinical features exhibited by VEEV infected humans (Charles *et al.*, 2001; Gardner *et al.*, 2008; Jackson *et al.*, 1991). Brain lesions develop 24 hours after virus infection of the area and spread to give a necrotizing panencephalitis. The olfactory bulb demonstrates lesions first, these then spread to more caudal regions. Lesions are characterized by damaged endothelia, perivascular oedema, neuronal death, astrocytosis, cytokine expression and leucocyte influx (Ryzhikov *et al.*, 1991; Steele *et al.*, 1998, 2006; Vogel *et al.*, 1996). During end-stage disease, lesions contain thicker perivascular cuffs, a substantial mononuclear infiltrate and severe necrosis. In NHPs, the most characteristic features are lesions of multifocal perivascular cuffs, composed of lymphocytes and gliosis. This peak 14 to 21 days post infection, with the thalamus most overtly affected (Gleiser *et al.*, 1962).

Finally, it should be noted that in the majority of cases, mice succumb to infection with virulent strains due to an overwhelming CNS infection (Steele and Twenhafel, 2010). However, there are a few strains in which extraneural processes most probably account for the cause of death. For example, infection of mice with the B-629 strain of WEEV in suckling mice results in death within 2 days of infection, most likely due to an acute and overwhelming peripheral inflammatory process, rather than a result of CNS pathology (Monath *et al.*, 1978). This is analogous to the situation in hamsters infected with a virulent strain of VEEE, in which overwhelming peripheral infection of lymphoid tissue precipitates an early death in the absence of neuroinvasion or CNS lesions (Gorelkin and Jahrling, 1975; Jackson *et al.*, 1991; Steele and Twenhafel, 2010; Walker *et al.*, 1976).

CNS immune response to alphaviruses

There are three phases of infection where the immune system has an impact; initial clearance of virus from the circulation, acute inflammatory encephalitis and late events after levels of virus have been reduced. Following inoculation, virus is rapidly disseminated to the draining LN where it triggers both innate and adaptive immune responses. Virus rapidly replicates in peripheral tissues giving rise to a high titre plasma viraemia. Within a few days this is brought under control by the developing innate and then adaptive immune responses. For neuroinvasive viruses, the transient viraemia is sufficient to disseminate virus to neural tissue. Events are then determined by whether or not the virus is able to rapidly spread in the brain or whether it is restricted. In the mouse, virulent strains, such as SFV L10, spread rapidly in the brain producing a panencephalitis within a few days. This rapid spread is not controlled by immune responses and infected mice die within a few days. With avirulent strains which spread slowly, or are confined to the perivascular foci around the blood vessels where they enter the brain, immune responses are activated and either eliminate or control the infection (Fazakerley, 2004; Griffin, 2010).

Innate recognition, cytokines and chemokines

The cellular mechanisms mediating recognition of alphavirus infections in the CNS remain unclear. Whilst the CNS is a immunospecialized environment, devoid of most immune processes, there is extensive expression of TLRs *in vivo* (Carpentier *et al.*, 2008; McKimmie *et al.*, 2005). Microglia and astrocytes most likely account for this expression (Bsibsi *et al.*, 2006; Farina *et al.*, 2005; Jack *et al.*, 2005; McKimmie and Fazakerley, 2005), although neuronal expression has also been suggested (McKimmie and Fazakerley, 2005; Préhaud *et al.*, 2005). In the brain, TLR3 is the most highly expressed TLR. Following infection with SFV, TLR3 expression is significantly up-regulated in a type I IFN dependent manner (McKimmie *et al.*, 2005). Interestingly, microglia can utilize their PRRs to distinguish between virus and bacterial infections and initiate pathogen specific responses (McKimmie *et al.*, 2006). Expression of RIG-I, MDA5 and other innate sensors are yet to be determined for neural cells. SFV CNS infection triggers the expression of pro-inflammatory mediators, including a large number of cytokines such as TNF-α, several interleukins and both type I and II IFNs (Fazakerley, 2004; Fragkoudis *et al.*, 2007; Ryman and Klimstra, 2008; Schoneboom *et al.*, 1999, 2000a,b). TNF-α and IFNα reduce the integrity of the BBB, enabling the entry of neutralizing antibody and antiviral leucocytes. Similar results have been found during VEEV infection in which pro-inflammatory cytokines, chemokines and other inflammatory mediators such as inducible nitric oxide synthase can be present (Grieder *et al.*, 1997; Schoneboom *et al.*, 1999, 2000a,b; Sharma and Maheshwari, 2009; Sharma *et al.*, 2008).

Astrocytes are innate immune sentinels that ensheath parenchymal blood vessels and sit at the gateway to the CNS parenchyma (Farina *et al.*, 2007). They express TLRs, can distinguish between bacterial and viral pathogens and express relevant chemokines that relate to the microbe recognized (Carpentier *et al.*, 2004; McKimmie and Fazakerley, 2005; McKimmie and Graham, 2010). Thus astrocytes can respond to infection in a distinct and appropriate manner, suggesting they have the capacity to attract appropriate sets of leucocytes into the brain parenchyma. Whilst the mechanisms that control the entry of leucocytes into the alphavirus-infected brain are little understood, in SFV A7(74) infection, the vascular adhesion protein CD49d (VLA-4) is up-regulated on cerebral endothelial cells. Interestingly, in this infection antibodies that block CD49d reduce leucocyte infiltration in the late phase of this infection (Smith *et al.*, 2000). Similarly, during SINV infection, glial cells quickly become activated and produce chemokines and there is increased expression of adhesion molecules on cerebral capillary endothelial cells (Griffin, 2010). Mice deficient in chemokine receptors that control leucocyte entry into tissues are highly susceptible to infection with neuroinvasive strains of West Nile virus (Glass, 2005; Lim *et al.*, 2011; Zhang *et al.*, 2008). In particular, CCR5 null mice demonstrate 100% mortality, whilst humans with a non-functional CCR5D32 allele are predisposed to more severe neurological disease (Glass, 2005; Glass *et al.*, 2006). The chemotactic cues that drive leucocytes into alphavirus infected brains are yet to be clarified.

CNS infection results in leucocyte influx

Whilst the exact pathways used by CNS tissue to attract leucocytes are yet to be clarified, the leucocytes that are recruited have been characterized. Infection with virulent strains of SFV, SINV, VEEV or EEEV leads to massive influx of leucocytes, mononuclear cells and substantial numbers of neutrophils. The latter are most likely responding to the tissue damage that results from these widespread infections (Jackson *et al.*, 1991; Schoneboom *et al.*, 2000a; Steele and Twenhafel, 2010). Studies with lymphocytic choriomeningitis virus suggest that this neutrophil influx is the most damaging and leads to death (Kim *et al.*, 2008).

During SFV A7(74) infection, perivascular MHC-I+ mononuclear cells are observed in areas of infection by day 3 post infection and influx of mononuclear cells into the cerebrospinal fluid (CSF) is greatest between days 5 and 10 (Morris *et al.*, 1997; Parsons and Webb, 1982b). During this period, brain virus infectivity titres are reduced to undetectable and numbers of cells immunostaining positive for virus proteins or for virus RNA by *in situ* hybridization are greatly reduced (Fazakerley *et al.*, 1993). However, virus RNA remains detectable in the brain for many months (Fragkoudis *et al.*, 2008), raising the possibility that replication competent virus may persist. Interestingly, numbers of inflammatory cells in the CSF do not return to the resting state but remain elevated; these cells are principally antibody secreting B cells (Morris *et al.*, 1997; Parsons and Webb, 1989).

The first leucocytes to enter the SINV-infected brain are CD8 T cells, followed by CD4 T cells and then CD19 B cells that can secrete antibodies *in situ* (Metcalf and Griffin, 2011; Rowell and Griffin, 1999). Together by day 7 these cells reduce the levels of infectious SINV to undetectable. Virus RNA can however be detected for many months (Levine and Griffin, 1992). After clearance of infectious virus, most T cells are CD4+ and the number of SINV specific antibody secreting cells steadily increases until they constitute 90% of all B cells. SINV-specific B cells are not present in the circulation at high levels, suggesting that they are specifically retained in the CNS (Metcalf and Griffin, 2011).

Role of antibodies in viral clearance

The function of antibodies during alphavirus infection has been characterized in many studies. Using SFV as an example, studies have been undertaken in *nu/nu* mice, severe combined immunodeficient (SCID) mice and μMT mice. The *nu/nu* strain lacks T cells, and as a result cannot provide cytokine help to B cells, without which they are unable to produce high affinity IgG, although B cells are able to produce IgM (Suckling *et al.*, 1982). SCID mice in comparison lack both T and B cells, and hence both cell-mediated and humoral immunity. μMT mice lack mature B-cells and cannot generate any antibodies but have normal T-cell responses. These mice cannot clear infectious SFV A7(74) virus from the blood or the brain (Fragkoudis *et al.*, 2008). During infection with avirulent SFV A7(74), adult wild type mice generate a rapid neutralizing IgM antibody response that is followed by an IgG2a and a later IgG1 response (Fazakerley *et al.*, 1993). IgM responses alone are sufficient to control the viraemia, since T cell deficient *nu/nu* mice successfully clear virus from the blood (Suckling *et al.*, 1982; Amor *et al.*, 1996). Furthermore, the transfer of anti-SFV IgM-containing serum from *nu/nu* mice clears the viraemia upon transfer to infected SCID mice, which as with μMT mice, otherwise suffer a persistent viraemia (Amor *et al.*, 1996). Despite the permeability of the BBB during CNS infection, IgM antibody is not sufficient to reduce brain virus titres to undetectable (Parsons and Webb, 1982a, 1992; Suckling *et al.*, 1982). There is a requirement for T cells to provide cytokine help to trigger immunoglobulin isotype switching in B cells. *Nu/nu* mice cannot clear infectious virus from the brain, nor can IgM containing serum from infected *nu/nu* mice clear infectious virus from the CNS of infected SCID mice (Amor *et al.*, 1996). Only serum from infected wild type mice that contains high affinity IgG antibodies is cable of clearing infectious CNS virus from either *nu/nu* or SCID mice (Amor *et al.*, 1996). Furthermore, mice given prophylactic monoclonal antibodies to the E2 glycoprotein are completely protected from infection with even virulent strains (Boere *et al.*, 1983).

Similar results have been obtained using SINV (Binder and Griffin, 2001; Metcalf and Griffin, 2011; Tyor *et al.*, 1992). μMT mice are unable to clear infectious virus (Burdeinick-Kerr *et al.*, 2007). Passive transfer of antibody neutralizes infectious virus, but also inhibits virus replication in CNS cells through a non-cytolytic mechanism (Griffin, 2010). This mechanism does not require the presence of either complement or leucocytes (Levine *et al.*, 1991; Ubol *et al.*, 1995). Antibody effectively blocks viral replication and budding and restores host protein synthesis and the ability of cells to respond to type I IFN (Despres *et al.*, 1995a,b). The relative contribution that plasma antibody and antibody from CNS resident B cells make towards antibody-mediated clearance is still being clarified (Griffin, 2010).

Persistence of CNS virus

Replicating virus cannot be completely removed from the CNS by antibody alone. If infected *nu/nu* mice that have been given antiviral antibodies are left for several months, the level of passively transferred antibody decays and viral replication eventually resumes (Levine and Griffin, 1992; Levine *et al.*, 1991). This indicates that low levels of replication competent RNA persist despite the presence of neutralizing IgG antibody. Whilst immunocompetent mice effectively eliminate infectious virus from the CNS, as measured by plaque assay (Fazakerley *et al.*, 1993; McFarland *et al.*, 1972), RT-PCR data indicate that viral RNA can be detected for many months after this (Donnelly *et al.*, 1997; Fragkoudis *et al.*, 2008; Levine and Griffin, 1992; Levine *et al.*, 1991), indicating that the additional presence of cell-mediated immunity is also insufficient to completely remove viral RNA and raising the possibility that, at least in the CNS, these viruses may be persistent infections (Fragkoudis *et al.*, 2008; Metcalf and Griffin, 2011). The long-term persistence of CNS inflammatory cells, principally B-cells, following SFV or SINV infections is consistent with this (Metcalf and Griffin, 2011; Parsons and Webb, 1989; Rowell and Griffin, 1999).

Role of T cells in viral clearance

Whereas T cells provide essential cytokine help to B cells in lymphoid tissue, their role within the CNS is controversial. A series of studies, including those discussed above, suggest that in the presence of high affinity virus-specific antibodies, T cells are not required to reduce virus infectivity titres to undetectable. Furthermore, T cells in the CNS may have a detrimental effect on CNS physiology. CD4 T cells can make pro-inflammatory cytokines, whilst CD8 T cells can kill virally infected cells; both may contribute to neuronal dysfunction. Despite evidence that IFN-γ can assist antibody-mediated clearance of SINV (Burdeinick-Kerr and Griffin, 2005), several studies have shown that depletion of either CD4 or CD8 T cells does not affect viral clearance and reduces inflammation. Together they suggest CNS T cells not only have a non-redundant role in viral clearance, but also unnecessarily contribute towards excessive immunopathology (Amor *et al.*, 1996; Fazakerley, 2004; Subak-Sharpe *et al.*, 1993). Interestingly, SJL mice and BALB/c mice exhibit divergent inflammatory responses despite having a similar ability to clear CNS virus. SJL mice exhibit exaggerated CNS inflammatory responses to SINV infection, compared to BALB/c mice. This difference may be due to significantly higher numbers of CD4 cells present during the second week of infection onwards, although other strain differences could also be involved (Rowell and Griffin, 1999). Further evidence for a pathogenic role for T cells is provided by SCID and T cell-depleted mice that, despite higher CNS viral titres, have extended survival times following infection with virulent VEEV (Charles *et al.*, 2001; Woodman *et al.*, 1975).

Demyelinating disease

In immunocompetent adult mice, following clearance of infectious SFV A7(74) from the brain from day 7 to 10, focal lesions of inflammatory demyelination develop throughout the CNS (Kelly *et al.*, 1982; Suckling *et al.*, 1978). In contrast, despite persistently high virus titres in the CNS of infected *nu/nu* or SCID mice, no demyelinating lesions are present (Amor *et al.*, 1996; Fragkoudis *et al.*, 2008; Jagelman *et al.*, 1978). Demyelination can be restored by transfer of splenocytes, but not by transfer of antibody (Amor *et al.*, 1996; Fazakerley and Webb, 1987a). Studies in mice immunosuppressed with total body irradiation, cyclophosphamide, cyclosporine, or cycloleucine also demonstrate that T cell responses are pathogenic and are required for demyelination (Amor and Webb, 1987; Fazakerley and Webb, 1987b,c). B cells are present in the demyelinating lesions but are dispensable for their formation since they develop in mMT mice (Fragkoudis *et al.*, 2008). Depletion of CD8+ but not CD4+ T cells prevents demyelination (Subak-Sharpe *et al.*, 1993). SFV A7(74) infects oligodendrocytes (Fragkoudis *et al.*, 2008) and it is most likely that the lesions of demyelination result from CD8+ T cell mediated destruction of virus infected oligodendrocytes. Unlike neurons, oligodendrocytes are IFN inducible for MHC-I expression and can therefore become targets of cytotoxic CD8+ T cells. That demyelination is only apparent late in infection may reflect a requirement for macrophages to remove sufficient myelin for this loss to be observed by histological staining.

Taken together, the results with avirulent strains of SFV and SINV in mice indicate that virus replication is partially restricted in at least some CNS neurons. IgG antibody is necessary and sufficient to reduce brain virus infectivity titres to undetectable, although antibody alone cannot eliminate these viruses; T cells are also required. T cells produce IFN-γ which contributes to the clearance of virus from neurons (Binder and Griffin, 2001) and T cells destroy at least some virus infected cells. In the case of SFV, this includes virus-infected oligodendrocytes, which generates lesions of demyelination. Given the restricted nature of the infection in neurons and their low propensity to undergo apoptosis, it is possible that even the combination of antibodies, T cells and cytokines is unable to eradicate these viruses and that a low-level infection and a low-grade immune response persist.

References

Abbott, N.J., Rönnbäck, L., and Hansson, E. (2006). Astrocyte–endothelial interactions at the blood–brain barrier. Nat. Rev. Neurosci. 7, 41–53.

Allsopp, T.E., and Fazakerley, J.K. (2000). Altruistic cell suicide and the specialized case of the virus-infected nervous system. Trends Neurosci. 23, 284–290.

Allsopp, T.E., Scallan, M.F., Williams, A., and Fazakerley, J.K. (1998). Virus infection induces neuronal apoptosis: A comparison with trophic factor withdrawal. Cell Death Differ. *5*, 50–59.

Alsharifi, M., Lobigs, M., Simon, M.M., Kersten, A., Müller, K., Koskinen, A., Lee, E., and Müllbacher, A. (2006). NK cell-mediated immunopathology during an acute viral infection of the CNS. Eur. J. Immunol. *36*, 887–896.

Amor, S., and Webb, H.E. (1987). The effect of cycloleucine on SFV A7(74) infection in mice. Br. J. Exp. Pathol. *68*, 225–235.

Amor, S., Scallan, M.F., Morris, M.M., Dyson, H., and Fazakerley, J.K. (1996). Role of immune responses in protection and pathogenesis during Semliki Forest virus encephalitis. J. Gen. Virol. 77(Pt 2), 281–291.

Arpino, C., Curatolo, P., and Rezza, G. (2009). Chikungunya and the nervous system: what we do and do not know. Rev. Med. Virol. *19*, 121–129.

Atkins, G.J., Sheahan, B.J., and Liljestrom, P. (1999). The molecular pathogenesis of Semliki Forest virus: a model virus made useful? J. Gen. Virol. *80*(Pt 9), 2287–2297.

Ayres, J.C., and Feemster, R. (1949). The sequelae of eastern equine encephalomyelitis. N. Engl. J. Med. *240*, 960–962.

Berglund, P., Smerdou, C., Fleeton, M.N., Tubulekas, I., and Liljestrom, P. (1998). Enhancing immune responses using suicidal DNA vaccines. Nat. Biotechnol. *16*, 562–565.

Binder, G.K., and Griffin, D.E. (2001). Interferon-gamma-mediated site-specific clearance of alphavirus from CNS neurons. Science *293*, 303–306.

Boere, W.A., Benaissa-Trouw, B.J., Harmsen, M., Kraaijeveld, C.A., and Snippe, H. (1983). Neutralizing and non-neutralizing monoclonal antibodies to the E2 glycoprotein of Semliki Forest virus can protect mice from lethal encephalitis. J. Gen. Virol. *64*(Pt 6), 1405–1408.

Bowen, G.S., Fashinell, T.R., Dean, P.B., and Gregg, M.B. (1976). Clinical aspects of human Venezuelan equine encephalitis in Texas. Bull. Pan. Am. Health Organ. *10*, 46–57.

Bradish, C.J., Allner, K., and Maber, H.B. (1971). The virulence of original and derived strains of Semliki forest virus for mice, guinea-pigs and rabbits. J. Gen. Virol. *12*, 141–160.

Brown, A., and Officer, J.E. (1975). An attenuated variant of Eastern encephalitis virus: biological properties and protection induced in mice. Arch. Virol. *47*, 123–138.

Brown, A.R., Rebus, S., McKimmie, C.S., Robertson, K., Williams, A., and Fazakerley, J.K. (2005). Gene expression profiling of the preclinical scrapie-infected hippocampus. Biochem. Biophys. Res. Commun. *334*, 86–95.

Bsibsi, M., Persoon-Deen, C., Verwer, R.W.H., Meeuwsen, S., Ravid, R., and van Noort, J.M. (2006). Toll-like receptor 3 on adult human astrocytes triggers production of neuroprotective mediators. Glia *53*, 688–695.

Burdeinick-Kerr, R., and Griffin, D.E. (2005). Gamma interferon-dependent, noncytolytic clearance of sindbis virus infection from neurons *in vitro*. J. Virol. *79*, 5374–5385.

Burdeinick-Kerr, R., Wind, J., and Griffin, D.E. (2007). Synergistic roles of antibody and interferon in noncytolytic clearance of Sindbis virus from different regions of the central nervous system. J. Virol. *81*, 5628–5636.

Burt, F.J., Rolph, M.S., Rulli, N.E., Mahalingam, S., and Heise, M.T. (2012). Chikungunya: a re-emerging virus. Lancet *379*, 662–671.

Calisher, C.H. (1994). Medically important arboviruses of the United States and Canada. Clin. Microbiol. Rev. *7*, 89–116.

Carpentier, P.A., Begolka, W.S., Olson, J.K., Elhofy, A., Karpus, W.J., and Miller, S.D. (2004). Differential activation of astrocytes by innate and adaptive immune stimuli. Glia *49*, 360–374.

Carpentier, P.A., Duncan, D.S., and Miller, S.D. (2008). Glial toll-like receptor signaling in central nervous system infection and autoimmunity. Brain Behav. Immun. *22*, 140–147.

Charles, P.C., Walters, E., Margolis, F., and Johnston, R.E. (1995). Mechanism of neuroinvasion of Venezuelan equine encephalitis virus in the mouse. Virology *208*, 662–671.

Charles, P.C., Trgovcich, J., Davis, N.L., and Johnston, R.E. (2001). Immunopathogenesis and immune modulation of Venezuelan equine encephalitis virus-induced disease in the mouse. Virology *284*, 190–202.

Danes, L., Rychterová, V., Kufner, J., and Hrusková, J. (1973). The role of the olfactory route on infection of the respiratory tract with Venezuelan equine encephalomyelitis virus in normal and operated Macaca rhesus monkeys. II. Results of histological examination. Acta Virol. *17*, 57–60.

Dent, J. (1955). Pathological changes in the brains of children infected with the virus of eastern equine encephalomyelitis. Bull. Tulane Univ. Med. Fac. *14*, 85–96.

Deresiewicz, R.L., Thaler, S.J., Hsu, L., and Zamani, A.A. (1997). Clinical and neuroradiographic manifestations of eastern equine encephalitis. N. Engl. J. Med. *336*, 1867–1874.

Despres, P., Griffin, J.W., and Griffin, D.E. (1995a). Antiviral activity of alpha interferon in Sindbis virus-infected cells is restored by anti-E2 monoclonal antibody treatment. J. Virol. *69*, 7345–7348.

Despres, P., Griffin, J.W., and Griffin, D.E. (1995b). Effects of anti-E2 monoclonal antibody on sindbis virus replication in AT3 cells expressing bcl-2. J. Virol. *69*, 7006–7014.

Donnelly, S.M., Sheahan, B.J., and Atkins, G.J. (1997). Long-term effects of Semliki Forest virus infection in the mouse central nervous system. Neuropathol. Appl. Neurobiol. *23*, 235–241.

Ehrenkranz, N.J., and Ventura, A.K. (1974). Venezuelan equine encephalitis virus infection in man. Annu. Rev. Med. *25*, 9–14.

Farina, C., Krummel, M.F., Giese, T., Hartmann, G., Aloisi, F., and Meinl, E. (2005). Preferential expression and function of Toll-like receptor 3 in human astrocytes. J. Neuroimmunol. *159*, 12–19.

Farina, C., Aloisi, F., and Meinl, E. (2007). Astrocytes are active players in cerebral innate immunity. Trends Immunol. *28*, 138–145.

Fazakerley, J.K. (2001). Neurovirology and developmental neurobiology. Adv. Virus Res. *56*, 73–124.

Fazakerley, J.K. (2002). Pathogenesis of Semliki Forest virus encephalitis. J. Neurovirol. *8*, 66–74.

Fazakerley, J.K. (2004). Semliki forest virus infection of laboratory mice: a model to study the pathogenesis of viral encephalitis. Arch. Virol. Suppl. 179–190.

Fazakerley, J.K., and Allsopp, T.E. (2001). Programmed cell death in virus infections of the nervous system. Curr. Top. Microbiol. Immunol. *253*, 95–119.

Fazakerley, J.K., and Buchmeier, M.J. (1993). Pathogenesis of virus-induced demyelination. Adv. Virus Res. *42*, 249–324.

Fazakerley, J.K., and Webb, H.E. (1987a). Semliki Forest virus-induced, immune-mediated demyelination: adoptive transfer studies and viral persistence in nude mice. J. Gen. Virol. *68*(Pt 2), 377–385.

Fazakerley, J.K., and Webb, H.E. (1987b). Cyclosporine enhances virally induced T-cell-mediated demyelination. The effect of cyclosporine on a demyelinating virus infection. J. Neurol. Sci. *78*, 35–50.

Fazakerley, J.K., and Webb, H.E. (1987c). Semliki Forest virus induced, immune mediated demyelination: the effect of irradiation. Br. J. Exp. Pathol. *68*, 101–113.

Fazakerley, J.K., Pathak, S., Scallan, M., Amor, S., and Dyson, H. (1993). Replication of the A7(74) strain of Semliki Forest virus is restricted in neurons. Virology *195*, 627–637.

Fleming, P. (1977). Age-dependent and strain-related differences of virulence of Semliki Forest virus in mice. J. Gen. Virol. *37*, 93–105.

Forrester, N.L., Kenney, J.L., Deardorff, E., Wang, E., and Weaver, S.C. (2008). Western Equine Encephalitis submergence: lack of evidence for a decline in virus virulence. Virology *380*, 170–172.

Fragkoudis, R., Breakwell, L., (null), Boyd, A., Barry, G., Kohl, A., Merits, A., and Fazakerley, J.K. (2007). The type I interferon system protects mice from Semliki Forest virus by preventing widespread virus dissemination in extraneural tissues, but does not mediate the restricted replication of avirulent virus in central nervous system neurons. J. Gen. Virol. *88*, 3373–3384.

Fragkoudis, R., Ballany, C.M., Boyd, A., and Fazakerley, J.K. (2008). In Semliki Forest virus encephalitis, antibody rapidly clears infectious virus and is required to eliminate viral material from the brain, but is not required to generate lesions of demyelination. J. Gen. Virol. *89*, 2565–2568.

Gardner, C.L., Burke, C.W., Tesfay, M.Z., Glass, P.J., Klimstra, W.B., and Ryman, K.D. (2008). Eastern and Venezuelan Equine encephalitis viruses differ in their ability to infect dendritic cells and macrophages: impact of altered cell tropism on pathogenesis. J. Virol. *82*, 10634–10646.

Glasgow, G.M., McGee, M.M., Sheahan, B.J., and Atkins, G.J. (1997). Death mechanisms in cultured cells infected by Semliki Forest virus. J. Gen. Virol. *78 (Pt 7)*, 1559–1563.

Glass, W.G. (2005). Chemokine receptor CCR5 promotes leukocyte trafficking to the brain and survival in West Nile virus infection. J. Exp. Med. *202*, 1087–1098.

Glass, W.G., McDermott, D.H., Lim, J.K., Lekhong, S., Yu, S.F., Frank, W.A., Pape, J., Cheshier, R.C., and Murphy, P.M. (2006). CCR5 deficiency increases risk of symptomatic West Nile virus infection. J. Exp. Med. *203*, 35–40.

Gleiser, C., Gochenour, W., Berge, T., and Tigertt, W. (1962). The comparative pathology of experimental Venezuelan equine encephalomyelitis infection in different animal hosts. J. Infect. Dis. *110*, 80–97.

Gorchakov, R., Frolova, E., Williams, B.R.G., Rice, C.M., and Frolov, I. (2004). PKR-dependent and -independent mechanisms are involved in translational shutoff during Sindbis virus infection. J. Virol. *78*, 8455–8467.

Gorelkin, L., and Jahrling, P.B. (1975). Virus-initiated septic shock. Acute death of Venezuelan encephalitis virus-infected hamsters. Lab. Invest. *32*, 78–85.

Gould, E., and Solomon, T. (2008). Pathogenic flaviviruses. Lancet *371*, 500–509.

Grieder, F.B., Davis, N.L., Aronson, J.F., Charles, P.C., Sellon, D.C., Suzuki, K., and Johnston, R.E. (1995). Specific restrictions in the progression of Venezuelan equine encephalitis virus-induced disease resulting from single amino acid changes in the glycoproteins. Virology *206*, 994–1006.

Grieder, F.B., Davis, B.K., Zhou, X.D., Chen, S.J., Finkelman, F.D., and Gause, W.C. (1997). Kinetics of cytokine expression and regulation of host protection following infection with molecularly cloned Venezuelan equine encephalitis virus. Virology *233*, 302–312.

Griffin, D.E. (1976). Role of the immune response in age-dependent resistance of mice to encephalitis due to Sindbis virus. J. Infect. Dis. *133*, 456–464.

Griffin, D.E. (2003). Immune responses to RNA-virus infections of the CNS. Nat. Rev. Immunol. *3*, 493–502.

Griffin, D.E. (2005). Neuronal cell death in alphavirus encephalomyelitis. Curr. Top. Microbiol. Immunol. *289*, 57–77.

Griffin, D.E. (2010). Recovery from viral encephalomyelitis: immune-mediated noncytolytic virus clearance from neurons. Immunol. Res. *47*, 123–133.

Hahn, C.S., Lustig, S., Strauss, E.G., and Strauss, J.H. (1988). Western equine encephalitis virus is a recombinant virus. Proc. Natl. Acad. Sci. U.S.A. *85*, 5997–6001.

Hanson, R.P., Sulkin, S.E., Beuscher, E.L., Hammon, W.M., McKinney, R.W., and Work, T.H. (1967). Arbovirus infections of laboratory workers. Extent of problem emphasizes the need for more effective measures to reduce hazards. Science *158*, 1283–1286.

Hart, M.K., Pratt, W., Panelo, F., Tammariello, R., and Dertzbaugh, M. (1997). Venezuelan equine encephalitis virus vaccines induce mucosal IgA responses and protection from airborne infection in BALB/c, but not C3H/HeN mice. Vaccine *15*, 363–369.

Hart, M.K., Caswell-Stephan, K., Bakken, R., Tammariello, R., Pratt, W., Davis, N., Johnston, R.E., Smith, J., and Steele, K. (2000). Improved mucosal protection against Venezuelan equine encephalitis virus is induced by the molecularly defined, live-attenuated V3526 vaccine candidate. Vaccine *18*, 3067–3075.

Hauser, G. (1948). Human equine encephalomyelitis, eastern type, in Louisiana. New Orleans Med. Surg. J. *100*, 551–558.

Havert, M.B., Schofield, B., Griffin, D.E., and Irani, D.N. (2000). Activation of divergent neuronal cell death pathways in different target cell populations during neuroadapted sindbis virus infection of mice. J. Virol. *74*, 5352–5356.

Jack, C.S., Arbour, N., Manusow, J., Montgrain, V., Blain, M., McCrea, E., Shapiro, A., and Antel, J.P. (2005). TLR signaling tailors innate immune responses in human microglia and astrocytes. J. Immunol. *175*, 4320–4330.

Jackson, A.C., and Rossiter, J.P. (1997). Apoptotic cell death is an important cause of neuronal injury in experimental Venezuelan equine encephalitis virus infection of mice. Acta Neuropathol. *93*, 349–353.

Jackson, A.C., SenGupta, S.K., and Smith, J.F. (1991). Pathogenesis of Venezuelan equine encephalitis virus infection in mice and hamsters. Vet. Pathol. *28*, 410–418.

Jagelman, S., Suckling, A.J., Webb, H.E., and Bowen, F.T. (1978). The pathogenesis of avirulent Semliki Forest virus infections in athymic nude mice. J. Gen. Virol. *41*, 599–607.

Johnson, K.M., Shelokov, A., Peralta, P.H., Dammin, G.J., and Young, N.A. (1968). Recovery of Venezuelan equine encephalomyelitis virus in Panama. A fatal case in man. Am. J. Trop. Med. Hyg. *17*, 432–440.

Johnston, R.E., and Peters, C.J. (1996). Alphaviruses. In Fields Virology, N. Fields, D.M. Knipe, and P.M. Howley, eds. (Lippincott Raven, Philadelphia, PA, USA), pp. 843–898.

Johnson, R.T., McFarland, H.F., and Levy, S.E. (1972). Age-dependent resistance to viral encephalitis: studies of infections due to Sindbis virus in mice. J. Infect. Dis. *125*, 257–262.

Junt, T., Moseman, E.A., Iannacone, M., Massberg, S., Lang, P.A., Boes, M., Fink, K., Henrickson, S.E., Shayakhmetov, D.M., Di Paolo, N.C., *et al.* (2007). Subcapsular sinus macrophages in lymph nodes clear lymph-borne viruses and present them to antiviral B cells. Nature *450*, 110–114.

Kam, Y.-W., Ong, E.K.S., Renia, L., Tong, J.-C., and Ng, L.F.P. (2009). Immuno-biology of Chikungunya and implications for disease intervention. Microbes Infect. *11*, 1186–1196.

Kelly, W.R., Blakemore, W.F., Jagelman, S., and Webb, H.E. (1982). Demyelination induced in mice by avirulent Semliki Forest virus. II. An ultrastructural study of focal demyelination in the brain. Neuropathol. Appl. Neurobiol. *8*, 43–53.

Kim, J.V., Kang, S.S., Dustin, M.L., and McGavern, D.B. (2008). Myelomonocytic cell recruitment causes fatal CNS vascular injury during acute viral meningitis. Nature *457*, 191–195.

Klimstra, W.B., Ryman, K.D., and Johnston, R.E. (1998). Adaptation of Sindbis virus to BHK cells selects for use of heparan sulfate as an attachment receptor. J. Virol. *72*, 7357–7366.

Klimstra, W.B., Ryman, K.D., Bernard, K.A., Nguyen, K.B., Biron, C.A., and Johnston, R.E. (1999). Infection of neonatal mice with sindbis virus results in a systemic inflammatory response syndrome. J. Virol. 73, 10387–10398.

Laine, M., Luukkainen, R., and Toivanen, A. (2004). Sindbis viruses and other alphaviruses as cause of human arthritic disease. J. Intern. Med. *256*, 457–471.

Levine, B., and Deretic, V. (2007). Unveiling the roles of autophagy in innate and adaptive immunity. Nat. Rev. Immunol. *7*, 767–777.

Levine, B., and Griffin, D.E. (1992). Persistence of viral RNA in mouse brains after recovery from acute alphavirus encephalitis. J. Virol. *66*, 6429–6435.

Levine, B., Hardwick, J.M., Trapp, B.D., Crawford, T.O., Bollinger, R.C., and Griffin, D.E. (1991). Antibody-mediated clearance of alphavirus infection from neurons. Science *254*, 856–860.

Levine, B., Huang, Q., Isaacs, J.T., Reed, J.C., Griffin, D.E., and Hardwick, J.M. (1993). Conversion of lytic to persistent alphavirus infection by the bcl-2 cellular oncogene. Nature *361*, 739–742.

Levitt, N.H., Miller, H.V., and Edelman, R. (1979). Interaction of alphaviruses with human peripheral leukocytes: *in vitro* replication of Venezuelan equine encephalomyelitis virus in monocyte cultures. Infect. Immun. *24*, 642–646.

Lewis, J., Wesselingh, S.L., Griffin, D.E., and Hardwick, J.M. (1996). Alphavirus-induced apoptosis in mouse brains correlates with neurovirulence. J. Virol. *70*, 1828–1835.

Liang, X.H., Kleeman, L.K., Jiang, H.H., Gordon, G., Goldman, J.E., Berry, G., Herman, B., and Levine, B. (1998). Protection against fatal Sindbis virus encephalitis by beclin, a novel Bcl-2-interacting protein. J. Virol. *72*, 8586–8596.

Lim, J.K., Obara, C.J., Rivollier, A., Pletnev, A.G., Kelsall, B.L., and Murphy, P.M. (2011). Chemokine receptor Ccr2 is critical for monocyte accumulation and survival in West Nile virus encephalitis. J. Immunol. *186*, 471–478.

Logue, C.H., Bosio, C.F., Welte, T., Keene, K.M., Ledermann, J.P., Phillips, A., Sheahan, B.J., Pierro, D.J., Marlenee, N., Brault, A.C., *et al.* (2009). Virulence variation among isolates of western equine encephalitis virus in an outbred mouse model. J. Gen. Virol. *90*, 1848–1858.

Lustig, S., Danenberg, H.D., Kafri, Y., Kobiler, D., and Ben-Nathan, D. (1992). Viral neuroinvasion and encephalitis induced by lipopolysaccharide and its mediators. J. Exp. Med. *176*, 707–712.

MacDonald, G.H., and Johnston, R.E. (2000). Role of dendritic cell targeting in Venezuelan equine encephalitis virus pathogenesis. J. Virol. *74*, 914–922.

McFarland, H.F., Griffin, D.E., and Johnson, R.T. (1972). Specificity of the inflammatory response in viral encephalitis. I. Adoptive immunization of immunosuppressed mice infected with Sindbis virus. J. Exp. Med. *136*, 216–226.

McKimmie, C.S., and Fazakerley, J.K. (2005). In response to pathogens, glial cells dynamically and differentially regulate Toll-like receptor gene expression. J. Neuroimmunol. *169*, 116–125.

McKimmie, C.S., and Graham, G.J. (2010). Astrocytes modulate the chemokine network in a pathogen-specific manner. Biochem. Biophys. Res. Commun. *394*, 1006–1011.

McKimmie, C.S., Johnson, N., Fooks, A.R., and Fazakerley, J.K. (2005). Viruses selectively upregulate Toll-like receptors in the central nervous system. Biochem. Biophys. Res. Commun. *336*, 925–933.

McKimmie, C.S., Roy, D., Forster, T., and Fazakerley, J.K. (2006). Innate immune response gene expression profiles of N9 microglia are pathogen-type specific. J. Neuroimmunol. *175*, 128–141.

Mathiot, C.C., Grimaud, G., Garry, P., Bouquety, J.C., Mada, A., Daguisy, A.M., and Georges, A.J. (1990). An outbreak of human Semliki Forest virus infections in Central African Republic. Am. J. Trop. Med. Hyg. *42*, 386–393.

Metcalf, T.U., and Griffin, D.E. (2011). Alphavirus-induced encephalomyelitis: antibody-secreting cells and viral clearance from the nervous system. J. Virol. *85*, 11490–11501.

Monath, T.P., Kemp, G.E., Cropp, C.B., and Chandler, F.W. (1978). Necrotizing myocarditis in mice infected with Western equine encephalitis virus: clinical, electrocardiographic, and histopathologic correlations. J. Infect. Dis. *138*, 59–66.

la Monte, de, S., Castro, F., Bonilla, N.J., Gaskin de Urdaneta, A., and Hutchins, G.M. (1985). The systemic pathology of Venezuelan equine encephalitis virus infection in humans. Am. J. Trop. Med. Hyg. *34*, 194–202.

Morris, M.M., Dyson, H., Baker, D., Harbige, L.S., Fazakerley, J.K., and Amor, S. (1997). Characterization of the cellular and cytokine response in the central nervous system following Semliki Forest virus infection. J. Neuroimmunol. *74*, 185–197.

Oliver, K.R., and Fazakerley, J.K. (1998). Transneuronal spread of Semliki Forest virus in the developing mouse olfactory system is determined by neuronal maturity. Neuroscience *82*, 867–877.

Oliver, K.R., Scallan, M.F., Dyson, H., and Fazakerley, J.K. (1997). Susceptibility to a neurotropic virus and its changing distribution in the developing brain is a function of CNS maturity. J. Neurovirol. *3*, 38–48.

Orvedahl, A., and Levine, B. (2008). Autophagy and viral neurovirulence. Cell. Microbiol. *10*, 1747–1756.

Parsons, L.M., and Webb, H.E. (1982a). Blood brain barrier disturbance and immunoglobulin G levels in the cerebrospinal fluid of the mouse following peripheral infection with the demyelinating strain of Semliki Forest virus. J. Neurol. Sci. *57*, 307–318.

Parsons, L.M., and Webb, H.E. (1982b). Virus titres and persistently raised white cell counts in cerebrospinal fluid in mice after peripheral infection with demyelinating Semliki Forest virus. Neuropathol. Appl. Neurobiol. *8*, 395–401.

Parsons, L.M., and Webb, H.E. (1989). Identification of immunoglobulin-containing cells in the central nervous system of the mouse following infection with the demyelinating strain of Semliki Forest virus. Br. J. Exp. Pathol. *70*, 247–255.

Parsons, L.M., and Webb, H.E. (1992). IgG subclass responses in brain and serum in Semliki Forest virus demyelinating encephalitis. Neuropathol. Appl. Neurobiol. *18*, 351–359.

Pathak, S., and Webb, H.E. (1974). Possible mechanisms for the transport of Semliki forest virus into and within mouse brain. An electron-microscopic study. J. Neurol. Sci. *23*, 175–184.

Pathak, S., and Webb, H.E. (1978). An electron-microscopic study of avirulent and virulent Semliki forest virus in the brains of different ages of mice. J. Neurol. Sci. *39*, 199–211.

Pathak, S., and Webb, H.E. (1983). Effect of myocrisin (sodium auro-thio-malate) on the morphogenesis of avirulent Semliki Forest virus in mouse brain: an electron microscopical study. Neuropathol. Appl. Neurobiol. *9*, 313–327.

Pathak, S., Webb, H., Oaten, S., and Bateman, S. (1976). An electron-microscopic study of the development of virulent and avirulent strains of Semliki Forest virus in mouse brain. J. Neurol. Sci. *28*, 289–300.

Powers, A.M., Brault, A.C., Shirako, Y., Strauss, E.G., Kang, W., Strauss, J.H., and Weaver, S.C. (2001). Evolutionary relationships and systematics of the alphaviruses. J. Virol. *75*, 10118–10131.

Pratt, W.D., Hart, M.K., Reed, D.S., and Steele, K.E. (2006). Alphaviruses. In Biodefense: Reserach Methodology and Animal Models, J.R. Swearengen, ed. (Taylor & Francis, Boca Raton, FL, USA), pp. 181–206.

Préhaud, C., Mégret, F., Lafage, M., and Lafon, M. (2005). Virus infection switches TLR-3-positive human neurons to become strong producers of beta interferon. J. Virol. *79*, 12893–12904.

Ransohoff, R.M., and Engelhardt, B. (2012). The anatomical and cellular basis of immune surveillance in the central nervous system. Nat. Rev. Immunol. *12*, 623–635.

Reboldi, A., Coisne, C., Baumjohann, D., Benvenuto, F., Bottinelli, D., Lira, S., Uccelli, A., Lanzavecchia, A., Engelhardt, B., and Sallusto, F. (2009). C-C chemokine receptor 6–regulated entry of TH-17 cells into the CNS through the choroid plexus is required for the initiation of EAE. Nat. Immunol. *10*, 514–523.

Reed, D.S., Larsen, T., Sullivan, L.J., Lind, C.M., Lackemeyer, M.G., Pratt, W.D., and Parker, M.D. (2005a). Aerosol exposure to western equine encephalitis virus causes fever and encephalitis in cynomolgus macaques. J. Infect. Dis. *192*, 1173–1182.

Reed, D.S., Lind, C.M., Lackemeyer, M.G., Sullivan, L.J., Pratt, W.D., and Parker, M.D. (2005b). Genetically engineered, live, attenuated vaccines protect nonhuman primates against aerosol challenge with a virulent IE strain of Venezuelan equine encephalitis virus. Vaccine 23, 3139–3147.

Reed, D.S., Lackemeyer, M.G., Garza, N.L., Norris, S., Gamble, S., Sullivan, L.J., Lind, C.M., and Raymond, J.L. (2007). Severe encephalitis in cynomolgus macaques exposed to aerosolized Eastern equine encephalitis virus. J. Infect. Dis. *196*, 441–450.

Rowell, J.F., and Griffin, D.E. (1999). The inflammatory response to nonfatal Sindbis virus infection of the nervous system is more severe in SJL than in BALB/c mice and is associated with low levels of IL-4 mRNA and high levels of IL-10-producing CD4+ T cells. J. Immunol. *162*, 1624–1632.

Rulli, N., Melton, J., Wilmes, A., Ewart, G., and Mahalingam, S. (2007). The molecular and cellular aspects of arthritis due to alphavirus infections. Ann. N. Y. Acad. Sci. *1102*, 96–108.

Ryman, K.D., and Klimstra, W.B. (2008). Host responses to alphavirus infection. Immunol. Rev. *225*, 27–45.

Ryman, K.D., Klimstra, W.B., Nguyen, K.B., Biron, C.A., and Johnston, R.E. (2000). Alpha/beta interferon protects adult mice from fatal Sindbis virus infection and is an important determinant of cell and tissue tropism. J. Virol. *74*, 3366–3378.

Ryzhikov, A.B., Tkacheva, N.V., Sergeev, A.N., and Ryabchikova, E.I. (1991). Venezuelan equine encephalitis virus propagation in the olfactory tract of normal and immunized mice. Biomed. Sci. 2, 607–614.

Ryzhikov, A.B., Ryabchikova, E.I., Sergeev, A.N., and Tkacheva, N.V. (1995). Spread of Venezuelan equine encephalitis virus in mice olfactory tract. Arch. Virol. *140*, 2243–2254.

Sammin, D.J., Butler, D., Atkins, G.J., and Sheahan, B.J. (1999). Cell death mechanisms in the olfactory bulb of rats infected intranasally with Semliki forest virus. Neuropathol. Appl. Neurobiol. 25, 236–243.

Sawicki, D.L., Silverman, R.H., Williams, B.R., and Sawicki, S.G. (2003). Alphavirus minus-strand synthesis and persistence in mouse embryo fibroblasts derived from mice lacking RNase L and protein kinase R. J. Virol. *77*, 1801–1811.

Scallan, M.F., and Fazakerley, J.K. (1999). Aurothiolates enhance the replication of Semliki Forest virus in the CNS and the exocrine pancreas. J. Neurovirol. *5*, 392–400.

Scallan, M.F., Allsopp, T.E., and Fazakerley, J.K. (1997). bcl-2 acts early to restrict Semliki Forest virus replication and delays virus-induced programmed cell death. J. Virol. *71*, 1583–1590.

Schlesinger, S. (2001). Alphavirus vectors: development and potential therapeutic applications. Expert Opin. Biol. Ther. *1*, 177–191.

Schoneboom, B.A., Fultz, M.J., Miller, T.H., McKinney, L.C., and Grieder, F.B. (1999). Astrocytes as targets for Venezuelan equine encephalitis virus infection. J. Neurovirol. 5, 342–354.

Schoneboom, B.A., Catlin, K.M., Marty, A.M., and Grieder, F.B. (2000a). Inflammation is a component of neurodegeneration in response to Venezuelan equine encephalitis virus infection in mice. J. Neuroimmunol. *109*, 132–146.

Schoneboom, B.A., Lee, J.S., and Grieder, F.B. (2000b). Early expression of IFN-alpha/beta and iNOS in the brains of Venezuelan equine encephalitis virus-infected mice. J. Interferon Cytokine Res. *20*, 205–215.

Sebastian, M.R., Lodha, R., and Kabra, S.K. (2009). Chikungunya infection in children. Indian J. Pediatr. *76*, 185–189.

Shabman, R.S., Morrison, T.E., Moore, C., White, L., Suthar, M.S., Hueston, L., Rulli, N., Lidbury, B., Ting, J.P.Y., Mahalingam, S., *et al.* (2006). Differential induction of type I interferon responses in myeloid dendritic cells by mosquito and mammalian-cell-derived alphaviruses. J. Virol. *81*, 237–247.

Sharma, A., and Maheshwari, R.K. (2009). Oligonucleotide array analysis of Toll-like receptors and associated signalling genes in Venezuelan equine encephalitis virus-infected mouse brain. J. Gen. Virol. *90*, 1836–1847.

Sharma, A., Bhattacharya, B., Puri, R.K., and Maheshwari, R.K. (2008). Venezuelan equine encephalitis virus infection causes modulation of inflammatory and immune response genes in mouse brain. BMC Genomics 9, 289.

Slepushkin, A.N. (1959). An epidemiological study on a case of Venezuelan equine encephalomyelitis in a laboratory. Vopr. Virusol. 4, 54–58.

Smith, J.P., Morris-Downes, M., Brennan, F.R., Wallace, G.J., and Amor, S. (2000). A role for alpha4-integrin in the pathology following Semliki Forest virus infection. J. Neuroimmunol. *106*, 60–68.

Soilu-Hanninen, M., Eralinna, J.P., Hukkanen, V., Roytta, M., Salmi, A.A., and Salonen, R. (1994). Semliki Forest virus infects mouse brain endothelial cells and causes blood–brain barrier damage. J. Virol. *68*, 6291.

Steele, K.E., and Twenhafel, N.A. (2010). Review paper: pathology of animal models of alphavirus encephalitis. Vet. Pathol. *47*, 790–805.

Steele, K.E., Davis, K.J., Stephan, K., Kell, W., Vogel, P., and Hart, M.K. (1998). Comparative neurovirulence and tissue tropism of wild-type and attenuated strains of Venezuelan equine encephalitis virus administered by aerosol in C3H/HeN and BALB/c mice. Vet. Pathol. *35*, 386–397.

Steele, K.E., Seth, P., Catlin-Lebaron, K.M.K., Schoneboom, B.A., Husain, M.M., Grieder, F., and Maheshwari, R.K. (2006). Tunicamycin enhances neuroinvasion and encephalitis in mice infected with Venezuelan equine encephalitis virus. Vet. Pathol. *43*, 904–913.

Styer, L.M., Kent, K.A., Albright, R.G., Bennett, C.J., Kramer, L.D., and Bernard, K.A. (2007). Mosquitoes inoculate high doses of west nile virus as they probe and feed on live hosts. PLoS Pathog. *3*, e132.

Styer, L.M., Lim, P.Y., Louie, K.L., Albright, R.G., Kramer, L.D., and Bernard, K.A. (2011). mosquito saliva causes enhancement of West Nile Virus infection in mice. J. Virol. *85*, 1517–1527.

Subak-Sharpe, I., Dyson, H., and Fazakerley, J. (1993). *In vivo* depletion of CD8+ T cells prevents lesions of demyelination in Semliki Forest virus infection. J. Virol. *67*, 7629–7633.

Suckling, A.J., Pathak, S., Jagelman, S., and Webb, H.E. (1978). Virus-associated demyelination. A model using avirulent Semliki Forest virus infection of mice. J. Neurol. Sci. *39*, 147–154.

Suckling, A.J., Jagelman, S., and Webb, H.E. (1982). Immunoglobulin synthesis in nude (nu/nu), nu/+ and reconstituted nu/nu mice infected with a demyelinating strain of Semliki Forest virus. Clin. Exp. Immunol. *47*, 283–288.

Tesh, R.B. (1982). Arthritides caused by mosquito-borne viruses. Annu. Rev. Med. *33*, 31–40.

Tesh, R.B., Watts, D.M., Russell, K.L., Damodaran, C., Calampa, C., Cabezas, C., Ramirez, G., Vasquez, B., Hayes, C.G., Rossi, C.A., *et al.* (1999). Mayaro virus disease: an emerging mosquito-borne zoonosis in tropical South America. Clin. Infect. Dis. *28*, 67–73.

Thach, D.C., Kimura, T., and Griffin, D.E. (2000). Differences between C57BL/6 and BALB/cBy mice in mortality and virus replication after intranasal infection with neuroadapted Sindbis virus. J. Virol. *74*, 6156–6161.

Tuittila, M.T., Santagati, M.G., Roytta, M., Määttä, J.A., and Hinkkanen, A.E. (2000). Replicase complex genes of Semliki Forest virus confer lethal neurovirulence. J. Virol. *74*, 4579–4589.

Tyor, W.R., Wesselingh, S., Levine, B., and Griffin, D.E. (1992). Long term intraparenchymal Ig secretion after acute viral encephalitis in mice. J. Immunol. *149*, 4016–4020.

Ubol, S., Levine, B., Lee, S.H., Greenspan, N.S., and Griffin, D.E. (1995). Roles of immunoglobulin valency and the heavy-chain constant domain in antibody-mediated downregulation of Sindbis virus replication in persistently infected neurons. J. Virol. *69*, 1990–1993.

Vernon, P.S., and Griffin, D.E. (2005). Characterization of an *in vitro* model of alphavirus infection of immature and mature neurons. J. Virol. *79*, 3438–3447.

Vogel, P., Abplanalp, D., Kell, W., Ibrahim, M.S., Downs, M.B., Pratt, W.D., and Davis, K.J. (1996). Venezuelan equine encephalitis in BALB/c mice: kinetic analysis of central nervous system infection following aerosol or subcutaneous inoculation. Arch. Pathol. Lab. Med. *120*, 164–172.

Vogel, P., Kell, W.M., Fritz, D.L., Parker, M.D., and Schoepp, R.J. (2005). Early events in the pathogenesis of eastern equine encephalitis virus in mice. AJPA *166*, 159–171.

Walker, D.H., Harrison, A., Murphy, K., Flemister, M., and Murphy, F.A. (1976). Lymphoreticular and myeloid pathogenesis of Venezuelan equine encephalitis in hamsters. AJPA *84*, 351–370.

Wang, T., Town, T., Alexopoulou, L., Anderson, J.F., Fikrig, E., and Flavell, R.A. (2004). Toll-like receptor 3 mediates West Nile virus entry into the brain causing lethal encephalitis. Nat. Med. *10*, 1366–1373.

Weaver, S.C., Hagenbaugh, A., Bellew, L.A., Netesov, S.V., Volchkov, V.E., Chang, G.J., Clarke, D.K., Gousset, L., Scott, T.W., and Trent, D.W. (1993). A comparison of the nucleotide sequences of eastern and western equine encephalomyelitis viruses with those of other alphaviruses and related RNA viruses. Virology *197*, 375–390.

Willems, W.R., Kaluza, G., Boschek, C.B., Bauer, H., Hager, H., Schütz, H.J., and Feistner, H. (1979). Semliki forest virus: cause of a fatal case of human encephalitis. Science *203*, 1127–1129.

Woodman, D.R., McManus, A.T., and Eddy, G.A. (1975). Extension of the mean time to death of mice with a lethal infection of Venezuelan equine encephalomyelitis virus by antithymocyte serum treatment. Infect. Immun. *12*, 1006–1011.

Zacks, M.A., and Paessler, S. (2010). Encephalitic alphaviruses. Vet. Microbiol. *140*, 281–286.

Zhang, B., Chan, Y.K., Lu, B., Diamond, M.S., and Klein, R.S. (2008). CXCR3 mediates region-specific antiviral T cell trafficking within the central nervous system during West Nile virus encephalitis. J. Immunol. *180*, 2641–2649.

Application of Alphavirus Vectors for Gene Therapy

9

Kenneth Lundstrom

Abstract

Alphavirus vectors based on Semliki Forest virus (SFV), Sindbis virus (SINV) and Venezuelan equine encephalitis virus (VEEV) have been engineered for efficient expression of heterologous proteins. In particular, SFV-based vectors have been frequently used for expression of membrane proteins in drug discovery. Likewise, SFV and SINV have been employed for numerous applications in neuroscience research due to their neurotrophic properties. Furthermore, all three alphaviruses have been used for vaccine development against infectious diseases and cancer. Alphavirus vectors have also found applications in gene therapy, especially in cancer therapy, due to their broad host range, high expression levels and transient nature. SINV vectors have demonstrated natural targeting to tumours whereas SFV vectors have been modified to enhance delivery to tumour tissue. Furthermore, oncolytic SFV vectors have been evaluated for increased therapeutic efficacy.

Introduction

Alphaviruses belong to the Togaviridae family (Luers *et al.*, 2005). The 11–12 kb single-stranded RNA genome is embedded in the icosahedral protein capsid structure, which is further enveloped by 240 heterotrimer glycoproteins (Strauss and Strauss, 1994). The alphavirus life cycle is presented below and schematically illustrated in Fig. 9.1. The spike proteins participate in receptor recognition and cell entry, which is mediated by endocytosis. Although specific alphavirus receptors have not been conclusively determined, laminin and heparin receptors have been suggested as targets (Wang *et al.*, 1992). After fusion of the viral envelope with the cell membrane, the viral RNA is released for immediate replication in the cytoplasm. A minus-strand RNA copy is generated from the plus-strand genome to act as a template for a large number of both full-length 42S RNA and subgenomic 26S RNA copies. Full-length RNA is packaged into nucleocapsids and transported to the plasma membrane as the envelope proteins are simultaneously processed through the rough endoplasmic reticulum and Golgi apparatus. The release of mature viral particles is highly efficient, resulting in titres of 10^9–10^{10} particles/ml within 24 hours.

Certain alphaviruses such as Eastern equine encephalitis virus (EEEV) and Western equine encephalitis virus (WEEV) can cause fatal encephalitis in humans (Griffin, 1986). However, the alphaviruses that have been engineered as viral expression vectors are generally considered as mild human pathogens despite their association with a couple of epidemics in Africa (Mathiot *et al.*, 1990; Niklasson *et al.*, 1988). All strains selected for engineering of expression vectors are avirulent to reduce the risk of pathogen effect. In this context, vectors based on Semliki Forest virus (SFV) (Liljestrom and Garoff, 1991), Sindbis virus (SINV) (Xiong *et al.*, 1989) and Venezuelan equine encephalitis virus (VEEV) (Davis *et al.*, 1989) have been applied. These vectors have provided high expression levels of topologically different proteins (Liljestrom and Garoff, 1991; Lundstrom, 2005). In particular, membrane proteins like G protein-coupled receptors (GPCRs) and ion channels have been expressed at extremely high levels (Lundstrom, 2003a). Alphaviruses

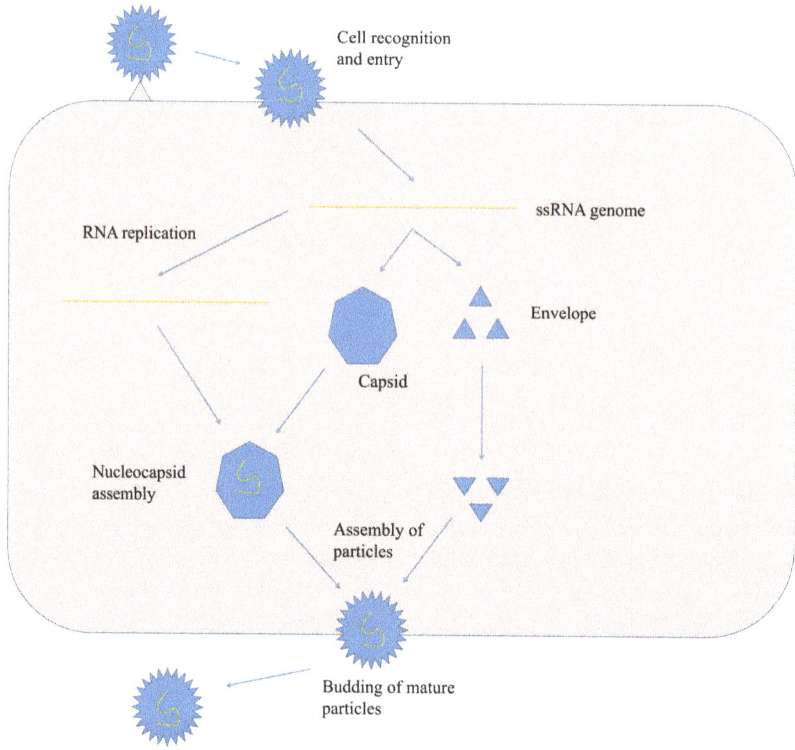

Figure 9.1 Life-cycle of alphaviruses. Infectious alphavirus particles recognize cell surface receptors and enter the cell through fusion and endocytosis. Immediate RNA replication occurs in the cytoplasm and capsid and envelope proteins are synthesized. Nucleocapsids and envelope proteins are transported to the surface and assembled particles released by budding.

have therefore provided excellent tools for drug discovery (Lundstrom, 2006) and in neuroscience (Ehrengruber *et al.*, 1999). As an example of their utility, SFV vectors were successfully used in one study to overexpress 101 GPCRs to levels suitable for structural genomics applications (Hassaine *et al.*, 2005).

Alphavirus expression vectors

Generally, alphavirus vectors come in three types, as described below and illustrated in Fig. 9.2: replication-deficient recombinant particles, replication-proficient recombinant particles and DNA layered vector system. To generate alphavirus particles, which are unable to produce virus progeny, the alphavirus genome is split into two vectors. The expression vector contains the non-structural genes responsible for highly efficient RNA replication, the strong subgenomic promoter, and the foreign gene of interest. The helper vector provides the structural genes responsible for capsid and envelope protein production. In this system, plasmid DNAs for the expression and helper vectors are subjected to in vitro RNA transcription followed by co-electroporation into baby hamster kidney (BHK) cells (Liljestrom and Garoff, 1991). As a packaging signal is present only on the RNA derived from the expression vector and no helper RNA is packaged into nucleocapsids, this results in release of mature viral particles lacking the genetic information for structural genes. These particles, however, are capable of efficiently infecting a broad range of host cells and of generating high expression of the gene of interest.

In replication-proficient recombinant particles, the gene of interest is added with an upstream second subgenomic promoter to the full-length alphavirus genome resulting in production of replication-proficient virus particles after in vitro transcription and electroporation. The resulting

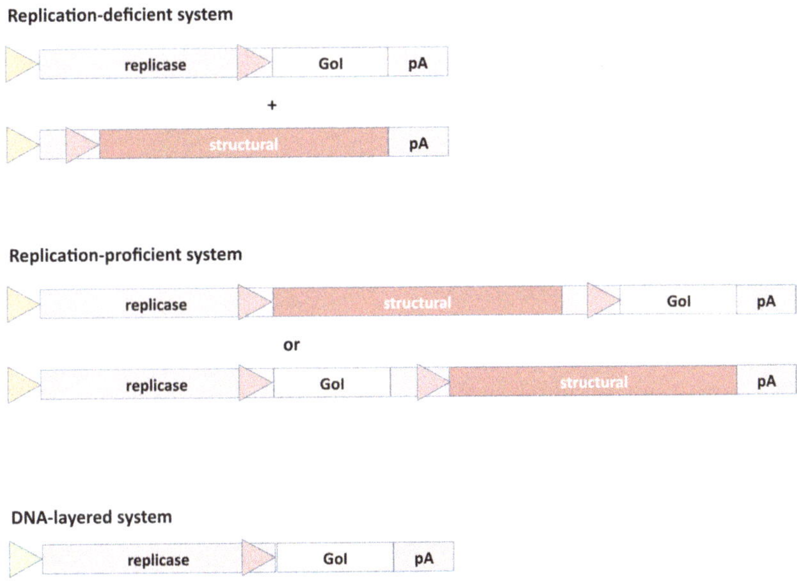

Figure 9.2 Alphavirus vector systems. The *replication-deficient system* applies RNA from an expression vector which is co-transfected with helper RNA providing structural alphavirus genes in trans to generate viral particles. The *replication-proficient system* includes a full-length genome and an additional subgenomic promoter and the gene of interest (GoI). The *DNA-layered system* employs an expression vector including a CMV promoter for transfection of DNA plasmids.

virus particles provide high levels of heterologous gene expression and simultaneous production of virus progeny.

In a DNA layered vector system, the SP6 RNA polymerase promoter in the alphavirus expression vector is replaced with a CMV promoter, which allows direct gene expression after plasmid DNA transfection of host cells. At no stage are viral particles generated using this system.

Each system has its own advantages and disadvantages. The DNA layered vectors provide high safety as no viral particles are generated at any stage of the process. However, their use is limited by and totally dependent on the availability of efficient transfection methods. Application of alphaviral particles allows efficient infection of a broad range of mammalian cell lines and primary cells, and *in vivo* delivery. The drawback in this case is the potential biohazard related to infectious viral particles. The replication-deficient particles are restricted to only one round of infection and have therefore been considered safer to use. However, especially with the first-generation helper vectors, a low frequency of homologous recombination was discovered between expression and helper vectors leading to replication-proficient particles. This issue was addressed by introduction of three point mutations in the pSFV-helper2 vector, which rendered it conditionally infectious and required an activation step of chymotrypsin treatment (Berglund *et al.*, 1993). Although this second-generation helper vector provided significantly reduced recombination, engineering of split helper vectors, where the capsid and envelope genes were placed on separate helper vectors, completely prevented production of replication-competent particles (Smerdou and Liljestrom, 1999). The drawback of replication-deficient vectors is the limited spread and transient nature of expression, which remains localized around the injection site and generally lasts less than 5 days (Lundstrom *et al.*, 1999). For improved delivery and prolonged expression time, replication-proficient vectors have been more suitable. However, *in vivo* use of these vectors requires caution to prevent unnecessary spread of infectious particles.

The question is then: which expression system to use? The use of each vector system is at least partly dictated by the required application. For example, in gene therapy, efficient delivery is most important. Depending on the indication, the duration of expression also needs to be addressed. Generally, even replication-proficient vectors will provide a relatively limited duration of expression, at least in comparison to other viral vectors such as retroviruses, which are capable of integration into the host genome (Lundstrom, 2003b).

In addition to engineering different types of expression systems for alphaviruses, the genomes themselves have also been subjected to modifications in attempts to improve the properties of the vectors. In this context, a number of point mutations with specific effects on cytopathogenicity and magnitude and duration of expression levels have been introduced in both SFV (Lundstrom *et al.*, 2003; Perri *et al.*, 2000) and SINV (Dryga *et al.*, 1997; Agapov *et al.*, 1998) vectors. The common feature of these mutant alphavirus vectors is the reduced cytotoxic effect in infected cells, which provides enhanced duration of gene expression. Certain SFV vectors, such as SFV-PD, also resulted in increased expression levels (Lundstrom *et al.*, 2003). Furthermore, some mutant SINV vectors demonstrated persistent infection in transduced host cells (Dryga *et al.*, 1997).

An alternative approach in vector development has been to engineer expression vectors carrying the viral non-structural genes from avirulent strains. In this context, vectors based on the avirulent SFV A7(74) demonstrated much reduced cytotoxicity and prolonged gene expression (Ehrengruber *et al.*, 2003). However, SFV A7(74)-based expression is highly temperature sensitive and provides high expression levels only at the permissive temperature (+31°C), which restricts its in vivo application. More recently, oncolytic alphavirus vectors have been engineered to improve delivery and gene expression profiles *in vivo* (Vähä-Koskela *et al.*, 2003; Heikkilä *et al.*, 2010). These vectors will be discussed in more detail below.

Alphavirus vectors in gene therapy

The attractive features of alphaviruses for gene therapy applications are the rapid generation of high titre viral particles, the broad host range and the high level of transgene expression. Rapid and simple virus production is also an asset, and the high ratio of RNA containing particles strongly supports therapeutic applications. Obviously, a high infection rate for a variety of cell types is important for achieving success in gene therapy, but it also raises concerns about potential damage to normal tissue, and this issue has been addressed by different approaches involving targeted delivery. High expression of therapeutic agents is also required for success. The transient nature of alphavirus-based expression has generally been considered a shortcoming, especially in gene therapy applications for chronic diseases. However, for cancer therapy the capacity of alphavirus vectors to generate short-term high-level expression of therapeutic gene sequences is considered to be an advantage.

Gene therapy in cancer

The most straightforward approach for applying alphaviruses in gene therapy has been intratumoral injection of vectors carrying reporter and/or therapeutic genes. In this context, significant tumour regression and inhibition of tumour blood vessel formation were observed in a murine B16 tumour model expressing the p40 and p35 subunits of interleukin-12 (IL-12) from replication-deficient SFV particles (Asselin-Paturel *et al.*, 1999). Similarly, intratumoral administration of SFV-GFP particles resulted in substantial tumour reduction in human lung tumour xenografts in immunodeficient mice (Murphy *et al.*, 2000). Furthermore, application of SFV vectors containing a translation enhancement signal provided a 10-fold enhancement of transgene expression and resulted in complete K-BALB tumour regression in BALB/c mice and therapeutic efficacy in CT26 tumours after delivery of SFV-IL-12 (Chikkanna-Gowda *et al.*, 2005) or SFV-IL-18 (Chikkanna-Gowda *et al.*, 2006). In another approach, reduced tumour growth and metastasis was observed after SFV-based expression of vascular endothelial growth factor receptor-2 (VEGFR-2) in mice (Lyons *et al.*, 2007).

Cancer vaccines

Alphaviruses administered as replication-deficient particles, naked RNA and layered DNA vectors have demonstrated great potential as

Table 9.1 Application of alphavirus vectors for generation of tumour vaccines

Target/tumour type	Vector/delivery	Response	Reference
IL-12/brain tumour	SFV/VPs	Immunogenicity	Yamanaka *et al.* (2002)
HPV E6-E7/cervix	SFV/VPs	Tumour protection	Daemen *et al.* (2002)
HPV E7/tumour	VEEV/VPs	Tumour protection	Velders *et al.* (2001)
HPV E7-VP22/tumour	SINV/VPs	CD8+ T-cell response	Cheng *et al.* (2002)
B16, 203/glioma	SFV/VPs	Tumour protection	Yamanaka *et al.* (2001)
Lac Z/tumour	SFV/RNA	Tumour protection	Ying *et al.* (1999)
P815A/tumour	SFV/VPs	Tumour protection	Colmenero *et al.* (2002)
P815/tumour	SFV/VPs	CTL, tumour protection	Colmenero *et al.* (1999)
MHC/tumour Ag	SFV/DNA, VPs	Immunogenicity	Ying *et al.* (2001)
Tyr rel-1/tumour Ag	SINV/DNA	Antitumour activity	Leitner *et al.* (2003)
MUC18/melanoma	SINV/DNA	Tumour protection	Leslie *et al.* (2007)
Neu/tumour	VEEV/VPs	Tumour protection	Moran *et al.* (2007)
PSMA/prostate	VEEV/VPs	Immunogenicity	Durso *et al.* (2007)

CTL, cytotoxic T-lymphocyte activity; HPV, human papilloma virus; IL, interleukin; MHC, major histocompatibility complex; MCAM, melanoma cell adhesion molecule; PSMA, prostate-specific membrane antigen; SFV, Semliki Forest virus; SINV, Sindbis virus; VEEV, Venezuelan equine encephalitis virus; VPs, virus particles.

cancer vaccines (Lundstrom, 2003c). A number of target genes (Table 9.1) such as IL-12 (Yamanaka *et al.*, 2003), human papilloma virus (HPV) E6 and E7 (Daemen *et al.*, 2002; Cheng *et al.*, 2002) and P815A tumour antigen (Colmenero *et al.*, 2002) have been evaluated in various animal models. For example, vaccination of mice with SFV particles expressing the P1A gene protected against P815A tumour challenges (Colmenero *et al.*, 1999). Similarly, protection was obtained after immunization with VEEV vectors expressing the HPV-E7 gene (Velders *et al.*, 2001). SFV particles expressing B16 and 203 tumour antigens provided protection and prolonged the survival of mice with established gliomas (Yamanaka *et al.*, 2001). Application of SINV vectors expressing the murine melanoma cell adhesion molecule (MCAM/MUC18) resulted in protection against lethal challenges with melanoma-expressing murine MUC18 in primary and metastatic tumour models (Leslie *et al.*, 2007). The efficacy of alphavirus particles was demonstrated by a single administration of dendritic cells transduced with VEEV expressing the truncated neu oncogene resulting in regression of large established tumours in mice (Moran *et al.*, 2007). Furthermore, mice vaccinated with VEEV particles expressing the prostate-specific membrane antigen (PSMA) elicited robust T-cell and B-cell responses, better than the immune responses against purified PSMA protein adjuvants (Durso *et al.*, 2007).

Although the majority of vaccine applications for alphaviruses involve administration of viral particles or layered DNA vectors, an interesting approach has been to employ naked replicon RNA. One of the pioneering studies involved vaccination of mice with naked SFV RNA replicons expressing bacterial β-galactosidase (Ying *et al.*, 1999). Immunization with only 1 μg of SFV-LacZ RNA showed therapeutic efficacy and complete tumour protection. Moreover, the survival rate in mice was extended by 10–20 days compared to control animals when tumours were injected 2 days prior to immunization. More recently, the efficacy of alphavirus-based RNA delivery for vaccine development was further enhanced by encapsulation of replicon RNA in lipid nanoparticles (Geall *et al.*, 2012). Application of secreted alkaline phosphatase (SEAP) as a reporter gene demonstrated superior expression levels of RNA-lipid nanoparticle complexes (RNA-LNPs) in comparison to naked RNA. Moreover, RNA-LNPs induced significantly better immune responses, as measured by RSV F-specific serum IgG titres.

Approaches for tumour targeting

Alphaviruses provide a broad host range and demonstrate a preference to target the central nervous system (Osten *et al.*, 2007), which has triggered investigations into targeting delivery specifically to the cell/tissue of therapeutic interest. In one approach, IgG binding domains of protein A were introduced into the SINV E2 envelope sequence to enhance cell targeting (Ohno *et al.*, 1997). The outcome was a 10^5-fold reduction in infection rates of BHK cells, whereas cells treated with a monoclonal antibody against surface proteins demonstrated enhanced infection efficacy. Unexpectedly, systemic delivery of SINV particles in general suggested a natural tumour-targeting property, as monitored by enhanced reporter gene expression in tumours (Tseng *et al.*, 2004). In this context, intratumoral injection demonstrated enhanced concentration of luciferase in tumour xenografts in mice. Furthermore, tumour load was reduced significantly after daily subcutaneous injections of SINV-IL-12. The tumour targeting property was further confirmed by intraperitoneal administration of SINV-Luc particles, which successfully targeted fibrosarcomas in the mouse-tail. In contrast to these findings with SINV, systemic delivery of SFV particles only resulted in tumour-specific infection after intratumoral injections (Rodriguez-Madoz *et al.*, 2007). Recently, SINV particles were evaluated in human xenograft tumour models in relation to tumour tropism (Huang *et al* 2012). The study indicated that SINV infection was not defined by the level of SINV receptors, but by the interferon (IFN) response in the tumours. Cells with defects in either IFN production or signalling were highly susceptible to SINV infection.

Another approach to enhance tumour targeting and delivery of alphavirus vectors has been to develop liposome formulations. The improved efficacy of replicon RNA-liposome nanoparticles has already been described in the section on 'Cancer vaccines' (Geall *et al.*, 2012). Encapsulation of alphavirus particles is a more complicated task as the particles themselves possess an envelope structure. Despite this challenge, formulations of SFV-LacZ particles demonstrated enhanced expression of the β-gal reporter gene in implanted tumours after systemic delivery (Lundstrom and Boulikas, 2002). Furthermore, extensive histology and toxicology studies confirmed the safety of systemic delivery of encapsulated SFV particles, which has set the ground for the first clinical trial in cancer patients, as described in the section 'Clinical trials'.

Oncolytic alphavirus vectors

In attempts to enhance the therapeutic efficacy of alphaviruses by prolonging transgene expression and extending *in vivo* distribution, oncolytic vectors capable of replication have been engineered (Vähä-Koskela *et al.*, 2006). In this context, the avirulent SFV strain A7(74) was shown to infect and lyse a number of cancer cell lines. Application of 10^6 viral particles resulted in significant tumour regression after a single injection (intratumoral, intraperitoneal or intravenous) in SCID mice with human melanoma grafts. However, virus was also found in normal tissue including the brain. For this reason, tissue-specific miRNA sequences were introduced into the SFV vector to alter the tissue tropism of viral replication (Ylösmäki *et al.*, 2012). Replicative SFV4 vectors carrying six tandem targets for the neuron-specific miR124 between the nsP3 and nsP4 genes displayed attenuated spread into the CNS in BALB/c mice and substantially prolonged survival of mice. In another study, oncolytic SFV A7(74) vectors expressing EGFP resulted in significant tumour regression in nude mice with human osteosarcoma xenografts (Ketola *et al.*, 2008). SFV treatment resulted in extended survival in animals with highly aggressive tumours although no final cure was achieved. Furthermore, oncolytic SFV-EGFP virus provided a significantly increased survival rate in comparison to conditionally replicating adenovirus (Ad5-Δ24TK-GFP) in nude mice with implanted orthotypic lung cancer xenografts after local administration (Määttä *et al.*, 2008). However, the effect was limited after systemic delivery.

Vectors for therapy in the CNS

Owing to the neurotropic nature of alphaviruses, attention has been paid to their potential use as therapeutic vectors in CNS-related diseases, although their neurovirulence has somewhat restricted their application in the CNS. For instance, *in vivo* administration of alphavirus vectors in rat (Lundstrom *et al.*, 1999) and mouse (Altman-Hamamdzic *et al.*, 1997) brain revealed transient local reporter gene expression after stereotactic injections into amygdala, striatum and substantia nigra, leading to some cell degeneration and axonopathy indicating neuronal loss. One solution has been to use avirulent strains (Atkins *et al.*, 1999). For example, the SFV-VA7-EGFP vector, which is based on the avirulent A7(74) strain, demonstrated reduced reporter gene expression in neurons and enhanced GFP fluorescence in glial cells (Ehrengruber *et al.*, 2003). Almost 20 years ago, SFV vectors were evaluated in experimental autoimmune encephalomyelitis (EAE) in the mouse CNS induced by injection of myelin as an animal model for multiple sclerosis (Mokhtarian *et al.*, 1994). Furthermore, SFV provides efficient delivery to the CNS via the olfactory bulb after intranasal administration (Sammin *et al.*, 1999). In this context, BALB/c mice were subjected to intranasal administration of SFV-EGFP particles, which resulted in reporter gene expression in the olfactory bulb as early as 1 day after administration (Jerusalmi *et al.*, 2003). The transient expression lasted for seven days. Intranasal administration of SFV particles expressing IL-10 significantly reduced EAE, while SFV-EGFP and empty SFV vectors exacerbated the disease. Furthermore, intranasal delivery of IFN β also inhibited EAE although the effect was dependent on the timing of the SFV administration and the number of injections (Quinn *et al.*, 2008). The SFV A7(74) strain has further been employed for the expression of transforming growth factor β (Vähä-Koskela *et al.*, 2007) and the inhibitor of metalloproteinase 2 (Nygårdas *et al.*, 2004) in mice. The SFV particles were administered intraperitoneally and were able to cross the blood–brain barrier and inhibit EAE. Additionally, treatment of SFV-infected δ-knock-out (KO) mice (incapable of remyelination) with SFV E2 peptide 2 in incomplete Freund's adjuvant (IFA) resulted in higher antibody production and earlier remyelination than in wild-type mice (Mokhtarian *et al.*, 2012). Likewise, in the EAE mouse model, treated mice showed significantly higher antibody titres and the average clinical disease scores were substantially lower, suggesting that the SFV E2 peptide 2 triggers a mechanism to promote remyelination.

Clinical trials

Alphaviruses have been used in a number of clinical trials, either in progress or already completed (http://clinicaltrials.gov/ct2/results?term=alphavirus). Most of the trials are phase I safety trials conducted in healthy volunteers. VEEV replicons were applied in a vaccine trial using the cytomegalovirus (CMV) proteins gB, pp65 and IE1b as antigens. In another study, influenza haemagglutinin (HA) was expressed from VEEV vectors to generate an influenza vaccine. Several studies have been initiated for HIV vaccines. A safety study sponsored by the National Institute for Allergies and Infectious Diseases (NIAID) applied a VEEV vector to evaluate an HIV gag vaccine in healthy uninfected volunteers. The safety of subcutaneous administration of a non-myristoylated form of gag was tested in two double-blind, randomized placebo-controlled trials (Wecker *et al.*, 2012). Owing to stability issues the first trial was prematurely halted. The second trial was also terminated early because of documentation issues with the contract manufacturer. Despite that, it was revealed that the vaccine was well-tolerated and presented only modest local reactogenicity. Five serious adverse events were reported, but none of these were considered to be related to the administered vaccine. Disappointingly, the immune responses in humans were limited, with only low levels of binding antibodies and T-cell responses. Another trial is currently recruiting for a study of VEEV vectors expressing carcinoembryonic antigen CEA (6D) in patients with advanced or metastatic CEA-expressing malignancies.

Liposome-encapsulated SFV particles expressing the p40 and p35 subunits of IL-12 have been used for systemic delivery in a phase I clinical trial in melanoma and kidney carcinoma patients (Lundstrom and Boulikas 2003). No liposome- or SFV-related toxicity was observed in any patient. The administration of SFV-IL12-p40p35 resulted in a 5-fold increase in IL-12 plasma levels for 4–7 days. The high IL-12 expression limited the maximum tolerated dose

(MTD) to 3×10^9 encapsulated particles per m^2. The liposome encapsulation prevented the SFV particles from being recognized by the host immune defence system and therefore allowed multiple repeated systemic injections. Furthermore, a protocol for a phase I/II trial for the treatment of recurrent glioblastoma multiforme applying encapsulated SFV particles was published in 2003 without the study being executed (Ren *et al.*, 2003).

Conclusions

Alphaviruses have been subjected to various applications for efficient gene delivery and expression *in vitro* and *in vivo*. Several studies have demonstrated that substantial tumour regression could be achieved after intratumoral injections of alphavirus vectors in animals implanted with tumour xenografts. Furthermore, particularly SINV-based vectors provide 'natural' tumour targeting. There has also been some success in targeting alphaviruses to tumours by application of RNA–liposome complexes or liposome-encapsulated particles. Moreover, the use of oncolytic vectors for prolonged expression of therapeutic genes has proven successful. Most importantly, alphavirus vectors have recently been used in clinical trials, which will certainly increase the interest in their application. As alphaviruses have clearly showed a strong preference for expression in the CNS, there is a genuine interest in targeting neurological disorders. However, certain chronic CNS diseases require continuous and long-term expression, which is not feasible with alphavirus-based gene delivery. On the other hand, where short-term and especially high-level expression is required, alphavirus vectors might be attractive candidates. In fact, short-term expression is an asset for cancer therapy where high concentrations of therapeutics for a short period of time are desired.

There is room for further development of more efficient alphavirus vectors by targeting the non-structural genes to reduce cytotoxicity, enhance expression levels and prolong expression time. Most importantly, engineering of high capacity packaging cell lines is necessary to further support the clinical applications of alphaviruses.

References

Agapov E.V., Frolov, I., Lindenbach, B.D., Pragai, B.M., Schlesinger, S., and Rice, C.M. (1998). Noncythopathic Sindbis virus RNA vectors for heterologous gene expression. Proc. Natl. Acad. Sci. U.S.A. *95*, 12989–12994.

Altman-Hamamdzic, S., Groseclose, C., Ma, J.X., Hamamdzic, D., Vrindavanam, N.S., Middaugh, L.D., Parratto, N.P., and Sallee, F.R. (1997). Expression of beta-galactosidase in mouse brain: utilization of a novel nonreplicative Sindbis virus vector as a neuronal gene delivery system. Gene Ther. *4*, 815–822.

Asselin-Paturel, C., Lassau, N., Guinebretière, J.M., Zhang, J., Gay, F., Bex, F., Hallez, S., Leclere, J., Peronneau, P., Mami-Chouaib, F., and Chouaib, S. (1999). Transfer of the murine interleukin-12 gene in vivo by a Semliki Forest virus vector induces B16 tumor regression through inhibition of tumor blood vessel formation monitored by Doppler ultrasonography. Gene Ther. *6*, 606–615.

Atkins, G.J., Sheahan, B.J., and Liljestrom, P. (1999). The molecular pathogenesis of Semliki Forest virus: a model virus made useful? J. Gen. Virol. *80*, 2287–2297.

Berglund, P., Sjöberg, M., Garoff, H., Atkins, G.J., Sheahan, B.J., and Liljestrom, P. (1993). Semliki Forest virus expression system: production of conditionally infectious recombinant particles. Biotechnology (NY) *11*, 916–920.

Cheng, W.F., Hung, C.F., Hsu, K.F., Chai, C.Y., He, L., Polo, J.M., Slater, L.A., Ling, M., and Wu, T.C. (2002). Cancer immunotherapy using Sindbis virus replicon particles encoding a VP22–antigen fusion. Hum. Gene Ther. *13*, 553–568.

Chikkanna-Gowda, C.P., Sheahan, B.J., Fleeton, M.N., and Atkins, G.J. (2005). Regression of mouse tumours and inhibition of metastases following administration of a Semliki Forest virus vector with enhanced expression of IL-12. Gene Ther. *12*, 1253–1263.

Chikkanna-Gowda, C.P., McNally, S., Sheahan, B.J., Fleeton, M.N., and Atkins, G.J. (2006). Inhibition of murine K-BALB and CT26 tumour growth using a Semliki Forest virus vector with enhanced expression of IL-18. Oncol. Rep. *16*, 713–719.

Colmenero, P., Liljestrom, P., and Jondal, M. (1999). Induction of P815 tumor immunity by recombinant Semliki Forest virus expressing the P1A gene. Gene Ther. *6*, 1728–1733.

Colmenero, P., Chen, M., Castaños-Velez, E., Liljeström, P., and Jondal, M. (2002). Immunotherapy with recombinant SFV-replicons expressing the P815A tumor antigen or IL-12 induces tumor regression. Int. J. Cancer *98*, 554–560.

Daemen, T., Regts, J., Holtrop, M., and Wilschut, J. (2002). Immunization strategy against cervical cancer involving an alphavirus vector expressing high levels of a stable fusion protein of human papillomavirus 16 E6 and E7. Gene Ther. *9*, 85–94.

Davis, N.L., Brown, K.W., and Johnston, R.E. (1989). In vitro synthesis of infectious Venezuelan equine encephalitis virus RNA from a cDNA clone: analysis of a viable deletion mutant. Virology *171*, 189–204.

Dryga, S.A., Dryga, O.A., and Schlesinger, S. (1997). Identification of mutations in a Sindbis virus variant able to establish a persistent infection in BHK cells: the importance of a mutation in the nsP2 gene. Virol. *228*, 74–83.

Durso, R.J., Andjelic, S., Gardner, J.P., Margitich, D.J., Donovan, G.P., Arrigale, R.R., Wang, X., Maughan, M.F., Talarico, T.L., Olmsted, R.A., Heston, W.D., Maddon, P.J., and Olson, W.C. (2007). A novel alphavirus vaccine encoding prostate-specific membrane antigen elicits potent cellular and humoral immune responses. Clin. Cancer Res. *13*, 3999–4008.

Ehrengruber, M.U., Lundstrom, K., Schweitzer, C., Heuss, C., Schlesinger, S., and Gahwiler, B.H. (1999). Recombinant Semliki Forest virus and Sindbis virus efficiently infect neurons in hippocampal slice cultures. Proc. Natl. Acad. Sci. U.S.A. *96*, 7041–7046.

Ehrengruber, M.U., Renggli, M., Raineteau, O., Hennou, S., Vähä-Koskela, M.J.V., Hinkkanen, A.E., and Lundstrom, K. (2003). Semliki Forest virus A7(74) transduces hippocampal neurons and glial cells in a temperature-dependent dual manner. J. Neurovirol. *9*, 16–28.

Geall A.J., Verma, A., Otten, G.R., Shaw, C.A., *et al.* (2012). Nonviral delivery of self-amplifying RNA vaccines. Proc. Natl. Acad. Sci. USA *109*, 14604–14609.

Griffin D.E. (1986). Alphavirus pathogenesis and immunity. In The Togaviridae and Flaviviridae, S. Schlesinger, and M.J. Schlesinger, eds. (Plenum PublishingCorp., New York, NY, USA), pp. 209–250.

Hassaine, G., Wagner, R., Kempf, J., Cherouati, N., Hassaine, N., Prual, C., André, N., Reinhart, C., Pattus, F., and Lundstrom, K. (2006). Semliki Forest virus vectors for overexpression of 101 G protein-coupled receptors in mammalian host cells. Prot. Expr. Purif. *45*, 343–351.

Heikkilä, J.E., Vähä-Koskela, M.J., Ruotsalainen, J.J., Martikainen, M.V., Stanford, M.M., McCart, J.A., Bell, J.C., and Hinkkanen, A.E. (2010). Intravenously administered alphavirus vector VA7 eradicates orthotopic human glioma xenografts in nude mice. PLOS ONE 5, e8603.

Huang, P.Y., Guo, J.H., and Hwang, L.H. (2012). Oncolytic Sindbis virus targets tumors defective in interferon response and induces significant bystander antitumor immunity *in vivo*. Mol. Ther. *20*, 298–305.

Jerusalmi, A., Morris-Downes, M.M., Sheahan, B.J., and Atkins, G.J. (2003). Effect of intranasal administration of Semliki Forest virus recombinant particles expressing reporter and cytokine genes on the progression of experimental autoimmune encephalomyelitis. Mol. Ther. *8*, 886–894.

Ketola, A., Hinkkanen, A., Yongabi, F., Furu, P., Määttä, A.M., Liimatainen, T., Pirinen, R., Björn, M., Hakkarainen, T., Mäkinen, K., Wahlfors, J., and Pellinen, R. (2008). Oncolytic Semliki forest virus vector as a novel candidate against unresectable osteosarcoma. Cancer Res. *68*, 8342–8350.

Leitner, W.W., Hwang, L.N., deVeer, M.J., Zhou, A., Silverman, R.H., Williams, B.R., Dubensky, T.W., Ying, H., and Restifo, N.P. (2003). Alphavirus-based DNA vaccine breaks immunological tolerance by activating innate antiviral pathways. Nature Med. *9*, 33–39.

Leslie, M.C., Zhao, Y.J., Lachman, L.B., Hwu, P., Wu, G.J., and Bar-Eli, M. (2007). Immunization against MUC18/MCAM, a novel antigen that drives melanoma invasion and metastasis. Gene Ther. *14*, 316–323.

Liljestrom, P., and Garoff, H. (1991). A new generation of animal cell expression vectors based on the Semliki Forest virus replicon. Nat. Biotechnol. *9*, 1356–1361.

Luers, A.J., Adams, S.D., Smalley, J.V., and Campanella, J.J. (2005). A phylogenomic study of the genus alphavirus employing whole genome comparison. Comp. Funct. Genomics *6*, 217–227.

Lundstrom, K. (2003a). Semliki Forest virus vectors for rapid and high-level expression of integral membrane proteins. Biochim. Biophys. Acta *1610*, 90–96.

Lundstrom, K. (2003b). Latest development in viral vectors for gene therapy. Trends Biotechnol. *21*, 117–122.

Lundstrom, K. (2003c). Alphavirus vectors for vaccine production and gene therapy. Exp. Rev. Vaccines 2, 447–459.

Lundstrom, K. (2005). Biology and application of alphaviruses in gene therapy. Gene Ther. *12* (Suppl. 1), S92–97.

Lundstrom, K. (2006). Latest development in drug discovery on G protein-coupled receptors. Curr. Protein Peptide Sci. *7*, 465–470.

Lundstrom, K., and Boulikas, T. (2002). Breakthrough in cancer therapy: encapsulation of drugs and viruses. Curr. Drug Discov. 2002, 19–23.

Lundstrom, K., Richards, J.G., Pink, J.R., and Jenck, F. (1999). Efficient *in vivo* expression of a reporter gene in rat brain after injection of recombinant replication-deficient Semliki Forest virus. Gene Ther. Mol. Biol. 3, 15–23.

Lundstrom, K., Abenavoli, A., Margaroli, A., and Ehrengruber, M.U. (2003). Novel Semliki Forest virus vectors with reduced cytotoxicity and temperature sensitivity for long-term enhancement of transgene expression. Mol. Ther. *7*, 202–209.

Lyons, J.A., Sheahan, B.J., Galbraith, S.E., Mehra, R., Atkins, G.J., and Fleeton, M.N. (2007). Inhibition of angiogenesis by a Semliki Forest virus vector expressing VEGFR-2 reduces tumour growth and metastasis in mice. Gene Ther. *14*, 503–513.

Määttä, A.M., Mäkinen, K., Ketola, A., Liimatainen, T., Yongabi, F.N., Vähä-Koskela, M., Pirinen, R., Rautsi, O., Pellinen, R., Hinkkanen, A., and Wahlfors, J. (2008). Replication competent Semliki Forest virus prolongs survival in experimental lung cancer. Int. J. Cancer *123*, 1704–1711.

Mathiot, C.C., Grimaud, G., Garry, P., Bouquety, J.C., Mada, A., Daguisy, A.M., and Georges, A.J. (1990). An outbreak of human Semliki Forest virus infection in Central African Republic. Am. J. Trop. Med. Hyg. *42*, 386–393.

Mokhtarian, F., Shi, Y., Zhu, P.F., and Grob, D. (1994). Immune responses, and autoimmune outcome, during virus infection of the central nervous system. Cell Immunol. *157*, 195–210.

Mokhtarian, F., Safavi, F., and Sarafraz-Yazdi, E. (2012). Immunization with a peptide of Semliki Forest virus peptide promotes remyelination in experimental autoimmune encephalomyelitis. Brain Res. *1488*, 92–103.

Moran, T.P., Burgents, J.E., Long, B., Ferrer, I., Jaffee, E.M., Tisch, R.M., Johnston, R.E., and Serody, J.S. (2007). Alphaviral vector-transduced dendritic cells are successful therapeutic vaccines against neu-overexpressing tumors in wild-type mice. Vaccine *25*, 6604–6612.

Murphy, A.M., Morris-Downes, M.M., Sheahan, B.J., and Atkins, G.J. (2000). Inhibition of human lung carcinoma cell growth by apoptosis induction using Semliki Forest virus recombinant particles. Gene Ther. *7*, 1477–1482.

Niklasson, B. (1988). Sindbis and Sindbis-like viruses. In The arboviruses: epidemiology and ecology. T.P. Monath, ed. (Boca Raton, Florida, CRC Press, Inc.), pp. 167–176.

Nygårdas, P.T., Grönberg, S.A., Heikkilä, J., Joronen, K., Sorsa, T., and Hinkkanen, A.E. (2004). Treatment of experimental autoimmune encephalomyelitis with a neurotropic alphavirus vector expressing tissue inhibitor of metalloproteinase-2. Scand. J. Immunol. *60*, 372–381.

Ohno, K., Sawai, K., Iijima, Y., Levin, B., and Meruelo, D. (1997). Cell-specific targeting of Sindbis virus vectors displaying IgG-binding domains of protein A. Nat. Biotechnol. *15*, 763–767.

Osten, P., Grinevich, V., and Cetin, A. (2007). Viral vectors:a wide range of choices and high levels of service. Handb. Exp. Pharmacol. *178*, 177–202.

Perri, S., Driver, D.A., Gardner, J.P., Sherill, S., Belli, B.A., Dubensky, T.W. Jr., and Polo, J.M. (2000). Replicon vectors derived from Sindbis virus and Semliki Forest virus that establish persistent replication in host cells. J. Virol. *74*, 9402–9407.

Quinn K. (2008). Effect of intranasal administration of Semliki Forest virus recombinant particles expressing interferon-β on the progression of experimental autoimmune encephalomyelitis. Mol. Med. Rep. *1*, 335–342.

Ren, H., Boulikas, T., Lundstrom, K., Söling, A., Warnke, P.C., and Rainov, N.G. (2003). Immunogene therapy of recurrent glioblastoma multiforme with a liposomally encapsulated replication-incompetent Semliki forest virus vector carrying the human interleukin-12 gene--a phase I/II clinical protocol. J. Neurooncol. *64*, 147–154.

Rodriguez-Madoz, J.R., Prieto, J., and Smerdou, C. (2007). Biodistribution and tumor infectivity of Semliki Forest virus vectors in mice: effect of re-administration. Mol. Ther. *15*, 2164–2171.

Sammin, D.J., Butler, D., Atkins, G.J., and Sheahan, B.J. (1999). Cell death mechanisms in the olfactory bulb of rats infected intranasally with Semliki forest virus. Neuropathol. Appl. Neurobiol. *25*, 236–243.

Smerdou, C., and Liljestrom, P. (1999). Two-helper RNA system for production of recombinant Semliki Forest virus particles. J. Virol. 73, 1092–1098.

Strauss, J.H., and Strauss, E.G. (1994). The Alphaviruses: gene expression, replication and evolution. Microbiol. Rev. *58*, 491–562.

Tseng, J.C., Levin, B., Hurtado, A., Yee, H., Perez de Castro, I., Jimenez, M., Shamamian, P., Jin, R., Novick, R.P., Pellicer, A., and Meruelo, D. (2004). Systemic tumor targeting and killing by Sindbis viral vectors. Nat. Biotechnol. *22*, 70–77.

Vähä-Koskela, M.J., Tuittila, M.T., Nygardas, P.T., Nyman, J.K., Ehrengruber, M.U., Renggli, M., and Hinkkanen, A.E. (2003). A novel neurotrophic expression vector based on the avirulent A7(74) strain of Semliki Forest virus. J. Neurovirol. *9*, 1–15.

Vähä-Koskela, M.J., Kallio, J.P., Jansson, L.C., Heikkilä, J.E., Zakhartchenko, V.A., Kallajoki, M.A., Kähäri, V.M., and Hinkkanen, A.E. (2006). Oncolytic capacity of attenuated replicative semliki forest virus in human melanoma xenografts in severe combined immunodeficient mice. Cancer Res. *66*, 7185–7194.

Vähä-Koskela, M.J., Kuusinen, T.I., Holmlund-Hampf, J.C., Furu, P.T., Heikkilä, J.E., and Hinkkanen, A.E. (2007). Semliki Forest virus vectors expressing transforming growth factor beta inhibit experimental autoimmune encephalomyelitis in Balb/c mice. Biochem. Biophys. Res. Commun. *355*, 776–781.

Velders, M.P., McElhiney, S., Cassetti, M.C., Eiben, G.L., Higgins, T., Kovacs, G.R., Elmishad, A.G., Kast, W.M., and Smith, L.R. (2001). Eradication of established tumors by vaccination with Venezuelan equine encephalitis virus replicon particles delivering human papillomavirus 16 E7 RNA. Cancer Res. *61*, 7861–7867.

Wang, K.S., Kuhn, R.J., Strauss, E.G., Ou, S., and Strauss, J.H. (1992). High-affinity laminin receptor is a receptor of Sindbis virus in mammalian cells. J. Virol. *66*, 4992–5001.

Wecker, M., Gilbert, P., Russell, N., Hural, J., Allen, M., Pensiero, M., Chulay, J., Chiu, Y.L., Abdool Karim, S.S., and Burke, D.S. (2012). Phase I safety and immunigenicity evaluations of an alphavirus replicon HIV-1 subtyoe C gag vaccine in healthy HIV-1-uninfected adults. Clin. Vaccine Immunol. *19*, 1651–1660.

Xiong, C., Levis, R., Shen, P., Schlesinger, S., Rice, C.M., and Huang, H.V. (1989). Sindbis virus: an efficient, broad host range vector for gene expression in animal cells. Science *243*, 1188–1191.

Yamanaka, R., Zullo, S.A., Tanaka, R., Blaese, M., and Xanthopoulos, K.G. (2001). Enhancement of antitumor response in glioma models in mice by genetically modified dendritic cells pulsed with Semliki Forest virus-mediated complementary DNA. J. Neurosurg. *94*, 478–481.

Yamanaka, R., Zullo, S.A., Ramsey, J., Yajima, N., Tsuchiya, N., Tanaka, R., Blaese, M., and Xanthopoulos, K.G. (2002). Marked enhancement of antitumor immune responses in mouse brain tumor models by genetically modified dendritic cells producing Semliki Forest virus-mediated interleukin-12. J. Neurosurg. 97, 611–618.

Ying, H., Zaks, T.Z., Wang, R.F., Irvine, K.R., Kammula, U.S., Marincola, F.M., Leitner, W.W., and Restifo, N.P. (1999). Cancer therapy using a self-replicating RNA vaccine. Nature Med. *5*, 823–827.

Ying, H., Zheng, G., and Black, K.L. (2001). Innovative cancer vaccine strategies based on the on the identification of tumor-associated antigens. Biodrugs *15*, 819–831.

Ylösmäki E., Martikainen, M., Hinkkanen, A., and Saksela, K. (2012). Attenuation of Semliki Forest virus neurovirulence by micro-RNA detargeting. J. Virol. *87*, 335–344.

Case Study: Chikungunya Virus

10

Lisa F.P. Ng

Abstract

The development of new and effective approaches to combat CHIKV infection is highly desirable. Molecular biology and genomic approaches could provide a brand new research paradigm. Nonetheless, to progress, potential drug candidates should also be extended to treat arthralgia and related chronic systemic inflammation induced by CHIKV. Whether these approaches will reach the clinic remains to be determined, but their success will rely on a better understanding of basic mechanisms of CHIKV infection and immunity. This chapter provides an overview of the cellular and molecular mechanisms of CHIKV pathogenesis and the various approaches leading to intervention.

Introduction

Infectious diseases are responsible for more than 20% of deaths worldwide, with a substantial fraction being caused by viruses (WHO Global status report on non-communicable diseases, 2010). Globalization, increased mobility and environmental change result in a highly increased chance of pandemics of new and re-emerging viral diseases that impose a tremendous threat for public health. Although more affluent countries are able to manage the spread and treatment of infectious viral diseases, it is also increasingly clear that they still face major challenges when dealing with diseases whose boundaries have been expanding due to warmer and wetter weather. A good example of such an abrupt increase in the incidence of disease is infections caused by arboviruses, whose expansion to new geographic areas has been facilitated by the establishment of new vectors. The re-emergence of chikungunya outbreaks in late 2005 represents an excellent example of how a virus of African origin and mosquitoes of Asian origin can meet in the Indian Ocean for re-emergence of the disease, and then spread to other parts of the world (Lumsden, 1955; Robinson, 1955; Kondekar and Gogtay, 2006; Lam *et al.*, 2001; Reiter *et al.*, 2006; Townson and Nathan, 2008; Her *et al.*, 2009; Morens and Fauci, 2014).

Chikungunya was first identified in the early 1950s in Tanzania, Africa. The word 'chikungunya' in the *Makonde* language translates to 'that which bends up', in reference to the stooped posture associated with the chronic and incapacitating arthralgia suffered by infected patients (Lumsden, 1955; Robinson, 1955). Chikungunya fever (CHIKF) is an acute illness caused by the chikungunya virus (CHIKV). It is an arbovirus and is transmitted primarily in a sylvatic cycle involving *Aedes* (*Ae. aegypti* and *Ae albopictus*) mosquitoes (Reiter *et al.*, 2006).

There are no approved antiviral drugs or vaccines. Patients are given analgesics, anti-inflammatory drugs and patient care (Parola *et al.*, 2006; Lemant *et al.*, 2008). The pathogenesis and viral tropism of CHIKV in humans remains an enigma. Efforts to develop effective antiviral drugs and vaccines have been greatly thwarted by the lack of information and tools (Kam *et al.*, 2009; Swartz and Albert, 2010). Despite being around for more than 50 years, CHIKV still remains elusive compared to other members of the alphavirus family. In light of the potential threat that CHIKV poses, encouraging efforts have been in place to understand CHIKV from different perspectives.

This chapter summarizes our current clinical, pathological, and immunological knowledge of CHIKV and examines how we could utilize our understanding of the host response for vaccine and therapeutic strategies.

Molecular biology of CHIKV

Like all members of the *Togaviridae* family, CHIKV and other alphaviruses are enveloped viruses with a positive-sense, single-stranded RNA genome (Strauss and Strauss, 1994; Chevillon *et al.*, 2008). The 12 kb viral genome consists of a 5′ cap-untranslated region (UTR), the coding regions for the non-structural (*nsP1–4*) and structural proteins (Capsid-E3-E2-6K-E1), and a 3′ terminal poly-A tail (Strauss and Strauss, 1994; Chevillon *et al.*, 2008). The RNA genome of the virus is encapsidated by the capsid proteins to form the nucleocapsid, which is enclosed by a lipoprotein envelope originating from the plasma membranes of the host cells formed during virus budding. The E1 and E2 glycoproteins form heterodimers that associate as trimeric spikes on the virion surface (Strauss and Strauss, 1994; Chevillon *et al.*, 2008; Li *et al.*, 2010; Voss *et al.*, 2010). The function of the E3 protein has yet to be fully defined, but it has been proposed to be involved in the process of envelope glycoprotein maturation (Voss *et al.*, 2010; Solignat *et al.*, 2009). Similarly, the function of 6K has not been studied, but it is thought to be involved in virus budding (Solignat *et al.*, 2009). X-ray crystallography data revealed that CHIKV virions have a typical icosahedral structure ($T = 4$ lattice) and a surface made up of 240 units of E1/E2 heterodimers, assembled into 80 glycoprotein spikes (Voss *et al.*, 2010). The CHIKV virion has also been determined by electron microscopy and atomic force microscopy to be between 70 to 100 nm in diameter (Strauss and Strauss, 1994; Her *et al.*, 2009).

Epidemiology of CHIKV infections

CHIKV infections have been documented in many parts of Southern and South-East Asia since the 1960s, but from 2001 to 2007, major outbreaks have occurred in islands in the Indian Ocean, including Mauritius, Mayotte, Madagascar and Reunion (Arankalle *et al.*, 2007; Powers and Logue, 2007; Powers *et al.*, 2007; Chretien *et al.*, 2007; Sergon *et al.*, 2008). The outbreaks in Reunion Island were the most severe, with one-third of its population infected (Schuffenecker *et al.*, 2006; Powers and Logue, 2007). Outbreaks then spread to several South-East Asian countries including Singapore (Arankalle *et al.*, 2007; Tandale *et al.*, 2008; Theamboonlers *et al.*, 2009; Sam *et al.*, 2010; Ng *et al.*, 2009; Her *et al.*, 2009), and also to European countries such as Italy and France from 2007 onwards (Rezza *et al.*, 2007). Sporadic outbreaks are still being reported constantly in several South-East Asian countries, and more recently, in the Americas (Morens and Fauci, 2014; Leparc-Goffart *et al.*, 2014).

Sequence analyses have suggested that CHIKV originated from Africa and was later introduced to Asia (Schuffenecker *et al.*, 2006; Powers and Logue, 2007). Based on partial E1 glycoprotein nucleotide sequences, phylogenetic analysis delineated CHIKV into three distinct clusters: East, Central and South African cluster (ECSA), Asian cluster, and West African cluster (Schuffenecker *et al.*, 2006). Phylogenetic analysis of CHIKV strains isolated from the Indian Ocean outbreaks indicated it to be more closely related to the ECSA cluster than to the Asian or West African clusters. However, 90% of the CHIKV strains isolated revealed an alanine to valine mutation at position 226 in E1 glycoprotein (Tsetsarkin *et al.*, 2007). This single amino acid change was of particular interest as it appeared to be exclusively found in CHIKV isolated from *Ae. albopictus*. This mutation has been shown to be associated with adaptation to *Ae. Albopictus*, which led to significantly increased fitness of CHIKV in *Ae. albopictus* that was attributable to the loss of cholesterol dependence for growth by CHIKV, thereby allowing the virus to adapt, replicate and disseminate more efficiently in *Ae. albopictus*. Subsequently, other evidence has shown the importance of another single mutation (leucine to glutamine) at position 210 in the E2 glycoprotein that contributed to the increased CHIKV dissemination in *Ae. albopictus* (Tsetsarkin and Weaker, 2011). More recent studies have further illustrated the importance of the A226V mutation in the E1 glycoprotein that give rise to sequential mutations in both the E1 and E2 glycoproteins and lead to virus fitness in *Ae. albopictus* (Tsetsarkin *et al.*, 2014; Stapleford *et al.*, 2014).

Clinical manifestations and pathology of CHIKV infections

CHIKF affects both genders and all ages, is usually self-limiting, and is rarely life-threatening (Deller and Russell, 1968; Edelman *et al.*, 2000; Powers and Logue, 2007; Parola *et al.*, 2006). CHIKV infection induced long-lasting protective immunity, and epidemic peaks dropped as an increasing percentage of the population gained immunity (Powers and Logue, 2007: Kam *et al.*, 2009). Clinical presentations are often similar to dengue fever, except that haemorrhagic or shock syndromes that occur in dengue infection are rarely seen in CHIKV infection (Powers and Logue, 2007: Kam *et al.*, 2009). In countries where dengue fever is endemic, CHIKF could readily be missed.

Non-complicated manifestations

The hallmark of CHIKV infection is polyarthralgia, which results in the stooped walking posture usually observed in patients (Lumsden, 1955; Robinson, 1955; Powers and Logue, 2007; Her *et al.*, 2009). Other classical symptoms of CHIKV infection include febrile illness (temperature usually >38.9°C), maculopapular rashes, myalgia, headache, oedema of the extremities, and gastrointestinal complaints (Lumsden, 1955; Robinson, 1955; Powers and Logue, 2007; Her *et al.*, 2009). Clinical examinations have revealed that polyarthralgia is often associated with synovitis at the metacarpophalangeal, proximal interphalangeal, and metatarsophalangeal joints, as well as at the ankles and wrists (Lumsden, 1955; Robinson, 1955; Powers and Logue, 2007; Her *et al.*, 2009). Tenosynovitis and enthesitis at the insertions of both Achilles tendons have also been reported, where inflammation of the periosteum was observed in several patients (Jaffar-Bandjee *et al.*, 2009, 2010).

Magnetic resonance imaging (MRI) and histology of muscle biopsies from CHIKV-infected patients have also revealed the presence of myositis, tenosynovitis, atrophy, collagenosis, vacuolisation and necrosis of muscle fibres at the sites of inflammation (Jaffar-Bandjee *et al.*, 2009, 2010). Concordant with these findings, biochemical analyses of urine samples from CHIKV patients have indicated increased mucopolysaccharide, proline and hydroxyproline excretion, suggesting increased metabolism and damage of the cartilage and connective tissues. In most cases, acute symptoms resolved within two weeks, but in about 12% of the patients, chronic arthritis persisting up to three years after the onset of illness has been documented (Jaffar-Bandjee *et al.*, 2010). After the CHIKV outbreaks in 2005, cohorts of patients followed up in Reunion Island reported more than 50% of the patients suffering from chronic arthralgia (Suhrbier *et al.*, 2012). Rheumatoid syndrome was suggested to be responsible for chronic joint pain as high antibody titres were detected against CHIKV in synovial fluids of these chronic patients (Gérardin *et al.*, 2013).

Complicated manifestations

Besides the classical CHIKV-associated pathologies, acute nephritis, cardiac myocarditis, pericarditis, myopericarditis, ophthalmological retrobulbar neuritis and neurological complications have also been reported (Lemant *et al.*, 2008). However, current knowledge on the mechanisms associated with these rare complications is largely speculative and unclear.

Of interest, increased incidences of neurological complications and associated lethality were reported for the CHIKV outbreaks in Réunion Island and India during 2005–2006 (Lemant *et al.*, 2008; Soumahoro *et al.*, 2009). In adults, neurological complications requiring hospitalization (incidence of 15–25%) included encephalopathy, encephalitis, Guillian–Barré syndrome and encephalomyeloradiculitis (Lemant *et al.*, 2008). In children, neurological manifestations occurred at an incidence rate of 61% in children aged 3–18 years old, but only 16% in children younger than three years of age (Lemant *et al.*, 2008). In addition, neurological signs were less severe in children (Ramful *et al.*, 2014), and these include encephalitis, febrile seizures, meningeal syndrome and acute encephalopathy (Ramful *et al.*,

2014). These observations strongly suggest age to be an important predisposing factor affecting neurological pathogenesis of CHIKV infection. However, mechanisms associated with these complications remain largely unknown. Current clinical studies on CHIKV-associated neurological manifestations are largely limited to MRI and cerebral spinal fluid (CSF) examination. Owing to the inherent risk involved in obtaining brain tissue biopsies from CHIKV patients, plus the lack of autopsy specimens, the understanding of neurotropism and viral pathogenesis in the brain is limited.

Immune responses against CHIKV infections

Current understanding of CHIKV-elicited immune responses suggests that host innate immunity plays a dual role in controlling viral dissemination and mediating arthritic manifestations of the disease (Kam *et al.*, 2009; Schwartz and Albert, 2010; Suhrbier *et al.*, 2012). In patients, viral elimination occurs before the host can produce IgGs and elicit any T-cell responses. High levels of pro-inflammatory cytokines were detected during acute CHIKV infections, further suggesting the involvement of immune cells and mediators in immune-pathogenesis (Ng *et al.*, 2009; Chow *et al.*, 2010; Wauquier *et al.*, 2011; Kelvin *et al.*, 2011).

Innate host response and the involvement type I interferons (IFNs) induction in CHIKV infection

The production of type I interferons, IFN-a and IFN-b, is the signature of an antiviral state in vertebrate hosts. They are essential to the functioning of innate immunity against the replication and spread of virus (Akira and Takeda, 2004; Stetson and Medzhitov, 2006). Interplay between type I IFNs and CHIKV in target cells during infection has been demonstrated (Schilte *et al.*, 2010; Her *et al.*, 2010; Fros *et al.*, 2010). Although the specific pathogen recognition receptors (PRRs) responsible for recognizing CHIKV and other alphaviruses remain to be fully identified, IPS-1 has been demonstrated to function as an essential downstream component of the unknown PRRs to mediate type I IFN production and limit CHIKV infection (White *et al.*, 2011). Following CHIKV infection of human fibroblasts, IPS-1-mediated signalling is required to activate IRF3, which subsequently induces the transcription of ISGs and IFN-b (Schilte *et al.*, 2012). Notably, CHIKV is able to avert the induction of the innate host response by inhibiting host cell translation in a PKR-independent pathway to block the expression of the antiviral proteins. Interestingly, CHIKV has also been shown to circumvent the induction of type I IFN-mediated innate immune responses independent of host shutdown (Fros *et al.*, 2010). It was shown that CHIKV could partially suppress type I IFN signalling by targeting the JAK/STAT pathway, in which the non-structural protein 2 (nsP2) of CHIKV specifically blocks the activation and nuclear translocation of STAT1 to possibly prevent expression of type I IFNs and ISGs. Together, these studies illustrate the various strategies employed by CHIKV to evade the innate immune response. In light of the importance of type I IFNs in controlling CHIKV infection, the identity and mechanism of the ISGs involved are still unknown, although ISG15 was demonstrated to be critical in controlling CHIKV infection (Werneke *et al.*, 2012). OAS has been shown to be involved in limiting CHIKV replication *in vitro* (Brehin *et al.*, 2009), while viperin, an ISG, has been demonstrated to inhibit CHIKV replication *in vitro* and control pathology *in vivo* (Teng *et al.*, 2012).

In recent years, there has been an increasing trend of incorporating computational and bioinformatics tools into bench-side research to facilitate the discovery of new knowledge (Cline *et al.*, 2007; Ramirez *et al.*, 2007). Therefore, one could harness the vast amount of information gathered from the studies of other alphaviruses and extrapolate into a model for investigation to elucidate the pathways and molecular mechanisms involved in the innate host response during CHIKV pathogenesis.

Adaptive immune response against CHIKV

Although the adaptive immune response against CHIKV has been reported, it remains to be fully characterized. It is well reported that the adaptive immune response becomes effective only after several days post-infection (Ross, 1956; Panning *et al.*, 2008). This is the usual time required for the viral antigens to be taken up by dendritic cells and presented in secondary lymphoid organs, before they are presented to naive T and B cells. Naive T and B cells will later proliferate and differentiate into effector cells that execute their antiviral functions for viral clearance.

The role of antibodies in CHIKV infection and pathology

The establishment of an anti-CHIKV immune response after a primary infection can confer complete protection against re-infection (Ross, 1956; Panning *et al.*, 2008), and this hypothesis has provided the basis to explain the time lapse between CHIKF epidemics (Powers and Logue, 2007). Anti-CHIKV IgM antibodies could be detected in the sera of infected patients during the acute phase of the infection during the first week of infection (Kam *et al.*, 2012a). On the other hand, anti-CHIKV IgGs were usually detected after the virus had been cleared from the blood and could persist for several months after infection (Kam *et al.*, 2012b). The ability of anti-CHIKV antibodies to neutralize virus infectivity was first demonstrated by Ross using sera from convalescent patients (Ross, 1956). Furthermore, it was also demonstrated that inoculation of CHIKV-containing sera from humans into monkeys could also induce the production of neutralizing antibodies (Ross, 1956). These findings suggest that anti-CHIKV antibodies could be used as a potential prophylactic strategy against CHIKF. In support of this notion, studies have showed that anti-CHIKV immunoglobulins (CHIKVIgs) exhibited high neutralizing activity *in vitro*, and administration of purified CHIKVIgs protected two mouse models from lethal dosage challenge *in vivo* (Couderc *et al.*, 2009). More recently, several reports showed efficient virus neutralization with antibodies from CHIKV-infected patients both *in vitro* and in *in vivo* animal models (Kam *et al.*, 2012a–c; Lum *et al.*, 2013). Epidemiological studies have also shown that in many parts of Africa where CHIKV is circulating, antibody prevalence increased with age (Osterrieth *et al.*, 1960). These data provide further evidence that anti-CHIKV antibodies would make good therapeutics for prophylactic treatment against CHIKV infection.

Antibody-mediated protection against alphaviruses has been studied extensively and surface viral glycoproteins have been speculated to be key targets for protective neutralizing antibodies against alphaviruses (Griffin *et al.*, 1997). Furthermore, it was shown that immunization with a CHIKV virus-like particle (VLP) vaccine comprising key surface viral glycoproteins induced production of neutralizing antibodies and protected against viral challenge in non-human primates (Akahata *et al.*, 2010) and, more recently, in a human trial (Lee *et al.*, 2014). In a step forward, the reports on the three-dimensional crystal structures of the CHIKV E3-E2–E1 glycoprotein complexes (Voss *et al.*, 2010), and that of the Sindbis virus E2–E1 heterodimer (Li *et al.*, 2010), further strengthened the understanding of the conformational changes of the virus during entry mechanism. Visualization of these three-dimensional structures also facilitated the identification of potential B-cell epitopes that are targeted by neutralizing antibodies (Kam *et al.*, 2012b,c). Such wealth of information will be valuable for the design of antibody-based vaccine strategies for protection against CHIKV and other alphaviruses.

The role of T cells in CHIKV infection and pathology

T cells are important effector cells during viral infection (Kagi and Hengartner, 1996). Both $CD4^+$ and $CD8^+$ T cells can eliminate virus-infected cells. Although the functions of T cells in CHIKV immunopathology have not been intensively explored, CD4+ and, rarely, CD8+ T cells were found in the synovial tissues of a CHIKV infected patient (Hoarau *et al.*, 2010). Separately, circulating T lymphocytes were found to exhibit a CD8+ T cell mediated response during early infection, before changing to a CD4+ T cell-mediated response later in the infection (Wauquier *et al.*, 2010). $CD4^+$ T cells are also important for the production of antibodies and were recently demonstrated to play a role determining anti-CHIKV antibody production (Teo *et al.*, 2013).

Up to now, only limited T-cell epitopes have been identified and validated for humans (Hoarau *et al.*, 2013), although in mice multiple T-cell epitopes have been identified in E1 and E2 glycoproteins proteins and in the capsid (Teo *et al.*, 2013).

Vaccines and therapy

Management and control of CHIKV infections involves several measures. In the absence of effective treatment, it is crucial to develop successful approaches in immunoprophylaxis and vaccine-mediated protection against CHIKV. Furthermore, identification of drug targets can provide CHIKV-specific therapy in the short term, while vaccine-mediated protection contributes to long-term health management.

Antiviral compounds and therapeutics against CHIKV

Antiviral compounds could be used as a basic therapeutic strategy against viral infection. Currently, there is no specific antiviral treatment for CHIKF. Treatment is usually symptomatic, and pain-relieving medicine is the only available option during the disease onset. Previous studies have showed a beneficial effect of the lysosomotropic agent chloroquine in improving the clinical condition of patients (Brighton, 1984). However, controversial results were observed from treatment of CHIKV-infected patients with this compound during the outbreaks in Reunion Island (de Lamballerie *et al.*, 2008). As a result, other compounds such as IFN-α, ribavirin and pharmacologically active compounds in natural products have been proposed as antiviral drugs against alphaviruses; but the clinical results are still under investigation (Briolant *et al.*, 2004). Nevertheless, further studies are needed to substantiate the beneficial effect of these potential compounds against CHIKV in clinically achievable conditions and to define the role of the immune response in CHIKF.

In terms of development of therapeutics against alphaviruses, it was shown for the first time in a viable mouse model that Bindarit, an inhibitor of MCP-1 could be used to ameliorate CHIKV infections (Rulli *et al.*, 2011). Bindarit functions by inhibiting the production of MCPs, the main components responsible for the recruitment of macrophages to the sites of inflammation and for controlling inflammation and disease pathology (Rulli *et al.*, 2011).

Development of CHIKV vaccines

Potential CHIKV vaccine candidates have been tested in humans and animals with varying success over the years (Harrison *et al.*, 1967; Eckels *et al.*, 1970; Eckels and Putnak, 2003; Edelman *et al.*, 2000; Harrison *et al.*, 1971; White *et al.*, 1972; Levitt *et al.*, 1986). Among these, many trials were eventually discontinued due to lack of funding and limited commercial potential. The re-emergence of CHIKV has highlighted the need for vaccine re-development and investigation of new vaccine candidates (Wang *et al.*, 2008; Muthumani *et al.*, 2008; Tiwari *et al.*, 2009; Akahata *et al.*, 2010; Mallilankaraman *et al.*, 2011; Plante *et al.*, 2011; Partidos *et al.*, 2011; Gorchakov *et al.*, 2012; Hallengärd *et al.*, 2014; García-Arriaza *et al.*, 2014; Roy *et al.*, 2014; Lee *et al.*, 2014).

First generation CHIKV vaccines involved mainly inactivated whole-virus formulations (either formalin-inactivated or live attenuated) that contained all the structural and non-structural proteins serving as antigens to induce immunological responses. Such formulations are attractive due to the ease of preparation and ability to satisfy the immediate needs in a short period of time in the event of an outbreak. Inactivated whole-virus vaccines have been studied for various alphaviruses but the results have been controversial. Therefore, inactivated vaccine formulations should be replaced once the immuno-reactive components have been identified. However, development is still focused on inactivated or live attenuated strains of CHIKV. Immunogenicity of these vaccine candidates has been tested with various animal models, and one of the promising candidates, TSI-GSD-218, has been tested for safety and immunogenicity in human phase II trials. However, transient arthralgia has still been observed in vaccinated volunteers (5 of 59 CHIKV vaccinees). In recent years, two new strategies using a CHIKV DNA chimeric vaccine and a VLP CHIKV vaccine have been reported (Muthumani *et al.*, 2008; Akahata *et al.*, 2010).

Epitope-based vaccines have been gaining prominence as an alternative approach that could elicit protective immune response without causing adverse effects. The potential advantages of this approach include improved safety, the ability to rationally design epitopes for increased vaccination efficiency, and the opportunity to design vaccines with higher population coverage (Sette and Fikes, 2003). Recent efforts have primarily focused on the discovery of (1) B-cell epitopes that could be the target of neutralizing antibodies; and (2) T-cell epitopes that could be the target of protective cytotoxic T cells (both $CD4^+$ and $CD8^+$ T cells) or recognized by $CD4^+$ helper T cells for optimal anti-CHIKV antibody production (Teo *et al.*, 2013).

Of interest, the existence of abnormal T cell responses recognizing particular T cell epitopes has stressed the necessity to eliminate the pathogenic T-cell epitopes from vaccine formulations (Welsh and Fujinami, 2007). Therefore, identification of both immunogenic and pathogenic epitopes will be necessary for the design and construction of new CHIKV vaccines in order to increase their prophylactic potential while minimizing the risk of side-effects from vaccine use.

In the last decade, a wide variety of epitope mapping tools have been reported, making computational pre-screening of antigens a standard approach in the discovery of vaccine candidates (Sette and Fikes 2003; Welsh and Fujinami, 2007). Computational modelling and prediction of B-cell epitopes is still at its formative stages. Much emphasis has been placed on the design of predictors for continuous epitopes, in part, due to lower complexity in system development, and also because the experimental design of discontinuous epitopes is difficult. Methods that can predict promiscuous human leucocyte antigen (HLA)-binding peptides are useful for the design of peptide vaccines with improved global coverage by ensuring that HLA alleles that are present in most individuals from all major ethnic groups may bind to at least one of the peptides in the vaccine (Bian and Hammer, 2004; Zhange *et al.*, 2009). Computational systems that simulate the cell-mediated immune system have also been reported, which involve integrating the different sub-components of the antigen processing and presentation pathway such as TAP, proteasome, and MHC (Tenzer *et al.*, 2005). Such tools would be particularly useful for screening large sets of protein antigens, such as those encoded by complete viral genomes.

Antibody intervention

Studies have demonstrated the efficacy of human and monkey anti-CHIKV antibodies in mouse models (Ross, 1956; Couderc *et al.*, 2009). Therefore, it is enticing to propose that passive antibody transfer might be useful. Pathogen-specific polyclonal immunoglobulins purified from immunized donors have been used for prophylaxis of several infectious diseases (Couderc *et al.*, 2009). The main advantage of these preparations is their broad reactivity; however, batch-to-batch variations in composition and activity remain a concern for clinical applications. As an alternative, human monoclonal antibodies have been used in prophylaxis as well as in therapeutic settings (Pal *et al.*, 2013). In addition to these, the generation of vaccines that could target multiple conserved epitopes and induce long-lasting antibody responses using a repeat-boosting strategy will be highly desirable.

References

Akahata, W., Yang, Z.Y., Andersen, H., Sun, S., Holdaway, H.A., Kong, W.P., Lewis, M.G., Higgs, S., Rossmann, M.G., Rao, S., and Nabel, G.J. (2010). A virus-like particle vaccine for epidemic chikungunya virus protects non-human primates against infection. Nat. Med. *16*, 334–338.

Akira, S., and Takeda, K. (2004). Toll-like receptor signalling. Nat. Rev. Immunol. *4*, 499–511.

Arankalle, V.A., Shrivastava, S., Cherian, S., Gunjikar, R.S., Walimbe, A.M., Jadhav, S.M., Sudeep, A.B., and Mishra, A.C. (2007). Genetic divergence of Chikungunya viruses in India (1963–2006) with special reference to the 2005–2006 explosive epidemic. J. Gen. Virol. *88*, 1967–1976.

Bian, H., and Hammer, J. (2004). Discovery of promiscuous HLA-II-restricted T cell epitopes with TEPITOPE. Methods *34*, 468–475.

Brehin, A.C., Casademont, I., Frenkiel, M.P., Julier, C., Sakuntabhai, A., and Despres, P. (2009). The large form of human 2',5'-oligoadenylate synthetase (OAS3) exerts antiviral effect against chikungunya virus. Virology *384*, 216–222.

Brighton, S.W. (1984). Chloroquine phosphate treatment of chronic chikungunya arthritis. An open pilot study. S. Afr. Med. J. *66*, 217–218.

Briolant, S., Garin, D., Scaramozzino, N., Jouan, A., and Crance, J.M. (2004). *In vitro* inhibition of chikungunya and semliki forest viruses replication by antiviral compounds: synergistic effect of interferon-[alpha] and ribavirin combination. Antiviral Res. *61*, 111–117.

Chang, L.J., Dowd, K.A., Mendoza, F.H., Saunders, J.G., Sitar, S., Plummer, S.H., Yamshchikov, G., Sarwar, U.N., Hu, Z., Enama, M.E., *et al.*; the VRC 311 Study Team. (2014). Safety and tolerability of chikungunya virus-like particle vaccine in healthy adults: a phase 1 dose-escalation trial. Lancet *384*, 2046–2052.

Chevillon, C., Briant, L., Renaud, L., and Devaux, C. (2008). The Chikungunya threat: an ecological and evolutionary perspective, Trends Microbiol. *16*, 80–88.

Chow, A., Her, Z., Ong, E.K.S., Chen, J.M., Dimatatac, F., Kwek, D.J., Barkham, T., Yang, H., Rénia, L., Leo, Y.S., and Ng, L.F.P. (2010). Persistent arthralgia induced by chikungunya virus infection is associated with interleukin-6 and granulocyte macrophage colony-stimulating factor. J. Infect. Dis. *203*, 149–157.

Chretien, J.P., Anyamba, A., Bedno, S.A., Breiman, R.F., Sang, R., Sergon, K., Powers, A.M., Onyango, C.O., Small, J., Tucker, C.J., and Linthicum, K.J. (2007). Drought-Associated Chikungunya Emergence Along Coastal East Africa. Am. J. Trop. Med. Hyg. *76*, 405–407.

Cline, M.S., Smoot, M., Cerami, E., Kuchinsky, A., Landys, N., Workman, C., Christmas, R., Avila-Campilo, I., Creech, M., *et al.* (2007). Integration of biological networks and gene expression data using Cytoscape. Nat. Protoc. *2*, 2366–2382.

Couderc, T., Khandoudi, N., Grandadam, M., Visse, C., Gangneux, N., Bagot, S., Prost, J.F., and Lecuit, M. (2009). Prophylaxis and therapy for Chikungunya virus infection. J. Infect. Dis. *200*, 516–523.

Deller, J.J., and Russell, P.K. (1968). Chikungunya disease. Am. J. Trop. Med. Hyg. *17*, 107–111.

Eckels, K.H., and Putnak, R. (2003). Formalin-inactivated whole virus and recombinant subunit flavivirus vaccines. Adv. Virus Res. *61*, 395–418.

Eckels, K.H., Harrison, V.R., and Hetrick, F.M. (1970). Chikungunya virus vaccine prepared by Tween-ether extraction. Appl. Microbiol. *19*, 321–325.

Edelman, R., Tacket, C.O., Wasserman, S.S., Bodison, S.A., Perry, J.G., and Mangiafico, J.A. (2000). Phase II safety and immunogenicity study of live chikungunya virus vaccine TSI-GSD-218. Am. J. Trop. Med. Hyg. *62*, 681–685.

Fros, J.J., Liu, W.J., Prow, N.A. Geertsema, C., Ligtenberg, M., Vanlandingham, D.L., Schnettler, E., Vlak, J.M., Suhrbier, A., Khromykh, A.A., and Pijlman, G.P. (2010). Chikungunya virus nonstructural protein 2 inhibits type I/II interferon-stimulated JAK-STAT signaling. J. Virol. *84*, 10877–10887.

García-Arriaza, J., Cepeda, V., Hallengärd, D., Sorzano, C.Ó., Kümmerer, B.M., Liljeström, P., and Esteban, M. (2014). J. Virol. *88*, 3527–3547.

Gérardin, P., Fianu, A., Michault, A., Mussard, C., Boussaïd, K., Rollot, O., Grivard, P., Kassab, S., Bouquillard, E., Borgherini, G., *et al.* (2013). Predictors of Chikungunya rheumatism: a prognostic survey ancillary to the TELECHIK cohort study. Arthritis Res. Ther. 15, R9.

Gorchakov, R., Wang, E., Leal, G., Forrester, N.L., Plante, K., Rossi, S.L., Partidos, C.D., Adams, A.P., Seymour, R.L., Weger, J., *et al.* (2012). Attenuation of Chikungunya virus vaccine strain 181/clone 25 is determined by two amino acid substitutions in the E2 envelope glycoprotein. J. Virol. *86*, 6084–6096.

Griffin, D., Levine, B., Tyor, W., Ubol, S., and Despres, P. (1997). The role of antibody in recovery from alphavirus encephalitis. Immunol. Rev. *159*, 155–161.

Hallengärd, D., Kakoulidou, M., Lulla, A., Kümmerer, B.M., Johansson, D.X., Mutso, M., Lulla, V., Fazakerley, J.K., Roques, P., Le Grand, R., Merits. A., and Liljeström, P. (2014). Novel attenuated Chikungunya vaccine candidates elicit protective immunity in C57BL/6 mice. J. Virol. *88*, 2858–2866.

Harrison, V.R., Binn, L.N., and Randall, R. (1967). Comparative Immunogenicities of chikungunya vaccines prepared in avian and mammalian tissues. Am. J. Trop. Med. Hyg. *16*, 786–791.

Harrison, V.R., Eckels, K.H., Bartelloni, and P.J., and Hampton, C. (1971). Production and evaluation of a formalin-killed Chikungunya vaccine. J. Immunol. *107*, 643–647.

Her, Z., Kam, Y., Lin, R.T.P., and Ng, L.F.P. (2009). Chikungunya: a bending reality. Microbes Infect. *11*, 1165–1176.

Her, Z., Malleret, B., Chan, M., Ong, E.K., Wong, S.C., Kwek, D.J., Tolou, H., Lin, R.T., Tambyah, P.A., Rénia, L., and Ng, L.F.P. (2010). Active infection of human blood monocytes by Chikungunya virus triggers an innate immune response. J. Immunol. *184*, 5903–5913.

Hoarau, J.J., Jaffar-Bandjee, M.C., Krejbich-Trotot, P., Das, T., Li-Pat-Yuen G., Dassa, B., Denizot, M., Guichard, E., Ribera, A., Henni, T., *et al.* (2010). Persistent chronic inflammation and infection by Chikungunya arthritogenic alphavirus in spite of a robust host immune response. J. Immunol. *184*, 5914–5927.

Hoarau, J.J., Gay, F., Pellé, O., Samri, A., Jaffar-Bandjee, M.C., Gasque, P., and Autran, B. (2013). Identical strength of the T cell responses against E2, nsP1 and capsid CHIKV proteins in recovered and chronic patients after the epidemics of 2005–2006 in La Reunion Island. PLOS ONE *8*, e84695.

Ivanov, A.P., Ivanova, O.E., Lomonosov, N.N., Pozdnyakov, S.V., Konstantinov, O.K., and Bah, M.A. (1992). Serological investigations of Chikungunya virus in the Republic of Guinea, Ann. Soc. Belg. Med. Trop. 72, 73–74.

Jaffar-Bandjee, M.C., Das, T., Hoarau, J.J., Krejbich-Trotot, P., Denizot, M., Ribera, A., Roques, P., and Gasque, P. (2009). Chikungunya virus takes centre stage in virally induced arthritis: possible cellular and molecular mechanisms to pathogenesis. Microbes Infect. *11*, 1206–1218.

Jaffar-Bandjee, M.C., Ramful, D., Gauzere, B.A., Hoarau, J.J., Krejbich-Trotot, P., Robin, S., Ribera, A., Selambarom, J., and Gasque, P. (2010). Emergence and clinical insights into the pathology of Chikungunya virus infection. Expert Rev. Anti. Infect. Ther. *8*, 987–996.

Kagi, D., and Hengartner, H. (1996). Different roles for cytotoxic T cells in the control of infections with cytopathic versus noncytopathic viruses, Curr. Opin. Immunol. 8, 472–477.

Kam, Y.W., Ong, E.K., Rénia, L., Tong, J.C., and Ng, L.F.P. (2009). Immuno-biology of Chikungunya and implications for disease intervention. Microbes Infect. *11*, 1186–1196.

Kam, Y.W., Simarmata, D., Chow, A., Her, Z., Teng, T.S., Ong, E.K., Rénia, L., Leo, Y.S., and Ng, L.F.P. (2012a). Early appearance of neutralizing immunoglobulin G3 antibodies is associated with chikungunya virus clearance and long-term clinical protection. J. Infect. Dis. *205*, 1147–1154.

Kam, Y.W., Lee, W.W., Simarmata, D., Harjanto, S., Teng, T.S., Tolou, H., Chow, A., Lin, R.T., Leo, Y.S., Rénia, L., and Ng, L.F.P. (2012b). Longitudinal analysis of the human antibody response to Chikungunya virus infection: implications for serodiagnosis and vaccine development. J. Virol. *86*, 13005–13015.

Kam, Y.W., Lum, F.M., Teo, T.H., Lee, W.W., Simarmata, D., Harjanto, S., Chua, C.L., Chan, Y.F., Wee, J.K., Chow, A., *et al.* (2012c). Early neutralizing IgG response to Chikungunya virus in infected patients targets a dominant linear epitope on the E2 glycoprotein. EMBO Mol. Med. *4*, 330–343.

Kelvin, A.A., Banner, D., Silvi, G., Moro, M.L., Spataro, N., Gaibani, P., Cavrini, F., Pierro, A., Rossini, G., Cameron, M.J., *et al.* (2011). Inflammatory cytokine expression is associated with chikungunya virus resolution and symptom severity. PLoS Negl. Trop. Dis. *5*, e1279.

Kondekar, S., and Gogtay, N.J. (2006). Why Chikungunya is called Chikungunya, J. Postgrad. Med. *52*, 307–307.

Lam, S.K., Chua, K.B., Hooi, P.S., Rahimah, M.A., Kumari, S., Tharmaratnam, M., Chuah, S.K., Smith, D.W., and Sampson, I.A. (2001). Chikungunya infection – an emerging disease in Malaysia, Southeast Asian J. Trop. Med. Public Health 32, 447–451.

de Lamballerie, X., Boisson, V., Reynier, J.C., Enault, S., Charrel, R.N., Flahault, A., Roques, P., and Le Grand, R. (2008). On Chikungunya acute infection and chloroquine treatment. Vector Borne Zoonotic Dis. *8*, 837–839.

Lemant, J., Boisson, V., Winer, A., Thibault, L., André, H., Tixier, F., Lemercier, M., Antok, E., Cresta, M.P., Grivard, P., *et al.* (2008). Serious acute Chikungunya virus infection requiring intensive care during the Reunion Island outbreak in 2005–2006. Crit. Care Med. *36*, 2536–2541.

Leparc-Goffart, I., Nougairede, A., Cassadou, S., Prat, C., and de Lamballerie, X. (2014). Chikungunya in the Americas. Lancet *383*: 514.

Levitt, N.H., Ramsburg, H.H., Hasty, S.E., Repik, P.M., Cole, F.E., and Lupton, H.W. (1986). Development of an attenuated strain of chikungunya virus for use in vaccine production. Vaccine *4*, 157–162.

Li, L., Jose, J., Xiang, Y., Kuhn, R.J., and Rossmann, M.G. (2010). Structural changes of envelope proteins during alphavirus fusion. Nature. *468*, 705–708.

Lum, F.M., Teo, T.H., Lee, W.W., Kam, Y.W., Rénia, L., and Ng, L.F.P. (2013). An essential role of antibodies in the control of Chikungunya virus infection. J. Immunol. *190*, 6295–6302.

Lumsden, W.H.R. (1955). An epidemic of virus disease in Southern Province, Tanganyika territory, in 1952–1953 II. General description and epidemiology, Trans. R. Soc. Trop. Med. Hyg. *49*, 33–57.

Mallilankaraman, K., Shedlock, D.J., Bao, H., Kawalekar, O.U., Fagone, P., Ramanathan, A.A., Ferraro, B., Stabenow, J., Vijayachari, P., Sundaram, S.G., *et al.* (2011). A DNA vaccine against chikungunya virus is protective in mice and induces neutralizing antibodies in mice and nonhuman primates. PLoS. Negl. Trop. Dis. *5*, e928.

Morens, D.M., and Fauci, A.S. (2014). Chikungunya at the door – deja vu all over again? N. Engl. J. Med. *371*, 885–887.

Muthumani, K., Lankaraman, K.M., Laddy, D.J., Sundaram, S.G., Chung, C.W., Sako, E., Wu, L., Khan, A., Sardesai, N., Kim, J.J., *et al.* (2008). Immunogenicity of novel consensus-based DNA vaccines against chikungunya virus. Vaccine *26*, 5128–5134.

Ng, L.F.P., Chow, A., Sun, Y.J., Kwek, D.J., Lim, P.L., Dimatatac, F., Ng, L.C., Ooi, E.E., Choo, K.H., Her, Z., *et al.* (2009). IL-1beta, IL-6, and RANTES as biomarkers of chikungunya severity. PLOS ONE *4*, e4261.

Osterrieth, P., Deleplanque-Liegeois, P., and Renoirte, R. (1960). Research on the Chikungunya virus in the Belgian Congo. II. Serological investigation. Ann. Soc. Belg. Med. Trop. *40*, 205–213.

Pal, P., Dowd, K.A., Brien, J.D., Edeling, M.A., Gorlatov, S., Johnson, S., Lee, I., Akahata, W., Nabel, G.J., Richter, M.K., *et al.* (2013). Development of a highly protective combination monoclonal antibody therapy against Chikungunya virus. PLoS Pathog. *9*, e1003312.

Panning, M., Grywna, K., van Esbroeck, M., Emmerich, P., and Drosten, C. (2008). Chikungunya fever in travelers returning to Europe from the Indian Ocean region, 2006, Emerg. Infect. Dis. *14*, 416–422.

Parola, P., de Lamballerie, X., Jourdan, J., Rovery, C., Vaillant, V., Minodier, P., Brouqui, P., Flahault, A., Raoult, D., and Charrel, R.N. (2006). Novel Chikungunya virus variant in travelers returning from Indian Ocean Islands, Emerg. Infect. Dis. *12*, 1493–1499.

Partidos, C.D., Weger, J., Brewoo, J., Seymour, R., Borland, E.M., Ledermann, J.P., Powers, A.M., Weaver, S.C., Stinchcomb, D.T., and Osorio, J.E. (2011). Probing the attenuation and protective efficacy of a candidate chikungunya virus vaccine in mice with compromised interferon (IFN) signaling. Vaccine *29*, 3067–3073.

Plante, K., Wang, E., Partidos, C.D., Weger, J., Gorchakov, R., Tsetsarkin, K., Borland, E.M., Powers, A.M., Seymour, R., Stinchcomb, D.T., Osorio, J.E., Frolov, I., and Weaver, S.C. (2011). Novel chikungunya vaccine candidate with an IRES-based attenuation and host range alteration mechanism. PLoS Pathog. *7*, e1002142.

Powers, A.M., and Logue, C.H. (2007). Changing patterns of chikungunya virus: re-emergence of a zoonotic arbovirus. J. Gen. Virol. *88*, 2363–2377.

Powers, A.M., Brault, A.C., Tesh, R.B., and Weaver, S.C. (2007). Re-emergence of chikungunya and o'nyong-nyong viruses: evidence for distinct geographical lineages and distant evolutionary relationships, J Gen. Virol. *81*, 471–479.

Ramful, D., Sampériz, S., Fritel, X., Michault, A., Jaffar-Bandjee, M.C., Rollot, O., Boumahni, B., and Gérardin, P. (2014). Antibody kinetics in infants exposed to Chikungunya virus infection during pregnancy reveals absence of congenital infection. J. Infect. Dis. *209*, 1726–1730.

Ramirez, F., Schlicker, A., Assenov, Y., Lengauer, T., and Albrecht, M. (2007). Computational analysis of human protein interaction networks. Proteomics *7*, 2541–2552.

Reiter, P., Fontenille, D., and Paupy, C. (2006). *Aedes albopictus* as an epidemic vector of chikungunya virus: another emerging problem?, Lancet Infect. Dis. *6*, 463–464.

Rezza, G., Nicoletti, L., Angelini, R., Romi, R., Finarelli, A.C., Panning, M., Cordioli, P., Fortuna, C., Boros, S., Magurano, F., *et al.*; CHIKV study group (2007). Infection with chikungunya virus in Italy: an outbreak in a temperate region. Lancet *370*, 1840–1846.

Robinson, M.C. (1955). An epidemic of virus disease in Southern Province, Tanganyika territory, in 1952–1953, Trans. R. Soc. Trop. Med. Hyg. *49*, 28–32.

Ross, R.W. (1956). The Newala epidemic. III. The virus: isolation, pathogenic properties and relationship to the epidemic, J. Hyg. (Lond). *54*, 177–191.

Roy, C.J., Adams, A.P., Wang, E., Plante, K., Gorchakov, R., Seymour, R.L., Vinet-Oliphant, H., and Weaver, S.C. (2014). Chikungunya vaccine candidate is highly attenuated and protects nonhuman primates against telemetrically monitored disease following a single dose. J. Infect. Dis. *209*, 1891–1899.

Rulli, N.E., Rolph, M.S., Srikiatkhachorn, A., Anantapreecha, S., Guglielmotti, A., and Mahalingam, S. (2011). Protection from arthritis and myositis in a mouse model of acute chikungunya virus disease by bindarit, an inhibitor of monocyte chemotactic protein-1 synthesis. J. Infect. Dis. *204*, 1026–1030.

Sam, I.C., Kamarulzaman, A., Ong, G.S., Veriah, R.S., Ponnampalavanar, S., Chan, Y.F., and AbuBakar, S. (2010). Chikungunya virus-associated death in Malaysia. Trop. Biomed. *27*, 343–347.

Schilte, C., Couderc, T., Chretien, F., Sourisseau, M., Gangneux, N., Guivel-Benhassine, F., Kraxner, A., Tschopp, J., Higgs, S., Michault, A., *et al.* (2010). Type I IFN controls chikungunya virus via its action on nonhematopoietic cells. J. Exp Med. *207*, 429–442.

Schilte, C., Buckwalter, M.R., Laird, M.E., Diamond, M.S., Schwartz, O., and Albert, M.L. (2012). Cutting edge: independent roles for IRF-3 and IRF-7 in hematopoietic and nonhematopoietic cells during host response to Chikungunya infection. J. Immunol. *188*, 2967–2971.

Schuffenecker, I., Iteman, I., Michault, A., Murri, S., Frangeul, L., Vaney, M.C., Lavenir, R., Pardigon, N., Reynes, J.M., Pettinelli, F., *et al.* (2006). Genome microevolution of chikungunya viruses causing the Indian Ocean outbreak. PLoS Med. *3*, 0001–0013.

Schwartz, O., and Albert, M.L. (2010). Biology and pathogenesis of chikungunya virus. Nat. Rev. Microbiol. *8*, 491–500.

Sergon, K., Njuguna, C., Kalani, R., Ofula, V., Onyango, C., Konongoi, L.S., Bedno, S., Burke, H., Dumilla, A.M., Konde, J., *et al.* (2008). Seroprevalence of chikungunya virus (CHIKV) infection on Lamu Island, Kenya, October 2004. Am. J. Trop. Med. Hyg. *78*, 333–337.

Sette, A., and Fikes, J. (2003). Epitope-based vaccines: an update on epitope identification, vaccine design and delivery, Curr. Opin. Immunol. *15*, 461–470.

Solignat, M., Gay, B., Higgs, S., Briant, L., and Devaux, C. (2009). Replication cycle of chikungunya: a re-emerging arbovirus. Virology *393*, 183–197.

Soumahoro, M.K., Gérardin, P., Boëlle, P.Y., Perrau, J., Fianu, A., Pouchot, J., Malvy, D., Flahault, A., Favier, F., and Hanslik, T. (2009). Impact of Chikungunya virus infection on health status and quality of life: a retrospective cohort study. PLOS ONE. *4*, e7800.

Stapleford, K.A., Coffey, L.L., Lay, S., Bordería, A.V., Duong, V., Isakov, O., Rozen-Gagnon, K., Arias-Goeta, C., Blanc, H., Beaucourt, S., *et al.* (2014). Cell Host Microbe.*15*, 706–716.

Stetson, D.B., and Medzhitov, R. (2006). Type I interferons in host defense. Immunity *25*, 373–381.

Strauss, J.E., and Strauss, E.G. (1994). The alphaviruses: gene expression, replication, and evolution, Microbiol. Mol. Biol. Rev. *58*, 491–562.

Suhrbier, A., Jaffar-Bandjee, M.C., and Gasque, P. (2012). Arthritogenic alphaviruses--an overview. Nat. Rev. Rheumatol. *8*, 420–429.

Tandale, B.V., Sathe, P.S., Arankalle, V.A., Wadia, R.S., Kulkarni, R., Shah, S.V., Shah, S.K., Sheth, J.K., Sudeep, A.B., Tripathy, A.S., and Mishra, A.C. (2009). Systemic involvements and fatalities during chikungunya epidemic in India, 2006. J. Clin. Virol. *46*, 145–149.

Teng, T.S., Foo, S.S., Simamarta, D., Lum, F.M., Teo, T.H., Lulla, A., Yeo, N.K., Koh, E.G., Chow, A., Leo, Y.S., *et al.* (2012). Viperin restricts chikungunya virus replication and pathology. J. Clin. Invest. *122*, 4447–4460.

Tenzer, S., Peters, B., Bulik, S., Schoor, O., Lemmel, C., Schatz, M.M., Kloetzel, P.M., Rammensee, H.G., Schild, H., and Holzhutter, H.G. (2005). Modeling the MHC class I pathway by combining predictions of proteasomal cleavage, TAP transport and MHC class I binding, Cell. Mol. Life. Sci. *62*, 1025–1037.

Teo, T.H., Lum, F.M., Claser, C., Lulla, V., Lulla, A., Merits, A., Rénia, L., and Ng, L.F.P. (2013). An essential role of antibodies in the control of Chikungunya virus infection. J. Immunol. *190*, 6295–6302.

Theamboonlers, A., Rianthavorn, P., Praianantathavorn, K., Wuttirattanakowit, N., and Poovorawan, Y. (2009). Clinical and molecular characterization of Chikungunya virus in South Thailand. Jpn. J. Infect. Dis. *62*, 303–305.

Tiwari, M., Parida, M., Santhosh, S.R., Khan, M., Dash, P.K., and Rao, P.V. (2009). Assessment of immunogenic potential of vero adapted formalin inactivated vaccine derived from novel ECSA genotype of chikungunya virus. Vaccine *27*, 2513–2522.

Townson, H., and Nathan, M.B. (2008). Resurgence of chikungunya, Trans. R. Soc. Trop. Med. Hyg. *102*, 308–309.

Tsetsarkin, K.A., and Weaver, S.C. (2011). Sequential adaptive mutations enhance efficient vector switching by Chikungunya virus and its epidemic emergence. PLoS Pathog. 7, e1002412.

Tsetsarkin, K.A., Vanlandingham, D.L., McGee, C.E., and Higgs, S. (2007). A single mutation in chikungunya virus affects vector specificity and epidemic potential. PLoS Pathog. *4*, e201.

Tsetsarkin, K.A., Chen, R., Yun, R., Rossi, S.L., Plante, K.S., Guerbois, M., Forrester, N., Perng, G.C., Sreekumar, E., Leal, G., *et al.* (2014). Multi-peaked adaptive landscape for chikungunya virus evolution predicts continued fitness optimization in Aedes albopictus mosquitoes. Nat. Commun. *5*, 4084.

Voss, J.E., Vaney, M.C., Duquerroy, S., Vonrhein, C., Girard-Blanc, C., Crublet, E., Thompson, A., Bricogne, G., and Rey, F.A. (2010). Glycoprotein organization of Chikungunya virus particles revealed by X-ray crystallography. Nature *468*, 709–712.

Wang, E., Volkova, E., Adams, A.P., Forrester, N., Xiao, S.Y., Frolov, I., and Weaver, S.C. (2008). Chimeric alphavirus vaccine candidates for chikungunya. Vaccine *26*, 5030–5039.

Wauquier, N., Becquart, P., Nkoghe, D., Padilla, C., Ndjoyi-Mbiguino, A.l., and Leroy, E.M. (2010). The acute phase of chikungunya virus infection in humans is associated with strong innate immunity and T CD8 cell activation. J. Infect. Dis. *204*, 115–123.

Welsh, R.M., and Fujinami, R.S. (2007). Pathogenic epitopes, heterologous immunity and vaccine design, Nat. Rev. Microbiol. *5*, 555–563.

Werneke, S.W., Schilte, C., Rohatgi, A., Monte, K.J., Michault, A., Arenzana-Seisdedos, F., Vanlandingham, D.L., Higgs, S., Fontanet, A., Albert, M.L., and Lenschow, D.J. (2011). ISG15 is critical in the control of Chikungunya virus infection independent of UbE1L mediated conjugation. PLoS Pathog. 7, e1002322.

White, A., Berman, S., and Lowenthal, J.P. (1972). Comparative immunogenicities of chikungunya vaccines propagated in monkey kidney monolayers and chick embryo suspension cultures. Appl. Microbiol. 23, 951–952.

White, L.K., Sali, T., Alvarado, D., Gatti, E., Pierre, P., Streblow, D., and Defilippis, V.R. (2011). Chikungunya virus induces IPS-1-dependent innate immune activation and protein kinase R-independent translational shutoff. J. Virol. *85*, 606–620.

WHO (2010). Burden: mortality, morbidity and risk factors, Chapter 1, In Global status report on noncommunicable diseases, pp9–31.

Zhang, H., Lundegaard, C., and Nielsen, M. (2009). Pan-specific MHC class I predictors: a benchmark of HLA class I pan-specific prediction methods, Bioinformatics *25*, 83–89.

Index

www.ingramcontent.com/pod-product-compliance
Lightning Source LLC
LaVergne TN
LVHW070830220926
848486LV00030B/247

* 9 7 8 1 9 1 0 1 9 0 1 5 9 *